Handbook of Polypropylene

Handbook of Polypropylene

Edited by **Will Davis**

New York

Published by NY Research Press,
23 West, 55th Street, Suite 816,
New York, NY 10019, USA
www.nyresearchpress.com

Handbook of Polypropylene
Edited by Will Davis

International Standard Book Number: 978-1-63238-267-2 (Hardback)

Printed in the United States of America.

Contents

Preface

Polypropylene is extensively used in packaging, textiles, equipment manufacturing and other areas. This book is a compilation of researches based on polypropylene, accomplished by experts around the globe over five decades. It demonstrates the various developmental stages that polypropylene has undergone. This book also consists of some current polypropylene theories. The book encompasses the various applications of polypropylene in diverse scientific fields. It intends to provide concise and well-compiled information related to polypropylene to its readers.

This book is the end result of constructive efforts and intensive research done by experts in this field. The aim of this book is to enlighten the readers with recent information in this area of research. The information provided in this profound book would serve as a valuable reference to students and researchers in this field.

At the end, I would like to thank all the authors for devoting their precious time and providing their valuable contribution to this book. I would also like to express my gratitude to my fellow colleagues who encouraged me throughout the process.

Editor

Polypropylene in the Science

Polypropylene Nanocomposites

Azza M. Mazrouaa

Petrochemical Department, Polymer Laboratory,
Egyptian Petroleum Research Institute, Nasr City, Cairo,
Egypt

1. Introduction

The possibility of manufacturing nano-composites materials with tailored properties at low cost has gained much interest. In fact, there is already more than two decades of research on those materials. Particular interest has been paid to clay nano-platelets and their composites with non-polar thermoplastic polyolefin matrixes, namely polypropylene (PP).

Imagine an industrialist and his design team relatively aware of the developments in the research with nano-fillers asking themselves: What can we do with nano-composites and make a net profit up to the 'promises' of the current state of the art? Research announced potential areas of interest for practical applications include mechanical performance, toughness improvement, surface hardening, fire retardancy, or, solvent and permeability reduction. However there remains the problem of how a company could set up the facility for compounding, and guarantee proper dispersion and minimization of health hazards. One should bear in mind that for industrial dissemination conventional equipments should be used and compounding achieved through in-line mixing of virgin resins and nanoclay master batches. Since the seventies polypropylene has been seen as the wonder engineering-commodity material with widespread application in numerous technical applications.

Current masterbatches are mainly based on thermoplastic polyolefin and anhydride functionalized PP as a compatibilizer. In principle, filling with a low incorporation level of nanoclay (typically less than 5%), makes PP adequate to applications with engineering requirements. Nevertheless, only well-dispersed and well-exfoliated nanoparticles can lead to the expected improvement of properties. The nanoparticle dispersion and exfoliation is usually assumed to be achieved during masterbatching, but the suppliers of master batches request a relatively high price. Underlining these evidences poor exfoliation was a common feature in moldings obtained using industry achievable processing conditions. There is an evident interest of bringing the benefits of nanocomposites at the laboratory scale to cost competitive industrial products. However the first available information leaves a number of treads that research could well follow, for example. Which level of exfoliation should be required to viable master batches? Is there any scope for hybrid compounding, i.e. combining particulate nanoclays with fibre reinforcements? Are there only a few niches of application for nanocomposites? Have nanofillers any chance of being full exfoliated within non polar matrixes? Do these nanocomposites will require alternative routes of processing? Should novel compatibilizers be developed in order to avoid unavoidable reagglomeration during injection molding? (Frontini & Pouzada, 2011).

2. Nanotechnology growth predicted

Nanocomposites, defined as polymers bonded with nanoparticles to produce materials with enhanced properties, have been in existence for years but are recently gaining momentum in mainstream commercial packaging use (Butschli 2004).The United States is leading in nanotechnology research with over 400 research centers and companies involved with over $3.4 billion in funding. Europe has over 175 companies and organizations involved in nanoscience research with $1.7 billion in funding. Japan is also very involved in research with over 100 companies working with nanotechnologies (Anyadike, 2005) . Globally, the market for nanocomposites is expected to grow to $250 million by 2008, with annual growth rates projected to be 18-25% per year (Principia, 2004). The global market for nanotechnology   products was worth an estimated $11.7 billion in 2009. The market is projected to grow to more than $15.7 billion in 2010 and nearly $26.7 billion in 2015 at a compound annual growth rate (CAGR) of 11.1% from 2010 to 2015 (bccresearch, 2010).

3. What does nanocomposites really mean?

Perhaps it is necessary to make clear the terms "hybrids" and "nanocomposites" before the discussion of the nanocomposites, since it is somewhat ambiguous to identify whether materials fall into "nanocomposites" or not. The most wide-ranging definition of a hybrid is a material that includes two moieties blended on the molecular scale.

Commonly the term "hybrids" is more often used if the inorganic units are formed in situ by the sol-gel process (Kickelbick, 2007).Meanwhile, use of the word "nanocomposites" implies that materials consist of various phases with different compositions, and at least one constituent phase (for polymer/silica nanocomposites, that phase is generally silica) has one dimension less than 100 nm. A gradual transition is implied by the fact that there is no clear borderline between "hybrids" and "nanocomposites"(Kickelbick, 2007).

Expressions of "nanocomposites" seem to be very trendy, and although the size of the silica particles is above 100 nm, the composites are often called "nanocomposites" in some literature. Organic/inorganic composite materials have been extensively studied for a long time. When inorganic phases in organic/inorganic composites become nano sized, they are called nanocomposites. Organic/inorganic nanocomposites are generally organic polymer composites with inorganic nanoscale building blocks. They combine the advantages of the inorganic material (e.g., rigidity, thermal stability) and the organic polymer (e.g., flexibility, dielectric, ductility, and processibility). Moreover, they usually also contain special properties of nanofillers leading to materials with improved properties.

A defining feature of polymer nanocomposites is that the small size of the fillers leads to a dramatic increase in interfacial area as compared with traditional composites (Balazs et al., 2006; Caseri & Nalwa, 2004; Caseri,2006, 2007; Schadler, 2003; Schadler et al., 2007; Schaefer &Justice,2007; Winey &Vaia, 2007;, Krishnamoorti &Vaia, 2007).

The next time you look at a car, you could be looking at nanotechnology without even realizing it. For the past several years, car companies have been using nanocomposites instead of plastic to make certain car parts. In 2001, Toyota started using nanocomposites to make bumpers for their cars. In 2002, General Motors (GM) made nanocomposite "step-assists" – external running boards that help people get into and out of cars – an option on the 2002 Chevrolet Astra and the GMC Safari (The Future of Automotive Plastics, 2003). Nanocomposites are lighter, stiffer, less brittle, and more dent- and scratch-resistant than conventional plastics. Some nanocomposites are also more recyclable, more flame retardant, less porous, better conductors of electricity, and can be painted more easily (Leaversuch & Buchholz, 2003).

4. How nanocomposites work?

Polymer nanocomposites are constructed by dispersing a filler material into nanoparticles that form flat platelets. These platelets are then distributed into a polymer matrix creating multiple parallel layers which force gases to flow through the polymer in a "torturous path", forming complex barriers to gases and water vapour, as seen in **Figure 1**. As more tortuosity is present in a polymer structure, higher barrier properties will result. The permeability coefficient of polymer films is determined using two factors: diffusion and solubility coefficients:

$$P = D \times S.$$

Effectively, more diffusion of nanoparticles throughout a polymer significantly reduces its permeability. According the Natick Soldier Center of the United States Army, "the degree of dispersion of the nanoparticles within the polymer relates to improvement in mechanical and barrier properties in the resulting nanocomposite films over those of pure polymer films". Nanoparticles allow for much lower loading levels than traditional fillers to achieve optimum performance. Usually addition levels of nanofillers are less than 5%, which significantly impact weight reduction of nanocomposite films. This dispersion process results in high aspect ratio and surface area causing higher performance plastics than with conventional fillers (Brody ,2003).

Idealized Oriented
Layered Nanoparticles

Fig. 1. Idealized Oriented Layered Nanoparticle

5. The most commonly used nanoparticles

Different types of fillers are utilized, the most common is a nanoclay material called montmorillonite—a layered smectite clay. Clays, in a natural state, are hydrophilic while polymers are hydrophobic. To make the two compatible, the clay's polarity must be modified to be more "organic" to interact successfully with polymers (Hay & Shaw, 2000; Ryan , 2003). One way to modify clay is by exchanging organic ammonium cations for inorganic cations from the clay's surface (Sherman,1999). Additional nanofillers include carbon nanotubes, graphite platelets, carbon nanofibers, as well as other fillers being investigated such as synthetic clays, natural fibers (hemp or flax), and POSS (polyhedral oligomeric silsesquioxane). Carbon nanotubes, a more expensive material than nanoclay fillers which are more readily available, offer superb electrical and thermal conductivity properties. The major suppliers for nanoclays are Nanocor and Southern Clay. Inorganic nanoscale building nanoparticles of metals (e.g., Au, Ag), and metal oxides (e.g., TiO_2, Al_2O_3 which SiO_2 is viewed as being very important.

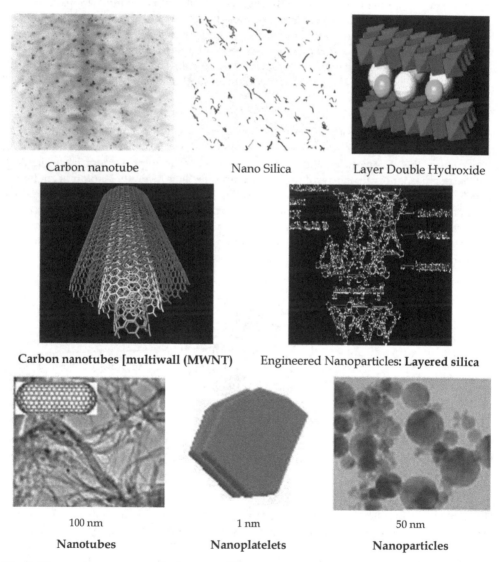

Fig. 2. The most common used of nanoparticle

Using polypropylene, polyethylene, and other polymers reinforced with natural minerals such as calcium carbonate, zeolite, mica, and talc combined with efficient coupling agents has proven to be a successful technology to obtain tailored reinforcement and optimize the cost to property balance. One important material used as filler is the talc mineral since it has unique reinforcing features as softness, lubricity, excellent wetting and dispersion in plastics, and other organics. Talc is a hydrated magnesium silicate mineral widely used in polymers as reinforcing filler. Its plate-like structure provide the talc-filled materials with tailored properties to be used in some industrial and commercial applications such as in refrigerators jackets, packaged components, blocking of infrared in agricultural films, and in automotive and appliance markets. Talc-filled polypropylene composite has low specific gravity and combines excellent chemical resistance with low cost (Zihlif & Ragosta, 2003; Tripathi & Pillia, 1994; Hijleh et al.,2000; Mahanwar et al.,2006; Xie et al.,2001; Chen, 2004; Hajji et al. 1999; Kojima et al., 1993). **Figure 2.**

6. Polymer nanocomposite synthesis

The synthesis of polymer nanocomposites is an integral aspect of polymer nanotechnology. By inserting the nanometric inorganic compounds, the properties of polymers improve and hence this has a lot of applications depending upon the inorganic material present in the polymers. Solvent casting is one of the easiest and less time consuming methods for the synthesis of polymer nanocomposites.

There are three common methods used to enhance polymers with nanofillers to produce nanocomposites: melt compounding, in-situ polymerization and the solvent method.

1. **Melt compounding – or processing –** of the nanofillers into a polymer is done simultaneously when the polymer is being processed through an extruder, injection molder, or other processing machine. The polymer pellets and filler (clay) are pressed together using shear forces to help with exfoliation and dispersion (Brody, 2003; Zihlif & Ragosta, 2003).
2. **In-situ polymerization,** the filler is added directly to the liquid monomer during the polymerization stage.
3. **The solution method,** fillers are added to a polymer solution using solvents such as toluene, chloroform and acetonitrile to integrate the polymer and filler molecules [19]. Since the use of solvents is not environmentally-friendly, melt processing and in-situ polymerization are the most widely used methods of nanocomposite production.

As pointed out(Tripathi &Pillia,1994) nanocomposite systems can be prepared by various synthesis routes, thanks to the ability to combine different ways to introduce each phase. *The organic component can be introduced as:*

i.   A precursor, which can be a monomer or an oligomer,
ii. A preformed linear polymer (in molten, solution, or emulsion states), or
iii. A polymer network, physically (e.g., semi crystalline linear polymer) or chemically (e.g., thermosets, elastomers) cross-linked.

The mineral part can be introduced as:

i.   A precursor (e.g., TEOS) or
ii. Preformed nanoparticles.

Organic or inorganic polymerization generally becomes necessary if at least one of the starting moieties is a precursor.

7. Polypropylene nanocomposites

Polypropylene (PP) is a versatile material its use has significantly penetrated numerous sectors of the manufacturing, medical, and packaging industries. Polymer clay nanocomposites are multiphase organic/inorganic hybrid materials pioneered by researchers at Toyota, (Kojima et al., Usuki et al., 1993)which may exhibit significantly improved mechanical, flammability, and permeability properties relative to the base polymer matrix at very low clay loading. Although first demonstrated for nylon, polymer clay nanocomposites have since been prepared for a range of thermoplastic and thermoset polymers. However, the development of PP/clay nanocomposites poses special challenges because of polypropylene's hydrophobicity.

The reinforcement of polypropylene and other thermoplastics with inorganic particles such as talc and glass is a common method of material property enhancement. Polymer clay nanocomposites extend this strategy to the nanoscale. The anisometric shape and approximately 1 nm width of the clay platelets dramatically increase the amount of interfacial contact between the clay and the polymer matrix. Thus the clay surface can mediate changes in matrix polymer conformation, crystal structure, and crystal morphology through interfacial mechanisms that are absent in classical polymer composite materials. For these reasons, it is believed that nanocomposite materials with the clay platelets dispersed as isolated, exfoliated platelets are optimal for end-use properties.

Recent research has generated advances in polypropylene nanocomposites that are sufficient to motivate new technological applications. For example, PP-based nanocomposites have been developed for application as exterior automotive components (Sherman, 1999) Cone calorimetry measurements of peak heat release rate from maleated/ PP nanocomposites with 4% loading are reduced by 75% relative to the pure polymer (Gilman et al.,2000). These improvements are relevant to applications requiring reduced flammability. Yet, relative to other thermoplastic nanocomposites, such as nylon 6, the improvement in end-use properties for polypropylene nanocomposites has been modest. In addition, noting that many synthesized PP nanocomposites are likely to exist as nonequilibrium structures, research into the aging and rejuvenation of these mesoscale structures is warranted. Furthermore, better methods to characterize the full distribution and hierarchy of structural states present in PP nanocomposites are required because, for example, rare aggregates can seriously compromise nonlinear mechanical properties such as toughness, yield stress, and elongation at break. Finally, the interaction between clay platelets and polymer crystallization requires further attention because these interactions are likely a significant determinant of the end-use properties of polypropylene nanocomposites.

8. What is organoclay?

Organoclays are manufactured by modifying bentonite with quaternary amines, a type of surfactant that contains a nitrogen ion. The nitrogen end of the quaternary amine, the hydrophilic end, is positively charged, and ion exchanges onto the clay platelet for sodium or calcium. The amines used are of the long chain type with 12-18 carbon atoms. After some 30 per cent of the clay surface is coated with these amines it becomes hydrophobic and, with certain amines, organophilic.

Fig. 3.

9. The nature of organoclays

The main component of organoclay is bentonite, a chemically altered volcanic ash that consists primarily of the clay mineral montmorillonite. The bentonite in its natural state can absorb up to seven times its weight in water, after treatment can absorb only 5 to 10 per cent of its weight in water, but 40 to 70 per cent in oil, grease, and other sparingly-soluble, hydrophobic chlorinated hydrocarbons. As the organoclay is introduced into water, the quaternary amine is activated and extends perpendicularly off the clay platelets into the water. A chlorine or bromine ion is loosely attached to the carbon chain. Since the sodium ions that were replaced by the nitrogen are positively charged, they bond with the chlorine ion, resulting in sodium salt that is washed away. The result is a neutral surfactant with a solid base, which is the organoclay. The hydrophilic end of the amine dissolves into the oil droplet because "like dissolves like," thus removing that droplet from water. Because the partition reaction takes place "outside" of the clay particle (in contrast to adsorption of oil by carbon, which takes place inside its pores), the organoclay does not foul quickly.

Organophilic clay can function is as a prepolisher to activated carbon, ion exchange resins, and membranes (to prevent fouling), and as a post polisher to oil/water separators, dissolved air flotation (DAF) units, evaporators, membranes, and skimmers. Organophilic clay powder can be a component or the main staple of a flocculent clay powder. They are excellent adsorbers for the removal of oil, surfactants, and solvents, including methyl ethyl ketone, t-butyl alcohol (TBA), and others.

10. Layered silicate / polypropylene nanocomposites

Layered silicate/polymer nanocomposites were first reported in 1950 as a patent literature (Carter et al.,1950) . However, it was not popular until Toyota researchers began a detailed experimentation in the year of 1996 on the nylon 6/clay nanocomposites (Kojima et al., 1993). In recent years, nanocomposites received a great interest in academic, governmental and industrial studies (Kojima et al., 1993).The improvements in thermal, mechanical and flammability properties of clay/polymer nanocomposites are significantly higher than those achieved in traditional filled polymers. Up to now, these systems have experienced some success for several kinds of polar polymers. However, for polymers with low polarity, such as polyolefins, the improvements are not very significant due to the low compatibility between the clay and the polyolefins.

One of the most commonly used organophilic layered silicates is derived from montmorillonite (MMT). Its structure is made of several stacked layers, with a layer thickness between 1.2-1.5 nm and a lateral dimension of 100– 200 nm (Marchant & Krishnamurthy ,2002; Moore & Reynolds ,1997) . These layers organize themselves to form the stacks with a regular gap between them, called interlayer or gallery. The sum of the single layer thickness and the interlayer represents the repeat unit of the multilayer material, called d-spacing or basal spacing (d001), and is calculated from the (001) harmonics obtained from X-ray diffraction patterns. The clay is naturally a hydrophilic material, which makes it difficult to exfoliate in a polymer matrix. Therefore, the surface treatment of silicate layers is necessary to render its surface more hydrophobic, which facilitates exfoliation. Generally, this can be done by ion-exchange reactions with cationic surfactants, including primary, secondary, tertiary and quaternary alkylammonium cations (Fornes et al., 2002; Le Pluart et al.,2002).This modification also leads to expand the basal spacing between the silicate layers due to the presence of alkyl chain intercalated in the interlayer and to obtain organoclay (OMMT).

Polypropylene (PP) is one of the most widely used plastics in large volume. To overcome the disadvantages of PP, such as low toughness and low service temperature, researchers have tried to improve the properties with the addition of nanoparticles that contains polar functional groups. An alkylammonium surfactant has been adequate to modify the clay surfaces and promote the formation of nanocomposite structure. Until now, two major methods, i.e., in-situ polymerization(Ma et al., 2001; Pinnavaia, 2000) and melt intercalation (Manias et al.,2001) have been the techniques to prepare clay/PP nanocomposites. In the former method, the clay is used as a catalyst carrier, propylene monomer intercalates into the interlayer space of the clay and then polymerizes there. The macromolecule chains exfoliate the silicate layers and make them disperse in the polymer matrix evenly. In melt intercalation, PP and organoclay are compounded in the molten state to form nanocomposites.

As the hydrophilic clay is incompatible with polypropylene, compatibilization between the clay and PP is necessary to form stable PP nanocomposites. There are two ways to compatibilize the clay and PP. In the first approach, the enthalpy of the interaction between the surfactant and the clay is reduced. In the second approach, a compatibilizer, such as maleic anhydride grafted PP (PPgMA) can be used(Manias et al.,2001). The clay is melt compounded with the more polar compatibilizer to form an intercalated master batch. The master batch is then compounded with the neat PP to form the PP nanocomposite.

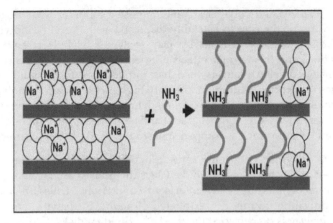

Fig. 4.a. Ion Exchange Reaction between Na-MMT and Alkyl Ammonium Molecules

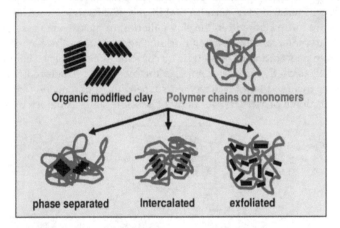

Fig. 4.b. Three main morphology achievable in nanocomposite structure

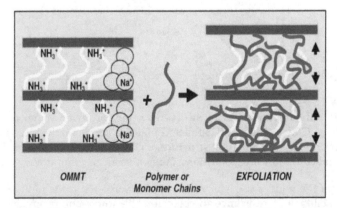

Fig. 4.c. Microstructural Development during Melt Intercalation Process

Mixtures of clay platelets and polymer chains compose a colloidal system. Thus in the melt state, the propensity for the clay to be stably dispersed at the level of individual disks (an exfoliated clay dispersion) is dictated by clay, polymer, stabilizer, and compatibilizer potential interactions and the entropic effects of orientational disorder and confinement. An isometric dimension of clay platelets also has implications for stability because liquid crystalline phases may form. In addition, the very high melt viscosity of polypropylene and the colloidal size of clay imply slow particulate dynamics, thus equilibrium structures may be attained only very gradually. Agglomerated and networked clay structures may also lead to nonequilibrium behavior such as trapped states, aging, and glassy dynamics.

Clay structure in polymer nanocomposites can be characterized as a combination of exfoliated platelets and intercalated tactoids. Clays themselves are layered silicate minerals with charged surfaces neutralized by interlayer counterions. Unless a liquid crystalline order disorder transition occurs, the exfoliated structure is spatially and orientationally disordered and the clay is dispersed at the level of individual disks. Intercalated clay retains interlayer ordering, at least within a particular tactoid; however, intergallery spacing is increased relative to natural clay because stabilizing surfactants, compatibilizers, and/ or matrix polymers are infiltrated within the clay galleries. In the extreme case of clay/polymer matrix immiscibility, intercalation spacing not much greater than the clay and its counterion indicates negligible penetration of polymeric or compatibilizing species between clay layers. Clay platelets or tactoids themselves comprise the mesoscale structure of nanocomposites. Possible structures include that of a dispersed suspension, a percolated network, or a liquid crystal with orientational order. The hierarchy of possible states is depicted in **Figure 5**.

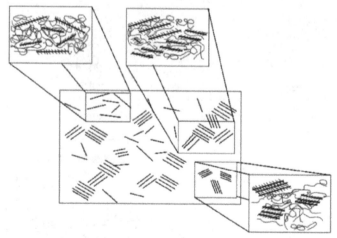

Fig. 5. Schematic of the hierarchy of clay structures in polypropylene nanocomposites of mixed morphology. Clay tactoids and exfoliated platelets comprise the mesoscale morphology. The internal intercalation structure of clay tactoids is determined by the compatibilizer and compounding conditions. (View this art in color at www.dekker.co

Polypropylene (PP) is widely used for many applications due to its low cost, low density, high thermal stability and resistance to corrosion. Blending polypropylene with clays to form nanocomposites is a way to increase its utility by improving its mechanical properties.

Layered silicates dispersed as a reinforcing phase in polymer matrix are one of the most important forms of hybrid organic-inorganic nanocomposites. MMT, hectorite, and saponite are the most commonly used layered silicates. Layered silicates have two types of structure: tetrahedral-substituted and octahedral substituted figure 6. In the case of tetrahedrally substituted layered silicates the negative charge is located on the surface of silicate layers, and hence, the polymer matrices can react interact more readily with these than with octahedrally-substituted material.

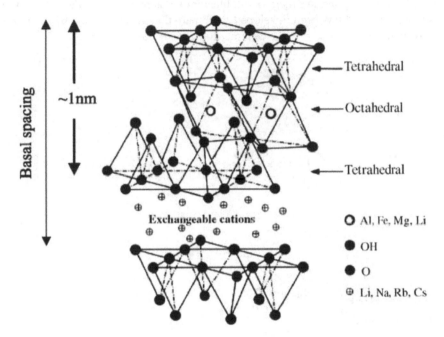

Fig. 6. Structure of Layered Silicates

Compared to conventional composites, polymer layered-silicate (PLS) nanocomposites have maximized polymer-clay interactions since the clay is dispersed on a nanometer scale.

There are three general methods for the preparation of polymer/ silica nanocomposites according to the starting materials and processing techniques: blending, sol-gel processes, and in situ polymerization.

Blending is generally just mixing of the silica nanoparticles into the polymer; *sol-gel process* can be done in situ in the presence of a preformed organic polymer or simultaneously during the polymerization of the monomer(s); and *in situ polymerization* involves the dispersion of nanosilica in the monomer(s) first and then polymerization is carried out.

Layered silicate/polypropylene nanocomposites were prepared by melt intercalation method. Homopolymers PP alone and maleic anhydride-grafted polypropylene (PPgMA) as a compatibilizer were used as the matrix. Clay (Na$^+$ montmorillonite, MMT) particles were used to obtain silicate nano-layers within the PP matrix. Structural modification of MMT using hexadecyltrimethyl ammonium chloride (HTAC) was applied to obtain organophilic silicates (OMMT) (Kıvanç, 2006).The most recent methods to prepare polymer-layered-silicate nanocomposites have primarily been developed by several other groups. In general these methods (shown in **Figure 7**) achieve molecular level incorporation of the layered silicate (e.g. montmorillonite clay or synthetic layered silicate) in the polymer by addition of a modified silicate either to a polymerization reaction (in situ method),   (Usuki et al.,1993,1997; Lan & Pinnavaia,1994) to a solvent-swollen polymer (solution blending), Jeon et al., 1998) or to a polymer melt (melt blending) (Giannelis,1996; Fisher et al.,1998). Additionally, a method has been developed to prepare the layered silicate by polymerizing silicate precursors in the presence of a polymer (Carrado & Langui, 1999)

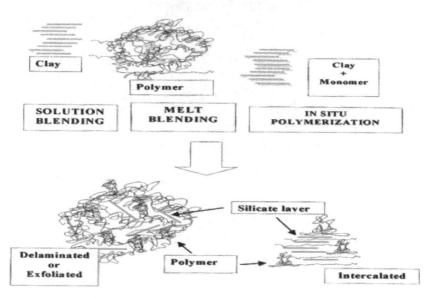

Fig. 7. Schematic representation of various methods (Solution blending, melt blending, and in situ polymerization). The delaminated (or exfoliated) and intercalated morphologies are shown.

Two terms (intercalated and delaminated) are used to describe the two general classes of nanomorphology that can be prepared. *Intercalated structures* are self assembled, well-ordered multilayered structures where the extended polymer chains are inserted into the gallery space between parallel individual silicate layers separated by 2-3 nm (see **Figure 8**). *The delaminated (or exfoliated)* structures result when the individual silicate layers are no longer close enough to interact with the adjacent layers' gallery cations (Lan & Pinnavaia, 1994). In the delaminated cases the interlayer spacing can be on the order of the radius of gyration of the polymer; therefore, the silicate layers may be considered to be well-dispersed in the organic polymer. The silicate layers in a delaminated structure may not be as well-ordered as in an intercalated structure. Both of these hybrid structures can also coexist in the polymer matrix; this mixed nanomorphology is very common for composites based on smectite silicates and clay minerals (Kroschurtz, 1993).

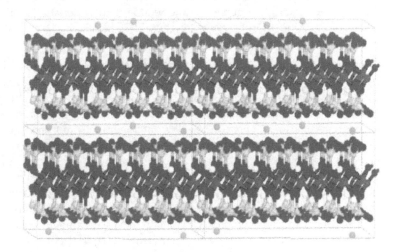

Fig. 8. Molecular representation of sodium montmorillonite, showing two aluminosilicate layers with the Na+ cations in the interlayer gap or gallery. The octahedral (O_h) alumina layer is shown as blue aluminum atoms surrounded by red oxygen atoms. The tetrahedral (Td) silicate layers are shown as yellow silicon atoms surrounded by red oxygen atoms. Hydrogen atoms are white and sodium (Na^+) cations are shown in green.

The very large commercial importance of polypropylene (PP) has also been driving an intense investigation of PP composites reinforced by particulates, fibers, and layered inorganic fillers (Karger, 1995;Karian,1999) .In particular, in the case of layered inorganic fillers, talc and mica had been traditionally attracting the most interest (Karian,1999). However, recent advances in polymer/clay and polymer/silicate nanocomposite material (Alexandre & Dubois, 2000; Giannelis et al., 1998) have inspired efforts to disperse montmorillonite-based fillers in PP. (Kato et al., 1997; Kawasumi et al., 1997; Hasegawa et al., 1998; Oya et al., 2000; Wolfet al., 1999; Reichert et al., 2000; Manias et al., 2000).

Although it has been long known that polymers can be mixed with appropriately modified clay minerals and synthetic clays, (Theng, 1979, 1974) the field of polymer/silicate nanocomposites has gained large momentum recently. Two were the major findings that pioneered the revival of these materials: First, the report of a nylon-6/montmorillonite material from Toyota research, (Kojima et al., 1993; Kojima et al.,1993) where very moderate inorganic loadings resulted in concurrent and remarkable enhancements of thermal and mechanical properties. Second, Giannelis et al. found that it is possible to melt-mix polymers with clays without the use of organic solvents (Vaia et al., 1993) .Since then, the high promise for industrial applications has motivated vigorous research, which revealed concurrent dramatic enhancements of many materials properties by the nanodispersion of inorganic silicate layers. Where the property enhancements originate from the nanocomposite structure, these improvements are generally applicable across a wide range of polymers (Alexandre & Dubois, 2000) .At the same time, there were also discovered property improvements in these nanoscale materials that could not be realized by conventional fillers, as for example a general flame retardant characteristic (Gilman et al., 2000) and a dramatic improvement in barrier properties (Strawhecker & Manias ,2000; Xu et al.,2001).

Montmorillonite (mmt) is a naturally occurring 2:1 phyllosilicate, which has the same layered and crystalline structure as talc and mica but a different layer charge (Theng, 1979, 1974). The mmt crystal lattice consists of 1-nmthin layers, with a central octahedral sheet of alumina fused between two external silica tetrahedral sheets (in such a way that the oxygens from the octahedral sheet also belong to the silica tetrahedra). Isomorphic substitution within the layers (for example, Al^{3+} replaced by Mg^{2+} or Fe^{2+}) generates a negative charges defined through the charge exchange capacity (CEC)sand for mmt is typically 0.9-1.2 mequiv /g depending on the mineral origin. These layers organize themselves in a parallel fashion to form stacks with a regular van der Waals gap between them, called interlayer or gallery. In their pristine form their excess negative charge is balanced by cations (Na^+, Li^+, Ca^{2+}) which exist hydrated in the interlayer. Obviously, in this pristine state mmt is only miscible with hydrophilic polymers, such as poly (ethylene oxide) and poly(vinyl alcohol) (Strawhecker & Manias ,2000; Vaia et al.,1995). To render mmt miscible with other polymers, one must exchange the alkali counterions with cationic-organic surfactants, such as alkylammoniums (Alexandre & Dubois, 2000; Giannelis et al., 1998).

11. Production of layered silicate/polypropylene nanocomposites

The production of polypropylene nanocomposites is shown in Figure 9.The homopolymer PP was fed into Haake two-roll mixer at 190 o C. After melting of the PP in 1 min, clay particles in the amounts of 3, 5 and 10 wt. % were added into molten PP and the mixing was continued for 10 min in the mixer. The blended samples were collected and left for cooling. After cooling, the blends were pressed into 100 mm x 100 mm samples having a

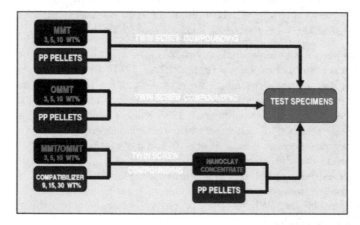

Fig. 9. Processing Stages for Clay/PP Nanocomposites

thickness of 1 mm using a hot press at 190°C. The tensile specimens were prepared by a pneumatic cutter and then the samples were left for two days to complete crystallization. Kıvanç I_IK 2006

12. Techniques used for the characterization of polypropylene/ nanocomposites

Generally, the structure of nanocomposites has typically been established using WAXD analysis and transmission electron micrographic (TEM) observation. Due to its easiness and availability WAXD is most commonly used to probe the nanocomposite structure (Giannelis, 1996; Giannelis et al., 1999; LeBaron et al.,1999; Vaia et al., 1999; Biswas & Sinha, 2001) and occasionally to study the kinetics of the polymer melt intercalation (Vaia et al.,1996) . By monitoring the position, shape, and intensity of the basal reflections from the distributed silicate layers, the nanocomposite structure (intercalated or exfoliated) may be identified. For example, in *an exfoliated nanocomposite*, the extensive layer separation associated with the delamination of the original silicate layers in the polymer matrix results in the eventual disappearance of any coherent X-ray diffraction from the distributed silicate layers. On the other hand, *for intercalated nanocomposites*, the finite layer expansion associated with the polymer intercalation results in the appearance of a new basal reflection corresponding to the larger gallery height. Although WAXD offers a convenient method to determine the interlayer spacing of the silicate layers in the original layered silicates and in the intercalated nanocomposites (within 1–4 nm), little can be said about the spatial distribution of the silicate layers or any structural non-homogeneities in nanocomposites.

Additionally, some layered silicates initially do not exhibit well-defined basal reflections. Thus, peak broadening and intensity decreases are very difficult to study systematically. Therefore, conclusions concerning the mechanism of nanocomposites formation and their structure based solely on WAXD patterns are only tentative. On the other hand, TEM allows a qualitative understanding of the internal structure, spatial distribution of the various phases, and views of the defect structure through direct visualization.

However, special care must be exercised to guarantee a representative cross-section of the sample. The WAXD patterns and corresponding TEM images of three different types of nanocomposites are presented in **Figure 10**. Both TEM and WAXD are essential tools (Morgan & Gilman, 2003) for evaluating nanocomposite structure. However, TEM is time-intensive, and only gives qualitative information on the sample as a whole, while low-angle peaks in WAXD allow quantification of changes in layer spacing. Typically, when layer spacing exceed 6–7 nm in intercalated nanocomposites or when the layers become relatively disordered in exfoliated nanocomposites, associated WAXD features weaken to the point of not being useful. However, recent simultaneous small angle X-ray scattering (SAXS) and WAXD studies yielded quantitative characterization of nanostructure and crystallite structure in N6 based nanocomposites (Mathias et al., 1999).

Very recently, (Bafna et al., 2003) developed a technique to determine the three-dimensional (3D) orientation of various hierarchical organic and inorganic structures in a PLS nanocomposite. They studied the effect of compatibilizer concentration on the orientation of various structures in PLS nanocomposites using 2D SAXS and 2D WAXD in three sample /camera orientations.

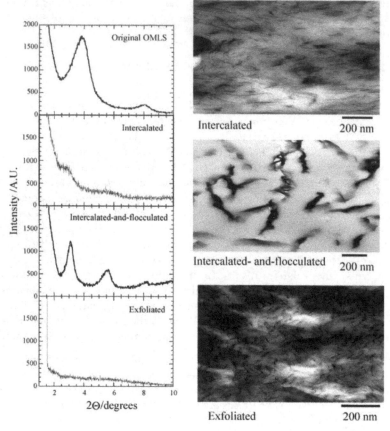

Fig. 10. (a) WAXD patterns and (b) TEM images of three different types of nanocomposites

13. The development and characterization of polypropylene-clay nanocomposites

The development and characterization of polymer-clay nanocomposites has been a subject of raising interest in the recent years (Vaia et al., 1996) . Polymer-layered silicate (PLS) nanocomposites exhibit outstanding properties that are synergistically derived from the organic and inorganic components. The enhanced properties are presumably due to the synergistic effects of nanoscale fillers within the polymer (Kojima et al., 1993). The delamination and dispersion of clays in the polymeric matrix is the key to design nanocomposites. In the ideal conditions, the delamination of the original clay structures, as well as the polymer intercalation in the clay can be achieved. Nanoparticles can significantly improve the stiffness, heat distortion temperature (HDT), dimensional stability, gas barrier properties, electrical conductivity and flame retardancy of polymer with only a 0.1–10 vol.% addition of dispersed nanophase (Wu et al., 2002; Tyan et al.,1999; Lan et al., 1996; Li et al., 2004; Ma et al.,2001; Li et al.,2001; Kawasumi et al.,1997; Usuki et al.,1997; Hasegawa et al., 2000). These performance improvements largely depend upon the spatial distribution, arrangement of intercalating polymer chains and interfacial interaction between the silicate layers and the polymer (Giannelies, 1996; Giannelies et al., 1999; LeBaron et al., 1999; Ray & Okamoto, 2003). Nanoclays like purified Na^+ or Ca^{+2} montmorillonites are inherently hydrophilic in nature, which leads to incompatibility with the hydrophobic polymer matrix with subsequent poor composite properties. Although, the modified clay is miscible with polar polymer like PS, Epoxy, Nylon etc., its dispersion within PE, PP, EVA, ABS etc is unsatisfactory (Komori & Kurada,2000; Cho & Paul, 2001). Grafting of polar functional groups onto polymer chains has been suggested to improve the properties in PE/clay nanocomposites (Fukushima & Inagaki, 1987).

Therefore, to obtain a nanocomposite with requisite properties, the inorganic clay must be modified with some organic surfactant, usually onium salt or an alkyl amine which, compatibilizes the surface chemistry of the clay and polymer matrix at the interface by replacing the inorganic cation and making the gallery space of the clay sufficiently organophilic to permit the entry of polymer matrix (Fukushima & Inagaki, 1987). PP nanocomposites have been the subject of research since several years (Boeing, 1997; Kurokawa et al., 1997; Kato et al., 1997; Reichert et al.,2000; Hasegawa et al.,1998; Zheng et al.,2001). Various organically modified nanoparticles have been prepared and incorporated within the PP matrix to enhance mechanical and thermal performance. However, the development of PP clay nanocomposites posses special challenges because of polypropylene's hydrophobicity. The dispersion of nanolayers strongly depends on the preparation techniques such as in-situ polymerization, solution blending or melt compounding (Okada et al., 1993; Akelah, 1995; Akelah et al., 1994; Giannelis, 1998; Zilg et al., 1998). Melt intercalation of inorganic clay mineral consisting of layered silicates with polymers is widely used, as it is environmental friendly and does not involve any solvent. Direct melt intercalation method offers convenient techniques for preparation of hybrids, which involve mixing the layered silicates with the polymer matrix above its softening point Rzaev et al., 2007). However, desired intercalation/ exfoliation of the clay galleries within PP based nanocomposites system posses host of technical issues which needs to explored and addressed.

14. Future of nanocomposites

By 2009, it is estimated that the flexible and rigid packaging industry will use five millionpounds of nanocomposites materials in the beverage and food industry. By 2011, consumption is estimated to be 100 million pounds. Beer is expected to be the biggest consumer by 2006 with 3 million pounds of nanocomposites until carbonated soft drinks bottles are projected to surpass that to use 50 million pounds of nanocomposites by 2011(Butschli,2004). Polymer nanocomposites are the future for the global packaging industry. Once production and materials cost are less, companies will be using this technology to increase their product's stability and survivability through the supply chain to deliver higher quality to their customers while saving money. The advantages that nanocomposites offer far outweigh the costs and concerns and with time the technology will be further refined and processes more developed. Research continues into other types of nanofillers (i.e., carbon nanotubes), allowing new nanocomposite structures with different improved properties that will further advance nanocomposite use in many diverse packaging applications.

15. Conclusion

The synthesis of polymer nanocomposites is an integral aspect of polymer nanotechnology. By inserting the nanometric inorganic compounds, the properties of polymers improve and hence this has a lot of applications depending upon the inorganic material present in the polymers. The improvements obtained in clay/PP nanocomposite structure can make this commercial thermoplastic polymer more suitable for automotive, construction and packaging applications. Different alkyl ammonium surfactants and compatibilizer was used to produce layered silicate/PP nanocomposites by the same melt intercalation technique.

Polypropylene nanocomposites are still challenging due to the lack of affinity of organophilic PP for hydrophilic clay. The reinforcement of polypropylene and other thermoplastics with inorganic particles such as talc and glass is a common method of material property enhancement. Polymer clay nanocomposites extend this strategy to the nanoscale. The anisometric shape and approximately 1 nm width of the clay platelets dramatically increase the amount of interfacial contact between the clay and the polymer matrix. Thus the clay surface can mediate changes in matrix polymer conformation, crystal structure, and crystal morphology through interfacial mechanisms that are absent in classical polymer composite materials. For these reasons, it is believed that nanocomposite materials with the clay platelets dispersed as isolated, exfoliated platelets are optimal for end-use properties. Yet, relative to other thermoplastic nanocomposites, such as nylon 6, the improvement in end-use properties for polypropylene nanocomposites has been modest. Thus research in the areas of synthesis and, especially, compounding, which are aimed at closing this performance gap, is necessary. Alternatively, improved fundamental understanding of the detailed interactions and chemistry between clays, amine surfactants, and maleic anhydride compatibilizers can help elucidate the complex thermodynamics of clay dispersion. In addition, noting that many synthesized PP nanocomposites are likely to exist as nonequilibrium structures, research into the aging and rejuvenation of these mesoscale structures is warranted. Furthermore, better methods to characterize the full distribution and hierarchy of structural states present in PP nanocomposites are required because, for example, rare aggregates can seriously compromise nonlinear mechanical properties such as toughness, yield stress, and elongation at break. Finally, the interaction between clay platelets and polymer crystallization requires further attention because these interactions are likely a significant determinant of the end-use properties of polypropylene nanocomposites.

16. References

Akelah, A. In: Prasad, P. E., Mark, N.J. and Fai, T.J. (1995). (eds), Polymers and Other Advanced Materials, pp. 625, *Plenum Press*, New York.

Akelah, A., Salahuddin, N., Hiltner, A., Baer, E. and Moet, A. (1994). Nanostruct. Mater., 4: 965.

Alexandre, M.; Dubois, P.(2000). *Mater. Sci., Eng. R: Reports* , 28, 1.

Anyadike.(2005). Nanotechnology in packaging. Retrieved on February 13, from Pira International at *http://pira.atalink.co.uk.packaging/130.html*

Bafna A, Beaucage G, Mirabella F, Mehta S.(2003). 3D hierarchical orientation in polymer-clay nanocomposite films. *Polymer* ;44:1103–15.

Balazs, A. C.; Emrick, T.; Russell, T. P.(2006). *Science*, 314, 1107.

Bccresearch.(2010).*wordpress.com/tag/nanotechnology* August 9.

Biswas M, Sinha Ray S. (2001).Recent progress in synthesis and evaluation of polymer-montmorillonite nanocomposites. *Adv Polym Sci .;*155:167–221.

Boeing, H.V.;(1997).Polyolefins, *Elsevier Reinhold,* New York.

Brody. (2003, December). "Nano, Nano" Food Packaging Technology. *Food Technology,* 52-54.

Buchholz, K. (2003).Nanocomposite debuts on GM vehicles. Automotive Engineering International Online. Accessed April 30. *URL: http://www.sae.org/automag/material/10-2001/index.htm*

Butschli. (2004, october). Nanotechnology in packaging. Retrieved on February 13, 2005 from Packaging World at *http://www.packworld.com/cds_print.html?rec_id=17883*

Carrado, K. A.; Langui, X.(1999). *Microporous Mesoporous Mater.,* 27, 87.

Carter L., Hendricks J.G., Bolley D.S., (1950).US 2,531,396; *[assigned to National Lead Co.].*

Caseri, W. Nalwa, H. S.,(2004) .(In Encyclopedia of Nanoscience and Nanotechnology; Ed.; *American Scientific Publishers: Stevenson Ranch,* CA ; Vol 6, pp 235-247. (b) Caseri, W. R. ,Kickelbick, G., (2006).*Mater. Sci. Technol.,* 22, 807. (c) Caseri, W.(2007). In Hybrid Materials. Synthesis, Characterization, and Applications; Ed.; *Wiley-VCH: Weinheim,* Germany ; Chapter 2.

Chen, B. (2004). Polymer-Clay nanocomposites: an overview with emphasis on interaction mechanisms. *British Ceramic Transactions:* Vol. 103, No. 6, pg 241.

Cho, J.W. and Paul, D.R.;(2001). Nylon 6 Nanocomposites by Melt Compounding, *Polymer,* 42(3): 1083–1094.

Fisher, H.; Gielgens. L.; Koster, T. (1998).Nanocomposites from Polymers and Layered Minerals; *TNO-TPD Report .*

Fornes T.D., Yoon P.J., Hunter D.L., Keskkula H., Paul D.R.,(2002). *Polymer ,* Vol.43, 5915–93.

Frontini , P. M. ; Pouzada, A. S. ; (2011). *eXPRESS Polymer Letters* Vol.5, No.8 , 661

Fukushima, Y. and Inagaki, S.; (1987). Synthesis of an Intercalated Compound of Montmorillonite and 6-polyamide, *J. Inclusion Phenom.,* 5(4): 473–482.

Giannelies, E.P. (1996). Polymer Layered Silicate Nanocomposites, *Adv. Mater.,* 8(1): 29–35.

Giannelies, E.P., Krishnamoorti, R. and Manias, E. (1999). Polymer-Silicate Nanocomposites: Model Systems for Confined Polymers and Polymer Brushes, *Adv. Polym. Sci.,* 138: 107–147.

Giannelis EP. (1996).Polymer layered silicate nanocomposites. *Adv Mater* ;8:29–35.

Giannelis, E. P. (1998). Krishnamoorti, R. K.; Manias, E. Adv. *Polym. Sci.,* 138, 107-148.

Giannelis, E.P. (1998). *Appl. Organomet. Chem.,* 3(5): 490.

Gilman, J.W.; Jackson, C.L.; Morgan, A.B.; Harris, R.; Manias, E.; Giannelis, E.P.; Wuthenow, M.; Hilton, D.; Phillips, S.H.(2000). Flammability properties of polymer–Layered-silicate nanocomposites. Polypropylene and polystyrene nanocomposites. *Chem. Mater.,* 12 (7), 1866–1873.

Hajji, P.; David, L.; Gerard, J. F.; Pascault, J. P.; Vigier, G.(1999). *J. Polym. Sci., Part B: Polym. Phys.,* 37, 3172.

Hasegawa, N., Kawasumi, M., Kato, M., Usuki, A. and Okada, A. (1998). Preparation and Mechanical Properties of Polypropylene-Clay Hybrids using a Maleic Anhydride-Modified Polypropylene Oligomer, *J. Appl. Polym. Sci.,* 67(1): 87–92.

Hasegawa, N., Okamoto, H., Kato, A. and Uauki, M. (2000). Preparation and Mechanical Properties of Polypropylene-clay Hybrids Based on Modified Polypropylene and Organophilic Clay, *J. Appl. Polym. Sci.*, 78(11): 1918-1922.

Hay, J. N. & Shaw, S. J. (2000). Nanocomposites – Properties and Applications. Abstracted from "A Review of Nanocomposites 2000". Retrieved on February 13, 2005 from Azom.com at *http://www.azom.com/details.asp?ArticleID=921*

Hijleh, M., Ramadin, Y. and Zihlif, A. (2000). *Inter. J. Polym. Mater.*, 40: 377-394. *http://specialchem4polymers.com/resources/latest/displaynews.aspx?id=1965*

Jeon, H. G.; Jung, H. T.; Lee, S. D.; Hudson, S.(1998). *Polymer Bulletin* , 41, 107.

Karger-Kocsis, J., (1995).Ed. Polypropylene: Structure, Blends and Composites, vol. 3; *Chapman and Hall*: London.

Karian, H. G., (1999).Ed. Handbook of Polypropylene and Polypropylene Composites; *Marcel Dekker*: New York,.

Kato, M., Usuki, A. and Okada, A. (1997). Synthesis of Polypropylene Oligomer – Clay Intercalation Compounds, *J. Appl. Polym. Sci.*, 66(9): 1781-1785.

Kawasumi, M., Hasegawa, N., Kato, M., Usuki, A. and Okada, A. (1997). Preparation and Mechanical Properties of Polypropylene-Clay Hybrids, *Macromol.*, 30(20): 6333-6338.

Kickelbick, G. (2007).In Hybrid Materials. Synthesis, Characterization, and Applications; Ed.; *Wiley-VCH: Weinheim*, Germany ; Chapter 1.

Kıvanç IŞIK(2006)." Layered Silicate / Polypropylene Nanocomposites" July,*Izmir*, Turkey

Kojima Y., Usuki A., Kawasumi M., Okada A., Kurauchi T., Kamigaito O.,(1993). "Synthesis of nylon 6-clay hybrid by montmorillonite intercalated with 3- caprolactam", *Journal of Polymer Science.*: Part A, Vol.31, p. 983.

Kojima, Y.; Usuki, A.; Kawasumi, M.; Okada, A.; Fukushima, Y.; Kurauchi, T.; Kamigaito, O. (1993).Mechanical properties of nylon 6-clay hybrid. *J. Mater. Res.*, 8 (5), 1185-1189.

Kojima, Y.; Usuki, A.; Kawasumi, M.; Okada, A.; Fukushima, Y.; Kurauchi, T. T.; Kamigaito. O. (1993). *J. Mater. Res.*, 8, 1179- 1185.

Komori, Y. and Kurada, K.;(2000). Layered silicate – Polymer Intercalation Compounds, In: Pinnavaia, T.J. and Beal, G.W. (eds), Polymer Layered Silicate Nanocomposites, *Wiley*, New York.

Kroschurtz, J. S., (1993). Ed.; For definitions and background on layered silicate and clay minerals, see: Kirk-Othmer *Encyclopedia of Chemical Technology*, 4th ed.; John Wiley and Sons: New York ; Vol. 6.

Kurokawa, Y., Yusuda, H. and Oya, A. (1997). *J. Matter. Sci. Lett.*, 1670.

Lan, T., Kaviratna, P.D. and Pinnavaia, T.J. (1996). Epoxy Self-Polymerization in Smectite Clays, *J. Phys. Chem. Solids,* 57(6-8): 1005-1010.

Lan, T.; Pinnavaia, T.(1994). *J. Chem. Mater.*, 6, 2216.

Le Pluart L., Duchet J., Sauterau H., Ge´rard J.F.,(2002). *Journal of Adhesives* ,Vol.78, pp.645-662.

Leaversuch, R.(2003). Nanocomposites Broaden Roles in Automotive, Barrier Packaging. Plastics Technology Online. Accessed April 30. *URL:* *http://www.plasticstechnology.com/articles/200110fa3.html*

LeBaron, P.C., Wang, Z. and Pinnavaia, T.J. (1999). Polymer-Layered Silicate Nanocomposites: An Overview, *Appl. Clay Sci.*, 15(1-2): 11-29.

Li, J., Chixing, Z., Wang, G., Yu, W., Tao, Y. and Liu, Q. (2004). Preparation and Linern Rheological behaviour of Polypropylene/MMT Nanocomposites, *Polymer Composites*, 24(3): 323–331.

Li, X.C., Kang, T., Cho, W.J., Lee, J.K. and Ha, Z.C.S. (2001). Brill Transition in Nylon 10 12 Investigated by Variable Temperature XRD and Real Time FT-IR, *Macromol. Rapid Commun.*, 21(15): 1040–1043.

Ma J.S., Qi Z.N., Hu Y.L.,(2001). "Synthesis and characterization of polypropylene/clay nanocomposites", *Journal of Applied Polymer Science*, Vol. 82, p.3611.

Ma, J., Zhang, S. and Qi, Z.N. (2001). Synthesis and Characterization of Elastomeric Polyurethane/Clay Nanocomposites, *J. Appl. Polym. Sci.*, 82(6): 1444–1448.

Mahanwar, P., Bose, S. and Raghu, H. (2006). *J. Thermoplast. Comp. Mater.*, 19: 491–506.

Manias E., Touny A., Wu L., Strawhecker K., Lu B., Chung T.C.,(2001). "Polypropylene/montmorillonite nanocomposites. Review of the synthetic routes and materials properties", *Chemical Materials*,Vol.10,p. 3516.

Manias, E.; Touny, A.; Wu, L.; Lu, B.; Strawhecker, K.; Gilman, J. W.; Chung, T. C. (2000).*Polym. Mater. Sci., Eng.*, 82, 282.

Marchant D., Krishnamurthy J.(2002). *Industrial Engineering Chemical Resources*, Vol.41, pp. 6402–6408.

Mathias LJ, Davis RD, Jarrett WL.(1999). Observation of a- and g-crystal forms and amorphous regions of nylon 6-clay nanocomposites using solid-state 15N nuclear magnetic resonance. *Macromolecules*;32:7958–60.

Moore D.M., Reynolds R.C.,(1997). X-Ray diffraction and the identification and analysis of clay minerals. *Oxford: Oxford University Press.*

Morgan AB, Gilman JW.(2003). Characterization of poly-layered silicate (clay) nanocomposites by transmission electron microscopy and X-ray diffraction: a comparative study. *J Appl Polym Sci .*;87:1329–38.

Okada, A., Kojima, Y., Kawasumi, M., Fukushima, Y., Kurauchi, T. and Kamigaito, O.(1993). *J. Mater. Res.*, 8: 1179.

Oya, A.; Kurokawa, Y.; Yasuda, H. J. Mater. Sci. 2000, 35, 1045- 1050.

Pinnavaia T.J., (Ed.),2000. Polymer–Clay Nanocomposite, *Wiley, London*, p. 151.

Principia Partners. (2004, December). Polymer Nanocomposites Create Exciting Opportunities in the Plastics Industry: Updated Study from Principia. Retrieved on February 13, (2005) from Special Chem at
http://specialchem4polymers.com/resources/latest/displaynews.aspx?id=1965

Ray, S.S. and Okamoto, M. (2003). Polymer/Layered Silicate Nanocomposites: A Review from Preparation to Processing, *Prog. Polym. Sci.*, 28(11): 1539–1641.

Reichert, P., Nitz, H., Klinke, S., Brandsch, R., Thomann, T. and Mulhaupt, R. (2000). Poly(propylene)/Organoclay Nanocomposite Formation: Influence of Compatibilizer Functionality and Organoclay Modification, *Macromol. Mater. Eng.*, 275(2): 8–17.

Ryan. (2003), January/February). Nanocomposites. *Polymer News*, Issue 8.

Rzaev, Z.M.O., Yilmazbayhan, A. and Alper, E. (2007). A one-step Preparation of Compatibilized Polypropylene-Nanocomposites by Reactive Extrusion Processing, *Adv. Polym. Tech.*, 26(1): 41–55.

Schadler, L. S.(2003). Nanocomposite Science and Technology; *Wiley- VCH: Weinheim*, Germany, Chapter 2. (b) Schadler, L. S.; Kumar, S. K.; Benicewicz, B. C.; Lewis, S. L.; Harton, S. E. (2007).*MRS Bull.*, 32, 335. (c) Schadler, L. S.; Brinson, L. C.; Sawyer, W. G.(2007). *JOM* , 59, 53.

Schaefer, D. W.; Justice, R. S.(2007). *Macromolecules* , 40, 8501.

Sherman, L.M. Nanocomposites: (1999). A little goes a long way. *Plast. Technol.*, 45 (6), 52–57.

Sherman, Lilli Manolis. (1999, June). Nanocomposites — A Little Goes a Long Way. Retrieved February 22, (2005) *from www.plasticstechnology.com/articles/articl_print1.cfm*

Strawhecker, K.; Manias, E.(2000). *Chem. Mater.*, 12, 2943-2949.

The Future of Automotive Plastics. (2003). *PR Newswire*. Accessed April 30.

Theng, B. K. G.(1974). Chemistry of clay-organic reactions; *Wiley: New York.*

Theng, B. K. G.(1979). Formation and properties of clay-polymer complexes; *Elsevier: Amsterdam.*

Tripathi, A. and Pillia, P. (1994). *J. Mater. Sci.: in Electronics*, 1: 143–147.

Tyan, H.L., Liu, Y.C. and Wei, K.H. (1999). Enhancement of Imidization of Poly(Amic Acid) through Forming Poly(Amic Acid)/Organoclay Nanocomposites, *Polymer*, 40(17): 4877–4876. URL: http://www.scprod.com/gm.html

Usuki, A., Kato, M., Okada, A. and Kurauchi, T. (1997). Synthesis of Polypropylene-Clay Hybrid, *J. Appl. Polym. Sci.*, 63(1): 137–138.

Usuki, A.; Kawasumi, M.; Kojima, Y.; Okada, A.; Kurauchi, T.; Kamigaito, O. (1993). Swelling behavior of montmorillonite cation exchanged for o-amino acids by e-caprolactam. *J. Mater. Sci.*, 8 (5), 1174– 1178.

Usuki, A.; Kojima, Y.; Kawasumi, M.; Okada, A.; Fukushima, Y.; Kurauchi, T.; Kamigaito, O.(1993). Synthesis of nylon 6–clay hybrid. *J. Mater. Res.* , 8 (5), 1179-1184.

Vaia RA, Jant KD, Kramer EJ, Giannelis EP.(1996). Microstructural evaluation of melt-intercalated polymer-organically modified layered silicate nanocomposites. *Chem Mater.*; 8: 2628–35.

Vaia RA, Price G, Ruth PN, Nguyen HT, Lichtenhan J.(1999). Polymer/layered silicate nanocomposites as high performance ablative materials. *Appl Clay Sci* .,15:67–92.

Vaia, R. A; Ishii, H.; Giannelis, E. P. (1993).*Chem. Mater.*, 5,1694-1696.

Vaia, R. A; Vasudevan, S.; Krawiec, W.; Scanlon, L. G.; Giannelis, E. P.(1995). *Adv. Mater.*, 7, 154.

Winey, K. I.; Vaia, R. A. (2007). *MRS Bull.* , 32, 314. (b) Krishnamoorti, R.; Vaia, R. A.(2007). *J. Polym. Sci.*, Part B: Polym. Phys., 45, 3252.

Wolf, D.; Fuchs, A.; Wagenknecht, U.; Kretzschmar, B.; Jehnichen, D.; Ha¨ ussler, L. (1999). *Proceedings of the Eurofiller 99*, Lyon- Villeurbanne; pp 6-9.

Wu, Z., Zhou, C. and Zhu, N. (2002). The Nucleating Effect of Montmorillonite on Crystallization of Nylon 1212/Montmorillonite Nanocomposite, Polym. Test., (4): 479–483.

Wu, Z.G., Zhou, C.X., Qi, R.R. and Zhang, H.B. (2002). Synthesis and Characterization of Nylon 1012/clay Nanocomposite, J. Appl. Polym. Sci., 83(11): 2403–2410.

Xie, X., Li, B. and Tjong, S. (2001). J. Appl. Polym. Sci., 80: 2105–2112.

Xu, R.; Manias, E.; Snyder, A. J.; Runt, (2001). J. Macromolecules,34, 337-339.

Zheng, L., Farris, R.J. and Coughlin, E.B. (2001). Novel Polyolefin Nanocomposites: Synthesis and Characterizations of Metallocene-Catalyzed Polyolefin Polyhedral Oligomeric Silsesquioxane Copolymers, Macromol., 34(23): 8034–8039.

Zihlif, A. and Ragosta, G. (2003). J. Thermoplast. Comp. Materials; 36: 273–283.

Zilg, C., Reichert, P., Dietsche, F., Engelhardt, T. and Mulhaupt, R. (1998). Kunststoffe, 88: 1812.

2

Organic Materials in Nanochemistry

Alireza Aslani

¹Nanobiotechnology Research Center,
Baqiyatallah University Medical of Science, Tehran,
²Department of Chemistry, Faculty of Basic Science,
Jundi Shapur University of Technology, Dizful,
Islamic Republic of Iran

1. Introduction

At the turn of twenty-first century, we entered nanoworld. These days, if our try to run a simple web search with the keyword "nano" more than thousands and thousands of references will come out: nanoparticles, nanowires, nanostructures, nanocomposite materials, nanoprobe microscopy, nanoelectronics, nanotechnology, nanochemistry, nanomaterials and so on. The list could be endless. When did this scientific nanorevolution actually happen? Perhaps, it was in the mid-1980s, when scanning tunneling microscopy (STM) was invented. Specialists in scanning electron microscopy (SEM) may strongly object to this fact by claiming decades of experience in observing features with nearly atomic resolution and later advances in electron-beam lithography. We should not omit molecular beam epitaxy, the revolutionary technology of the 1980s, which allows producing layered structures with the thickness of each layer in the nanometer range. Colloid chemists would listen to that with a wry smile, and say that in the 1960s and 1970s, they made Langmuir-Blodgett (LB) films with extremely high periodicity in nanometer scale. From this point of view, the nanorevolution was originated from the works of Irving Langmuir and Katherine Blodgett in 1930s. What is the point of such imaginary arguments? All parties were right. We cannot imagine modern nanotechnology without any of the abovementioned contributions. The fact is that we are in the nanoworld now, and the words with prefix "nano" suddenly have become everyday reality. Perhaps it is not that important how it happened. Hence nanotechnology is a very promising field for industrial applications. In fact, several products are already on the market for certain niche sectors with high added value, e.g., biomedical materials and analytic devices. The real revolution in nanomaterial applications, however, is expected to involve widely used bulk products. Polymers like polyolefins and polyvinylchloride (PVC), for example, are good candidates in this respect because of their large-scale use and versatility. Indeed, one of the first applications of nanotechnology was the production of nanofillers for the improvement of the mechanical properties of polymers. Polypropylene (PP) is particularly interesting because of its low cost and good mechanical properties. This polymer has been used in conventional composites for a long time and, in combination with nanofillers, shows better mechanical properties with even low amounts of filler. The main nanofillers used today are nanoclay (natural product) and other nanomaterials (synthetic). Synthetic carbon nanotubes are very expensive. Nanoclays (layered silicates), in contrast, are especially interesting for bulk applications because they are relatively inexpensive and they cause an improvement in the mechanical properties of polymers. Commonly used nanoclays include montmorillonite, hectorite, and saponite, all of which belong to the same general family of 2:1 layered or phyllosilicates. As a result of the material reduction, the environmental impact of PP nanocomposite products can be expected to be lower than that of products made out of conventional material unless the production of the nanoparticles is accompanied by particularly high environmental impacts. Nanocomposites are as multiphase materials, where one of the phases has nanoscale additives. They are expected to display unusual properties emerging from the combination of each component. According to their matrix materials, nanocomposites can be classified as ceramic matrix nanocomposites (CMNC), metal matrix nanocomposites (MMNC), and polymer matrix nanocomposites (PMNC). Polymers are now the most widely used in the field of technical textiles. The widespread use of common organic polymers such as polyolefins, polyesters and polyurethanes emanates from key features such as lightweight, easy fabrication, exceptional processability, durability and relatively low cost. A major challenge in polymer science is to broaden the application window of such materials by retaining the above features while enhancing particular characteristics such as modulus, strength, fire performance and heat resistance. However, polymers have relatively poor mechanical, thermal, and electrical properties as compared to metals and ceramics. Many types of polymers such as homopolymers, co-polymers, blended polymers and modified polymers are not sufficient enough to compensate various properties, which we have demanded. Alternative approaches to improve their properties are to reinforce polymers with inclusion of fiber, whisker, platelets or particles. The choice of the polymers is usually guided mainly by their mechanical, thermal, electrical, optical and magnetic behaviors. However, other properties such as hydrophobic-hydrophilic balance, chemical stability, bio-compatibility, opto-electronic properties and chemical functionalities have to be considered in the choice of the polymers. The polymers in many cases can also allow easier shaping and better processing of the composite materials. The inorganic particles not only provide mechanical and thermal stability, but also new functionalities that depend on the chemical nature, the structure, the size, and crystallinity of the inorganic nanoparticles (silica, transition metal oxides, metallic phosphates, nanoclays, nanometals and metal chalcogenides). Indeed, the inorganic particles can implement or improve mechanical, thermal, electronic, magnetic and redox properties, density, refractive index. Organic polymer-based inorganic nanoparticle composites have attracted increasing attention because of their unique properties emerging from the combination of organic and inorganic hybrid materials. The composites have been widely used in the various fields such as military equipments, safety, protective garments, automotive, aerospace, electronics, stabilizer and optical devices. However, these application areas continuously demand additional properties and functions such as high mechanical properties, flame retardation, chemical resistance, UV resistance, electrical conductivity, environmental stability, water repellency, magnetic field resistance and radar absorption. Moreover, the effective properties of the composites are dependent upon the properties of constituents, the volume fraction of components, shape and arrangement of inclusions and interfacial interaction between matrix and inclusion. With the recent development in the nanoscience and nanotechnology fields, the correlation of material properties with filler size has become a focal point of significant interest. On the other hand Polypropylene (PP) is one of the fastest growing commercial thermoplastics due to its attractive combination of low density and high heat distortion temperature. There are some limitations in physico-chemical properties that restrict PP applications. A typical illustration is in packaging, where PP has poor oxygen gas barrier resistance. No single polymer has shown the ideal combination of performance features. PP possesses good water vapor barrier properties, but it is easily permeated by oxygen, carbon dioxide, and hydrocarbons. The necessity of developing more effective barrier polymers has given rise to different strategies to incorporate and optimize the features from several components. Most schemes to improve PP gas barrier properties involve either addition of higher barrier plastics via a multilayer structure (co-extrusion) or by introducing filler with high aspect ratio in the polymer matrix. Co-extrusion allows tailoring of film properties through the use of different materials where each material component maintains its own set of properties, compared with blending of polymers in a mono-extrusion technique. Co-extrusion is used to generate multilayer laminate structures from separately extruded polymer films that are sandwiched together. Resulting films may comprise many layers, such as the PP-adhesive poly (ethylene-co-vinyl alcohol) (EVOH)-adhesive PP system: EVOH barrier sheet trapped between two layers of moisture resistant PP and two additional adhesive strata. However, by nature co-extrusion is a complex and expensive process. Alternatively, nano fillers with high aspect ratio can be loaded into the polymer matrix. Polymer nanocomposites are a better choice with significant property increments from some materials. Nanocomposite materials are one of the methods for improving gas barrier properties of polyolefin. Recent developments in polymer nanocomposites have attracted attention due to the possibilities offered by this technology to enhance the barrier properties of inexpensive commodity polymers. Many studies have demonstrated improvements in permeability reduction to gases, moisture and organic vapors resulting from the addition of low concentrations of layered some nanoparticles to various thermoplastic matrices. This is mainly due to their nanometer scale particle size and intraparticle distances. The desired properties are usually reached at low filler volume fraction, allowing the nanocomposites to retain macroscopic dispersion and low density of the polymer. The geometrical shape of the particle plays an important role in determining the properties of the composites. The improved nanocomposite barrier behavior illustrated by many examples has been explained by the tortuous path model, in which the presence of impermeable some platelets generates an overlapped structure that hinders penetrate diffusion and thus decreases the permeability of the material.

2. Nanotechnology

Nanotechnology is receiving a lot of attention of late across the globe. The term nano originates etymologically from the Greek, and it means "dwarf." The term indicates physical dimensions that are in the range of one-billionth (10^{-9} or $\frac{1}{10^9}$) of a meter. This scale is called colloquially nanometer scale, or also nanoscale. One nanometer is approximately the length of two hydrogen atoms. Nanotechnology relates to the design, creation, and utilization of materials whose constituent structures exist at the nanoscale; these constituent structures can, by convention, be up to 100 nm in size. Nanotechnology is a growing field that explores electrical, optical, and magnetic activity as well as structural behavior at the molecular and sub-molecular level. These questions should be answered: What is nanotechnology? What are the applications of nanotechnology? What is the market potential for nanotechnology? What are the global research activities in nanotechnology? Why would a practitioner, need to care?

Research and technology development at the atomic, molecular, or macromolecular levels, in the length scale of approximately 1 to 100 nm range called nanotechnology. Creating and using structures, devices, and systems that have novel properties and functions because of their small and/or intermediate size are application of nanomaterials. Hence, nanotechnology can be defined as the ability to work at the molecular level, atom by atom, to create large structures with fundamentally new properties and functions. Nanotechnology can be described as the precision-creation and precision-manipulation of atomic-scale matter; hence, it is also referred to as precision molecular engineering.

Nanotechnology is the application of nanoscience to control processes on the nanometer scale that is, between 1 to 100 nm or call better 2 to 50 nm. The field is also known as molecular engineering or molecular nanotechnology (MNT). MNT deals with the control of the structure of matter based on atom-by-atom and/or molecule-by-molecule engineering; also, it deals with the products and processes of molecular manufacturing. The term engineered nanoparticles describes particles that do not occur naturally; humans have been putting together different materials throughout time, and now with nanotechnology they are doing so at the nanoscale. As it might be inferred, nanotechnology is highly interdisciplinary as a field, and it requires knowledge drawn from a variety of scientific and engineering arenas: Designing at the nanoscale is working in a world where physics, chemistry, electrical engineering, mechanical engineering, and even biology become unified into an integrated field. "Building blocks" for nanomaterials include carbon-based components and organics, semiconductors, metals, and metal oxides; nanomaterials are the infrastructure, or building blocks, for nanotechnology. The term nanotechnology was introduced by Nori Taniguchi in 1974 at the Tokyo International Conference on Production Engineering. He used the word to describe ultrafine machining: the processing of a material to nanoscale precision. This work was focused on studying the mechanisms of machining hard and brittle materials such as quartz crystals, silicon, and alumina ceramics by ultrasonic machining. Years earlier, in a lecture at the annual meeting of the American Physical Society in 1959 (There's Plenty of Room at the Bottom) American Physicist and Nobel Laureate Richard Feynman argued (although he did not coin or use the word nanotechnology) that the scanning electron microscope could be improved in resolution and stability, so that one would be able to "see" atoms. Feynman proceeded to predict the ability to arrange atoms the way a researcher would want them, within the bounds of chemical stability, in order to build tiny structures that in turn would lead to molecular or atomic synthesis of materials. Based on Feynman's idea, K. E. Drexler advanced the idea of "molecular nanotechnology" in 1986 in the book Engines of Creation, where he postulated the concept of using nanoscale molecular structures to act in a machinelike manner to guide and activate the synthesis of larger molecules. Drexler proposed the use of a large number (billions) of robotic-like machines called "assemblers" (or nanobots) that would form the basis of a molecular manufacturing technology capable of building literally anything atom by atom and molecule by molecule. At this time, an engineering discipline has already grown out of the pure and applied science; however, nanoscience still remains somewhat of a maturing field. Nanotechnology can be identified precisely with the concept of "molecular manufacturing" (molecular nanotechnology) introduced above or with a broader definition that also includes laterally related sub-disciplines. The nanoscale is where physical and biological systems approach a comparable dimensional scale. A basic "difference" between systems biology and nanotechnology is the goal of the science: systems biology aims to uncover the fundamental operation of the cell in an effort to predict the exact response to specific stimuli and genetic variations (has scientific discovery focus); nanotechnology, on the other hand, does not attempt to be so precise but is chiefly concerned with useful design.

A nanometer is about the width of four silicon atoms (with a radius of 0.13 nm) or two hydrogen atoms (radius of 0.21 nm); .For comparison purposes, the core of a single-mode fiber is 10.000 nm in diameter, and a 10 nm nanowire is 1000 times smaller than (the core of) a fiber. The nanoscale exists at a boundary between the "classical world" and the "quantum mechanical world"; therefore, realization of nanotechnology promises to afford revolutionary new capabilities.

The nanoparticles are ultrafine particles in the size of nanometer order. "Nano" is a prefix denoting the minus 9th power of ten, namely one billionth. Here it means nanometer (nm) applied for the length. One nm is extremely small length corresponding to one billionth of 1 m, one millionth of 1 mm, or one thousandth of 1 μm. The definition of nanoparticles differs depending upon the materials, fields and applications concerned. In the narrower sense, they are regarded as the particles smaller than 10 to 20 nm, where the physical properties of solid materials themselves would drastically change. On the other hand, the particles in the three digit range of nanometer from 1 nm to 1μm could be called as nanoparticles. In many cases, the particles from 1 to 100 nm are generally called as nanoparticles, but here they will be regarded as the particles smaller than those called conventionally "submicron particles", and concretely less than the wavelength of visible light (its lower limit is about 400 nm) as a measure, which need to be treated differently from the submicron particles.

3. Features of nanoparticles: "Activation of particle surface"

All the solid particles consist of the atoms or the molecules. As they are micronized, they tend to be affected by the behavior of atoms or the molecules themselves and to show different properties from those of the bulk solid of the same material. It is attributable to the change of the bonding state of the atoms or the molecules constructing the particles. The diameter of the smallest hydrogen atom is 0.074 nm, and that of the relatively large lead atom (atomic number is 82) is 0.35 nm. From these sizes, it is estimated that the particle with a size of 2 nm consists of only several tens to thousands atoms. When the particle is constructed by larger molecules, the number decreases furthermore [M. Arakawa, 2005a and 1983b]. It is indicated that the fraction of surface atoms of a 20 μm cubic particle is only 0.006%, but it increases to 0.6% for a 200 nm particle and then it is estimated almost half of the atoms are situated at the surface of a 2 nm particle. On the other hand, as the micronization of solid particles, the specific surface area increases generally in reversal proportion to the particle size. In the above-mentioned case, when the particle of 1cm is micronized to 1 μm and 10 nm, the specific surface area becomes ten thousand times and million times, respectively. As the increase in the specific surface area directly influences such properties like the solution and reaction rates of the particles, it is one of major reasons for the unique properties of the nanoparticles different from the bulk material together with the change in the surface properties of the particles itself.

4. Evaluation of size of nanoparticles

In order to elucidate the change in properties and characteristics of nanoparticles with the particle size, it is essential first of all to measure the size of the nanoparticles accurately. The most basic method to measure the size of nanoparticles is the size analysis from the picture image using the transmission electron microscope, which could also give the particle size distribution. For this analysis, preparation of the well-dispersed particles on the sample mount is the key issue. The grain size of the particles can be obtained from peak width at half height in the X-ray diffraction analysis and it is regarded as an average primary particle size of particles. Meanwhile, the laser diffraction and scattering method, which is popular for the size analysis of micron-sized particles, would hardly measure the particle size of individual nanoparticles but that of the agglomerated particles. The photon correlation method often used for the particle analysis in the nanosized range might not give accurate results in many cases, when the particle size distribution is wide. Then the BET (Brunauer-Emmett-Teller) specific surface measurement based on the gas adsorption is often applied as a simple method to evaluate the size of nanosized primary particles. By this method, it is possible to estimate the particle size from the specific surface area under the assumption of spherical particle shape. This equivalent particle size based on the specific surface area is useful for the evaluation of nanoparticle size, though it may differ from the particle size observed by the electron microscope depending upon the surface state and the inner structure of the particles.

5. Properties of nanoparticles and size effect

As mentioned above, with the decreasing particle size, the solid particles generally tend to show different properties from the bulk material and even the physical properties like melting point and dielectric constant themselves which have been considered as specific properties may change, when the particles become in several nanometer size. These changes in the fundamental properties with the particle size are called "size effect" in a narrower sense. On the contrary, in a broader sense, it could also include the change in the various characteristics and behaviors of particles and powders with the particle size. The nanoparticles have various unique features in the morphological/structural properties, thermal properties, electromagnetic properties, optical properties, mechanical properties as described briefly in the following:

5.1 Morphological and structural properties

The ultrafine size of the nanoparticles itself is one of useful functions. For example, the finer particles are apt to be absorbed more easily through the biological membrane. It is known as the enhanced permeation and retention (EPR) effect [H. Maeda, 1992] that the particles having a particle size from about 50 to 100 nm, which would not be transferred to the normal cells through the vascular wall could be delivered selectively to a certain affected cells because of the enlarged cell gap of this part. As mentioned above, the large specific surface area of the nanoparticles is an important property to the reactivity, solubility, sintering performance etc. related with the mass and heat transfer between the particles and their surroundings from the morphological viewpoint apart from the control of the surface and inner structures of the nanoparticles. Furthermore, the crystal structure of the particles may change with the particle size in the nanosized range in some cases. Uchino et al. [K. Uchino, 1982] reported that from the X-ray diffraction analysis of the lattice constant of $BaTiO_3$ powder prepared by hydrothermal synthesis method, the c/a axis length ratio showing the tetragonal characteristics decreased to indicate the increasing symmetric property with the decreasing particle size from about 200 nm as shown in fig 1. This is considered to be attributable to the compressive force exerted on the particles as a result of the surface tension of the particle itself. For PbTiO₃, it is reported that the tetragonal crystals decreased and the cubical crystal increased in the particles from the particle size of about 18 nm [H. Suzuki, 2002]. In this way, the critical particle size for the crystal structure and the size effect differ with the materials concerned.

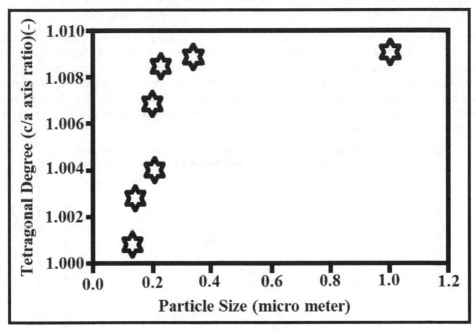

Fig. 1. Relationship between particle size and tetragonal degree (c/a axis ratio) of BaTiO₃ powder.

5.2 Thermal properties

As the atoms and molecules located at the particle surface become influential in the nanometer order, the melting point of the material decreases from that of the bulk material because they tend to be able to move easier at the lower temperature. For example, the melting point of gold is 1336 K as a bulk but starts to decrease remarkably below the particle size of about 20 nm and drastically below 10 nm and then becomes more than 500 degrees lower than that of the gold bulk around 2 nm. The reduction of the melting point of ultrafine particles is regarded as one of the unique features of the nanoparticles related with aggregation and grain growth of the nanoparticles or improvement of sintering performance of ceramic materials [N. Wada, 1984].

5.3 Electromagnetic properties

The nanoparticles are used as the raw material for a number of electronic devices. The electric properties and particle size of these nanoparticles play a great role for the improvement of the product performance [I. Matsui, 2005]. As an example, there is a strong demand for the materials with a high dielectric constant to develop small and thin electronic devices. For this purpose, it has been confirmed by the X-ray diffraction analysis for instance that the dielectric constant of $PbTiO_3$ tends to increase considerably as the particles become smaller than about 20 nm. Meanwhile, it is also known that when the dielectric constant is measured with a pellet prepared by pressing these nanoparticles, it shows a peak with the raw material around 100nm and decreases with the decreasing particle size, which is attributable to the influence of the grain boundary and void in the pellet [M. Takashige, 1981].

On the other hand, the minimum particles size to keep the ferroelectric property (critical size) differs depending upon the kind and composition of the materials. Summarizing the data of various kinds of materials, it varies from 7 nm for $PbTiO_3$ to 317 nm for Ba–Pb–Ti compounds. The Curie point defined as the point changing from the ferroelectric material to the paraelectric phase of $PbTiO_3$ reduces drastically with the decreasing particle size below 20–30 nm as shown in fig 2. As for the Curie point, some equations have been proposed for its estimation [K. Ishikawa, 2001].

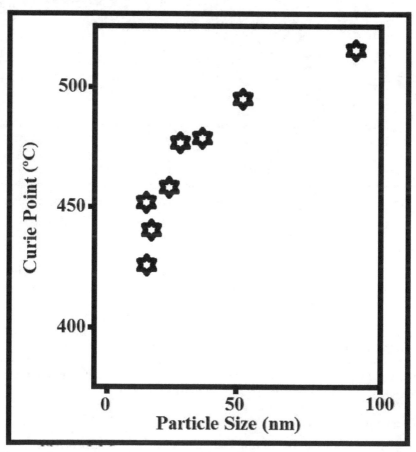

Fig. 2. Change of Curie point of $PbTiO_3$ with its particle size.

As for the magnetic property, ferromagnetic fine particles have a single magnetic domain structure as they become very small as in the order less than about 1 μm and show super-paramagnetic property, when they get further finer. In this case, although the individual particles are ferromagnetic with the single magnetic domain structure, the particles collectively behave as a paramagnetic. It is magnetized as a whole in the same direction of the external magnetic field but the magnetization disappears by the thermal fluctuation, when the external magnetic field is taken away. The time for disappearing of magnetization depends upon the particle size, namely the magnetization of the material responds promptly with the external magnetic field as a paramagnetic when the particles are small enough but it decreases gradually as the particle size becomes larger. As a result of such change in the electromagnetic properties of nanoparticles, it is known for instance that the gold which is a stable substance as a bulk shows unique catalytic characteristics as nanoparticles [K. Ishikawa, 1998 and M. Haruta, 1994].

5.4 Optical properties

As the size of particles becomes in the several nanometers range, they absorb the light with a specific wavelength as the plasmon absorption [Y. Kurokawa, 1996] caused by the plasma oscillation of the electrons and the transmitted light with different color depending upon the kind of metal and particle size is obtained [K. Kobayashi, 2004 and S. Sato, 1996].

In case of gold nanoparticles, it is reported that the maximum light absorption wavelength is 525 nm for the particles of 15nm but it is enlarged by about 50 nm for 45 nm particles. In this way, these gold and silver nanoparticles show the color phenomena with splendid tinting strength, color saturation and transparency compared with the conventional pigments for the paint in the submicron size and the tinting strength per unit volume of silver nanoparticles becomes about 100 times higher than that of organic pigments. Furthermore, since the nanoparticles are smaller than the wavelength of visible light and the light scattering by the particles becomes negligible, higher transparency can be obtained with the nanoparticles than the conventional pigment.

5.5 Mechanical properties

It is known that the hardness of the crystalline materials generally increases with the decreasing crystalline size and that the mechanical strength of the materials considerably increases by micronizing the structure of the metal and ceramic material or composing them in the nano range [K. Niihara, 1991 and T. Sekino, 2000]. Furthermore, with the ceramic material having crystalline size less than several hundred nanometers, the unique super-plastic phenomenon is seen that it is extended several to several thousand times from the original size at the elevated temperature over 50 % of the melting point [F. Wakai, 1990], which may provide the possibility of forming and processing of ceramics like metallic materials.

6. Existing conditions of particles and their properties

The nanoparticles usually exhibit collective functions. Therefore, the dispersing state and the surrounding conditions in addition to the physical properties of the particles themselves are important. In many cases, the nanoparticles exist as aggregates of the primary particles by the adhesion and bonding during the production process because of their high adhesiveness. The existing state of the nanoparticles is greatly influenced by the surrounding conditions if they are in gas, liquid, solid or in a vacuum and what sort of interaction they have with the surrounding materials. The nanoparticles are rarely used by themselves but dispersed in other materials or combined with them. The dispersing process of the nanoparticles is a key for the nanoparticle technology as well as their preparation methods, since the performance of the final products are affected by their dispersing conditions [T. Yokoyama, 2005]. In this way, it is expected with great possibility to develop various new materials and applications by the nanoparticle technology producing and processing the nanoparticles, which have different properties from the bulk material by the size effects as mentioned above and in the following sections.

7. Wet technologies for the formation of organic nanostructures

Chemical methods of material processing were known for years, existing in parallel with physical and other methods of film deposition. Recent advances in electron microscopy and scanning nanoprobe microscopy (STM, AFM) have revealed that some of the materials produced by the chemical methods have distinctive nanocrystalline structure. Furthermore, due to the achievements of colloid chemistry in the last 20 years, a large variety of colloid nanoparticles have become available for film deposition. This has stimulated great interest in further development of chemical methods as cost-effective alternatives to such physical methods as: thermal evaporation; magnetron sputtering; chemical and physical vapor deposition (CVD, PVD); and molecular beam epitaxy (MBE).

7.1 Formation of colloid nanoparticles

The most advanced chemical method for nano-structured materials processing is the deposition of colloid inorganic particles. Recent achievements in colloid chemistry have made a large variety of colloid compounds commercially available. The list of colloid nanoparticles with uniform (low-dispersed) dimensions in the range from 3 to 50 nm includes the noble metals (e.g., Au, Ag, Pt, Pd, and Cu), semiconductors (e.g., Si, Ge, III-V and II-VI, and metal oxides), insulators (e.g., mica, silica, different ceramic materials, polymers), and magnetic materials (e.g, Fe_2O_3, Ni, Co, and Fe). The growth of colloid particles is usually stabilized during synthesis by adding surfactants to the reagents [Edelstein. A. S. 1996]. Therefore, the stable nanoparticles produced are coated with a thin shell of functionalized hydrocarbons, or some other compounds. Typical examples of the chemistry of formation of colloid nanoparticles are shown below. Gold stable colloids can be prepared by the reduction of $AuCl_4$ with sodium borohydride in the presence of alkane-thiols [Brust. M, 1994]. Other colloids, such as Ag, CdS, CdSe, and ZnS, can be prepared in a similar way. InP nanocrystals can be synthesized by the following reaction, with temperatures ranging from 150 °C to 280 °C in the presence of either primary amines, tri-n-octylphosphine (TOP), or tri-n-octylphosphine oxide (TOPO) as stabilizing agents, preventing further InP aggregation [Talapin. D. V, 2002].

$$InCl_3 + [(CH_3)_3Si]_3P \rightarrow InP + 3(CH_3)_3SiCl$$

The particles appear to be mono-dispersed, with a mean cluster size varying from 2 to 6 nm depending on the stabilizer used. The particles show strong resonance luminescence after etching in HF. Cobalt mono-dispersed nanocrystals can be produced by rapid pyrolysis of the organic precursor $Co(CO)_8$ in an inert Ar-atmosphere, and in the presence of organic surfactants, such as oleic-acid and trioctylphosphonic acid at high temperatures [Puntes. V. F., 2001]. The particles appear to have ideal spherical, cubical, or rod-like shapes, with sizes in the range from 3 to 17 nm depending on surfactant concentration. The Co nanoparticles demonstrate superparamagnetic ferromagnetic transition. CdTe nanoparticle colloids can be prepared by the reaction of Na_2Te with CdI_2 in methanol at $-78\ °C$.

The diameter of CdTe colloid particles is in the range from 2.2 to 2.5 nm [Schultz. D. L, 1996]. An alternative method for the formation of stabilized colloid particles is to utilize self-assembled membranes, such as micelles, microemulsions, liposomes, and vesicles. Typical dimensions are from 3 to 6 nm for reverse micelles in aqueous solutions, from 5 to 100 nm for emulsions, and from 100 to 800 nm for vesicles. Liposomes are similar to vesicles, but they have bilayer membranes made of phospholipids. Such membranes may act as the reaction cage during the formation of nanoparticles, and may prevent their further aggregation. The idea of the formation of nanoparticles inside micelles is to trap respective cations there. This can be done by sonification of the mixture of required salts and surfactants. Since the permeability of the membrane for anions is about 100 times higher than for cations, the formation of nanoparticles takes place within micelles, with a constant supply of anions from outside. A number of different colloids, such as CdSe, Ag_2O, Fe_2O_3, Al_2O_3, and cobalt ferrite, were prepared using the above methods [Bhandarkar. S, 1990, Mann. S, 1983, Mann. S. J. P, 1986, Cortan. A. R, 1990, Yaacob. I. I, 1993, and Li. S, 2001].

7.2 Self-assembly of colloid nanoparticles

The deposition of colloid nanoparticles onto solid substrates can be accomplished by different methods, such as simple casting, electrostatic deposition, Langmuir-Blodgett, or spin coating techniques. However, the simplest method of nanoparticles deposition, which gives some remarkable results, is the so-called self-assembly or chemical self-assembly method. This method, which was first introduced by Netzer and Sagiv, is based upon strong covalent bonding of the adsorbed objects (i.e., monomer or polymer molecules and nanoparticles) to the substrate via special functional groups. It is known, for example, that the compounds containing thiol (SH) or amine (NH2) groups have strong affinity to gold. The silane group (SiH3) with silicon is another pair having very strong affinity.

The first work on the self-assembly of gold colloid particles capped with alkanethiols was done by Brust and coworkers [Netzer, L, 1983]. This routine has been adopted by other scientists for the deposition of self-assembled monolayers of different colloid nanoparticles (e.g., Ag, CdS, CdSe, and ZnS), which were prepared using mercapto-alcohols, mercaptocarboxylic acids, and thiophenols as capping agents. Self-assembled nanoparticles usually show well-ordered lateral structures, proved by numerous observations with SEM, STM, and AFM [Collier. C, 1997, Lover. T, 1997, Rogach. A. L, 1999 and Vogel. W, 2000].

Two-dimensional ordering in self-assembled nanoparticle monolayers can be substantially improved by thermal annealing at temperatures ranging from 100 °C to 200 °C, depending on the material used. The use of bi-functional $HS\text{-}(CH_2)_{10}\text{-}COOH$ bridging molecules, which combines both the affinity of thiol groups to gold and carboxylic group to titania, can provide more flexibility in the self-assembly. Both self-assembly routes were exploited for deposition of TiO_2 nanoparticles onto the gold surface [Rizza. R, 1997]. In the first one, unmodified TiO_2 nanoparticles were self-assembled onto the gold surface, coated with a monolayer of $HS-(CH_2)_{10}-COOH$; while in the second one, TiO_2 nanoparticles stabilized with $HS-(CH_2)_{10}-COOH$ were self-assembled onto the bare gold surface. For some time, chemical self-assembly was limited to the formation of organized monolayers. The use of bi-functional bridge molecules overcomes this relative disadvantage. For example, multi-layers of Au colloid particles can be deposited using di-thiol spacing layers. A similar routine was applied for the fabrication of Au/CdS super lattices [Sarathy. K. V, 1999 and Nakanishi. T, 1998].

7.3 Morphology and crystallography of nanostructured materials prepared by chemical routes

The structural study of materials was always of a high priority, because the physical properties of materials depend very much on their structure. There are several levels of structural study, which start with the investigation of the morphology of the material surfaces, closely related to their in-plane ordering. Many nano-structured materials prepared with the help of layer-by-layer deposition techniques, such as LB or electrostatic self-assembly, have a distinctive periodicity in the direction normal to the surface, which determines their main electrical and optical properties. This is why the study of the layer-by-layer structure of such materials is of crucial importance.

The materials consisting of colloid nanoparticles have a tendency to form two dimensional structures according to the close packing order. This trend can stimulate the formation of multilayered quasi-3D structures of closely packed nanoparticles. The final stage of structural study is the crystallography of individual nanoparticles, clusters, and grains of materials. This is a very interesting and important subject, since the crystallography of nanoclusters, which consist of several hundred to several thousand atoms, is very often different from that of their respective bulk materials.

The planar order of nanostructures deposited by chemical routes has become an important issue, because of the competition with solid-state nanotechnology capable of the fabrication of fine two-dimensional structures. The main concern is with the layers of nanoparticles produced by chemical self-assembly, because methods of electrostatic self-assembly and LB is not capable of producing two-dimensional ordered arrays of nanoparticles. The features of the lateral arrangement of particles, which are buried under layers of either closely packed amphiphilic compounds or polymers, are usually smeared and difficult to observe. In the case of relatively thick (quasi-3D) films, produced by electrodeposition and sol-gel techniques, the morphology study usually reveals polycrystallites. Therefore, the quality of these materials can be assessed by the size of the crystallites and by the presence of preferential orientation, which may cause anisotropy of the electrical and optical properties of materials.

7.4 Morphology and crystallography of chemically self-assembled nanoparticles

In contrast to the previous deposition techniques (i.e., LB and electrostatic self-assembly), nanoparticles, which are chemically self-assembled onto the solid substrates, tend to form regular two-dimensional structures, especially after annealing at moderate temperatures.

The type of two-dimensional structure, which usually follows the trend of close packing arrangement, depends on the particles' shapes. For example, a simple hexagonal pattern is formed by spherical nanoparticles. A classic example of such structures is gold colloid particles chemically self-assembled onto the surface of gold via thiol groups. The observation of such structures is possible with scanning nanoprobe microscopy, such as STM and AFM, as well as with TEM and high resolution SEM [49-73][Roberts.G. G, 1983[a], 1990[b] and 1985[c], Ross. J, 1986, Lvov. Y. M, 1987[a], 1989[b] and 1994[c], 2000[d], Petty. M. C, 1990[a] and 1995[b], Ulman. A, 1991[a] and 1995[b], Wegner. G, 1993, Yarwood. J, 1993, Tredgold. R. H, 1994, Tsukruk. V. V, 1997, Bliznyuk. V. N, 1998, Bonnell. D. A, 2001, Stefanis. A, 2001, Gabriel. B. L, 1985, Keyse. R. J, 1998, Greffet. J. J, 1997, Kitaigorodski. A. L, 1961, Nabok.A. V, 1998[a] and 2003[b]].

8. Elemental and chemical composition of organic/inorganic nanostructures

8.1 Stabilization of colloidal metal particles in liquids

Before beginning a description of synthetic methods, a general and crucial aspect of colloid chemistry should be considered, and that is the means by which the metal particles are stabilized in the dispersing medium, since small metal particles are unstable with respect to agglomeration to the bulk. At short inter-particle distances, two particles would be attracted to each other by van-der-Waals forces and in the absence of repulsive forces to counteract this attraction an unprotected sol would coagulate. This counteraction can be achieved by two methods, electrostatic stabilization and steric stabilization. In classical gold sols, for example, prepared by the reduction of aqueous $[AuCl_4]^-$ by sodium citrate, the colloidal gold particles are surrounded by an electrical double layer formed by adsorbed citrate and chloride ions and cations which are attracted to them. This results in a Columbic repulsion between particles. The weak minimum in potential energy at moderate interparticle distance defines a stable arrangement of colloidal particles which is easily disrupted by medium effects and, at normal temperatures, by the thermal motion of the particles.

Thus, if the electric potential associated with the double layer is sufficiently high, electrostatic repulsion will prevent particle agglomeration, but an electrostatically stabilized sol can be coagulated if the ionic strength of the dispersing medium is increased sufficiently. If the surface charge is reduced by the displacement of adsorbed anions by a more strongly binding neutral adsorbate, the colloidal particles can now collide and agglomerate under the influence of the van-der-Waals attractive forces [J. S. Bradley, 1993]. Even in organic media, in which electrostatic effects might not normally be considered to be important, the development of charge has been demonstrated on inorganic surfaces, including metals, in contact with organic phases such as solvents and polymers. For example, the acquisition of charge by gold particles in organic liquids has been demonstrated, and the sign and magnitude of the charge has been found to vary as a function of the donor properties of the liquid [M. E. Labib, 1984]. Thus, even for colloidal metals in suspension in relatively non-polar liquids, the possibility cannot be excluded that electrostatic stabilization contributes to the stability of the sol.

A second means by which colloidal particles can be prevented from aggregating is by the adsorption of molecules such as polymers, surfactants or ligands at the surface of the particles, thus providing a protective layer. Polymers are widely used, and it is obvious that the protectant, in order to function effectively, must not only coordinate to the particle surface, but must also be adequately solvated by the dispersing fluid - such polymers are termed amphiphilic. The choice of polymer is determined by consideration of the solubility of the metal colloid precursor. The solvent of choice and the ability of the polymer to stabilize the reduced metal particles in the colloidal state. Natural polymers such as gelatin and agar were often used before the advent of synthetic polymer chemistry, and related stabilizers such as cellulose acetate, cellulose nitrate [A. Duteil, 1993], and cyclodextrins [M. Komiyama, 1983] have been used more recently. Thiele [V. H. Thiele, 1965] proposed the Protective Value as a measure of the ability of a polymer to stabilize colloidal metal. It was defined, similarly to the older Gold Number of Zsigmondy. as the weight of the polymer which would stabilize 1 g of a standard red gold sol containing 50 mg/L gold against the coagulating effect of 1% sodium chloride solution. Several other studies have been performed on the relative ability of polymers to act as steric stabilizers [P. H. Hess, 1966, H. Hirai, 1979a and 1985b], and, despite the fact that these quite subjective studies focus on very specific (and quite different) sol systems, it seems that; of the synthetic polymers considered, vinyl polymers with polar side groups such as poly(vinylpyrrolidone) (PVP) and poly(vinyl alcohole) are especially useful in this respect. The use of copolymers introduces another degree of variability to colloidal stabilization. As the co-monomer ratio can be varied. For example, the use of vinyl-pyrrolidone-vinyl-alcohol copolymers is reported for the preparation of platinum and silver hydrosols [K. Megure, 1988]. The silver sols were stable only in the presence of the copolymer, and the size of the silver particles decreased with an increase in vinyl-pyrrolidone content of the copolymer. Electrostatic and steric stabilization are in a sense combined in the use of long chain alkyl-ammonium cations and surfactants, either in single-phase sols or in reverse micelle synthesis of colloidal metals. A new class of metal colloids has recently been established in which the surface of the particle is covered by relatively small ligand molecules such as sulfonated triphenylphosphine or alkane-thiols.

8.2 Synthetic methods for the preparation of colloidal transition metals

The synthetic methods which have been used include modern versions of established methods of metal colloid preparation such as the mild chemical reduction of solutions of transition metal salts and complexes and newer methods such as radiolysis and photochemical reduction, metal atom extrusion from labile organometallics. And the use of metal vapor synthesis techniques. Some of these reactions have been in use for many years, and some are the results of research stimulated by the current resurgence in metal colloid chemistry. The list of preparative methods is being extended daily, and, as examples of these methods are described below, the reader will quickly be made aware that almost any organometallic reaction or physical process which results in the deposition of a metal is in fact a resource for the metal colloid chemist. The acquisition of new methods requires only the opportunism of the synthetic chemist in turning a previously negative result into a synthetic possibility.

8.3 The role of surfactants in nanomarerials synthesis

For over 2000 years, humankind has used surfactants or surface-active ingredients in various aspects of daily life, for washing, laundry, cosmetics, and housecleaning.

However, the development of more economical processes for the manufacture of surfactants has contributed to an increased consumption of synthetic detergents. However, the major surfactants common (with respect to detergent) to all regions are linear alkyl benzene sulfonates (LASs), alcohol ether sulfates (AESs), aliphatic alcohols (AEs), alcohol sulfates (ASs), and soap. In the past decades, new surfactants have proliferated mainly as nonionic or non-soap surfactants offering unique properties and features to both industrial and household markets. Non-soap surfactants are widely used in diverse applications such as detergents, paints, and in the cosmetics, stabilized and synthesis of new materials and pharmaceutical industries. Since the 1960s, biodegradability and a growing environmental awareness have been the driving forces for the introduction of new surfactants. These forces continue to grow and influence the surfactant market and production. A new class of surfactants, carbohydrate-based surfactants, has gained significant interest and increased market share. Consequently, sugar-based surfactants, such as alkyl-poly-glycoside (APG), are used as a replacement for poly-oxy-ethylene alkyl phenols (APEs) where biodegradability is a concern. They represent a new concept in compatibility and care. Nonetheless, over 40 different types of surfactants are produced and used commercially in the formulation of home care, personal care, and industrial products. Contrary to many textbooks that elaborate on surfactant physical properties or formulation guidelines. Surfactants are primarily anionic, nonionic, cationic, and amphoteric.

8.4 "Surfactant systems"– Structure

The interactions between the hydrophilic head-groups and the hydrophobic tail of the surfactants, they tend to form aggregates spontaneously when placed into water, oil, or a mixture of the two. The aggregates are thermodynamically stable and therefore long lived. This interaction can be represented by a geometric packing parameter. Because the head-groups of surfactants are either hydrated or charged, they prefer to maintain a certain distance from their nearest neighbors, which is represented by the area, a_0. Tail interactions, from energy and enthalpy differences between the solvent and the environment of hydrophobic tails, lead to another area based on the length and volume of the tail, v/lc. The packing parameter is the ratio of these two areas ($P = v/a_0lc$). For certain values of P, the surfactant packs into different preferred geometries, which are shown in fig 3. Normal phases (fig 3 a-c), where water is continuous, have packing parameters less than one, and reverse phases (fig 3d and 3e), where oil is continuous, have packing parameters greater than one. When spheres (fig 3a) are in random order, the system is called a micellar solution. If the spheres close pack into a lattice, a discrete cubic liquid crystal can be formed. As the packing parameter is increased, cylinders (fig 3b) are formed. In dilute solution when the orientation of the cylinders is random (i.e., uncorrelated), they are called rodlike micelles, but when close packed into a lattice, they form a hexagonal liquid crystal. Bilayer sheets (fig 3c) can give many structures. When P is close to one, the bilayers are parallel to each other and form the lamellar liquid crystal. If the packing parameter is less than one, these sheets can no longer remain parallel to each other and may fold back on themselves, trapping solvent in a bilayer container called a vesicle. The bilayers may also maintain a curved shape and fill space by following an infinite periodic minimal surface, IPMS. IPMS are a geometry that is periodic in three directions and has constant mean curvature. When the bilayer conforms to this geometry, both the water and oil fractions are continuous in three directions, and because the IPMS give a cubic crystal pattern, these liquid crystals are known as bicontinuous cubic liquid crystals. The packing of bicontinuous cubics as well as all liquid crystals is thermodynamically controlled, so the structures conform to a regular array of fixed size. Once past the value $P = 1$, the geometries and ordered phase repeat in reverse order but with the oil and solvated surfactant tails becoming the continuous phase, i.e., reverse phases. Along with the ordered liquid crystal structures there are random geometries that are also thermodynamically stable. These systems are called micro-emulsions. They are related to regular emulsions in that they are dispersions of two immiscible liquids stabilized by surfactants. Unlike emulsions, micro-emulsions are thermodynamically stable and optically transparent. The basic structure of micro-emulsions is a swollen micellar solution. When water is the solvent, the systems are known as oil-in-water (O/W) micro-emulsions. Similarly, if the oil is the continuous phase and the water is within a reverse micelle, the system is water-in-oil (W/O) micro-emulsion. When the amounts of oil and water are nearly equal, a bicontinuous structure can be formed. These structures are the disordered analogues of the cubic and lamellar phases.

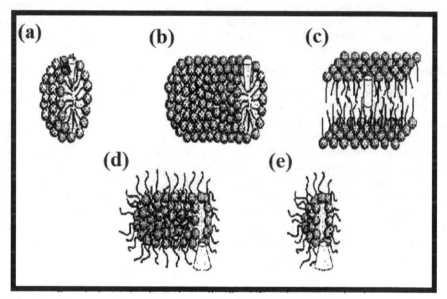

Fig. 3. Packing Parameter ($P=v/a_0l_c$). (a) Spherical Micelle ($P<1/3$) With a geographical Representation of the Three Values That Make up P. (b) A Cylindrical Micelle ($1/3<P<1/2$). (c) A Bilayer Sheet ($1/2<P<2$). (d) A Reverse Cylindrical Micelle ($2<P<3$). (e) A Reverse Spherical Micelle ($P>3$).

9. Polymers

Polymers are long chain-like giant molecules (macromolecules) made by the linkage of large numbers of small repeating molecules called monomers. Short chain lengths formed in the course of synthesis or degradation of polymers is called oligomers. The majority of polymers, and the only ones considered here, are compounds of carbon. Polymers are very widespread and can be synthetic (e.g. nylon) or natural (e.g. rubber). They form vital components of living organisms, and the most important molecule, DNA, is a polymer of amino acids. Colloquially, polymers are often called plastics. More precisely, plastics are sometimes defined as polymers that can be easily formed at low temperatures, and sometimes as a pure polymer together with a nonpolymeric additive, which may be solid, liquid or gas. There are two main divisions of polymeric materials: thermoplastic and thermosetting. Thermoplastic materials can be formed repeatedly; that is, they can be melted and reformed a number of times. Thermosetting materials can be formed only once; they cannot be re-melted. They are usually strong, and are typified by resins. A further group of polymers merits mention, elastomers. Elastomers can be deformed a considerable amount and return to their original size rapidly when the force is removed. The properties of polymers depend both on the details of the carbon chain of the polymer molecule and on the way in which these chains fit together. The chain form can be linear, branched or cross linked, and a great variety of chemical groups can be linked to the chain backbone. The chains can be carefully packed to form crystals, or they can be tangled in amorphous regions. Amorphous polymers tend not to have a sharp melting point, but soften gradually. These materials are characterized by a glass transition temperature, T_g, and in a pure state are often transparent. Although polymers are associated with electrically insulating behavior, the increasing ability to control both the fabrication and the constitution of polymers has led to the development of polymers that show metallic conductivity superior to that of copper and to polymers that can conduct ions well enough to serve as polymer electrolytes in batteries and fuel cells.

9.1 History of conducting polymers

Historically, polymers have been considered as insulators and found application areas due to their insulating properties. Infact, so far, any electrical conduction in polymers which is generally due to loosely bound ions was mostly regarded as an undesirable fact. However, emerging as one of the most important materials in the twentieth century, the use of polymers move from primarily passive materials such as coatings and containers to active materials with useful optical, electronic, energy storage and mechanical properties. Indeed, discovery and study of conducting polymers have already started this development [Freund. M. S, 2006, Epstein. A. J, 1999 and Inzelt. G, 2008]. Electrically conducting polymers are defined as materials with an extended system of conjugated carbon-carbon double bonds (fig 4) [Advani. S. G, 2007]. They are synthesized either by reduction or oxidation reaction, which is called doping process, giving materials with electrical conductivities up to 105 S/cm. Conducting polymers are different from polymers filled with carbon black or metals, since the latter are only conductive if the individual conductive particles are mutually in contact and form a coherent phase.

Although conducting polymers are known as new materials in terms of their properties, the first work describing the synthesis of a conducting polymer was published in the ninetieth century. In 1862, Henry Letheby prepared polyaniline by anodic oxidation of aniline, which was conductive and showed electrochromic behavior. However, electronic properties of so called aniline black were not determined.

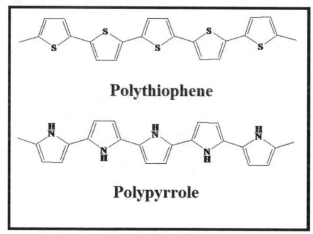

Fig. 4. Some Examples for Conducting Polymers.

In 1958, Natta et al. synthesized polyacetylene as a black powder which was found to be a semiconductor with conductivity in the range of 10-11 to 10-3 S/cm, depending on the process conditions of the polymer. In 1977, drawing attention on "conducting polymers", the first intrinsic electrically conducting organic polymer, doped polyacetylene, was reported. Intrinsically conducting polymers are a different class of materials than conducting polymers, which are a physical mixture of a non-conductive polymer with a conducting material such as metal or carbon powder. The preparation of polyacetylene by Sirakawa and coworkers and the discovery of the large increase in its conductivity after "doping" by the group led by MacDiarmid and Heeger actually launched this new field of research.

Electronically conducting polymers possess a variety of properties related to their electrochemical behavior and are therefore active materials whose properties can be altered as a function of their electrochemical potential. The importance and potential impact of this new class of material was recognized by the world scientific community when Hideki Shirakawa, Alan J. Heeger and Alan G. MacDiarmid were awarded the Nobel Prize in Chemistry in 2000 "for the discovery and development of electronically conductive polymers".

9.2 Composites

A composite is defined as a material created by combination of two or more components namely, selected filler or reinforcing agent and a compatible matrix binder. The combination of these component results in formation of a new material with specific characteristics and properties. The synthetic assemblage of the components does not occur as a dissolvation but rather like merging into each other to act in concert. Although the components act together as a single material, both the components and the interface between them can usually be physically identified. Generally, the behavior and the properties of the composite are controlled by the interface of the components. Since the composite is a totally new material having new and specific characteristics, its properties cannot be achieved by any of its components acting alone.

The classification of composites can be done in different ways. The composites can be classified on the basis of the form of their structural components: (i) fibrous where the composite is composed of fibers in a matrix, (ii) laminar where the composite is composed of layers in a matrix, and (iii) particulate where the composite is composed of particles in a matrix [Kricheldorf. H. R, 2005, Lubin. G, 1982].

Another type of classification can be done on the basis of filler or reinforcing agent used namely polymer matrix composites (PMCs), metal matrix composites (MMCs), ceramic matrix composites (CMCs), carbon-carbon matrix composites (CCCs), intermetallic composites (IMCs), or hybrid composites [Sanjay. K, 2002].

9.3 Polymer matrix composites

Composite materials have been utilized to solve technological problems for a long time. In 1960s with the introduction of polymeric-based composites, composites start capturing the attention of industries. Since then, composite materials have become common engineering materials. They are designed and manufactured for various applications including automotive components, sporting goods, aerospace parts, consumer goods, and in the marine and oil industries. Increasing awareness of product performance and competition in the global market for lightweight components also supported the growth in composite usage. Among all materials, composite materials have the potential to replace widely used steel and aluminum, and many times with better performance. Replacing steel components with composite components can save 60 to 80% in component weight, and 20 to 50% weight by replacing aluminum parts. Today, it appears that composites are the materials of choice for many engineering applications.

The matrix material used in polymer-based composites can either be thermoset (epoxies, phenolics) or thermoplastic resins (low density polyethylene, high density polyethylene, polypropylene, nylon, acrylics). The filler or reinforcing agent can be chosen according to the desired properties. The properties of polymer matrix composites are determined by properties, orientation and concentration of fibers and properties of matrix.

The matrix has various functions such as providing rigidity, shaping the structure by transferring the load to fiber, isolating the fiber to stop or slow the propagation of crack, providing protection to reinforcing fibers against chemical attack and mechanical damage (wear), and affecting the performance characteristics such as ductility, impact strength, etc. depending on its type. The failure mode is strongly affected by the type of matrix material used in the composite as well as its compatibility with the fiber. The important functions of fibers include carrying the load, providing stiffness, strength, thermal stability, and other structural properties in the composites and providing electrical conductivity or insulation, depending on the type of fiber used [Omastova. M, 1998].

Polypropylene (PP) is one of the fastest growing commercial thermoplastics due to its attractive combination of low density and high heat distortion temperature. There are some limitations in physicochemical properties that restrict PP applications. A typical illustration is in packaging, where PP has poor oxygen gas barrier resistance. No single polymer has shown the ideal combination of performance features. PP possesses good water vapor barrier properties, but it is easily permeated by oxygen, carbon dioxide, and hydrocarbons. The necessity of developing more effective barrier polymers has given rise to different strategies to incorporate and optimize the features from several components. Most schemes to improve PP gas barrier properties involve either addition of higher barrier plastics via a multilayer structure (co-extrusion) or by introducing filler with high aspect ratio in the polymer matrix. 1. Co-extrusion allows tailoring of film properties through the use of different materials where each material component maintains its own set of properties, compared with blending of polymers in a mono-extrusion technique. Co-extrusion is used to generate multilayer laminate structures from separately extruded polymer films that are sandwiched together [109 and 110]. Resulting films may comprise many layers, such as the PP-adhesivepoly (ethylene-co-vinyl-alcohol) (EVOH)-adhesive PP system: EVOH barrier sheet trapped between two layers of moisture resistant PP and two additional adhesive strata. However, by nature co-extrusion is a complex and expensive process.

9.4 Composites of polypyrrole

Maria Omastova and Ivan Chodak prepared conductive polypropylene/polypyrrole composites using the method of chemically initiated oxidative modification of polypropylene particles in suspension by pyrrole. In order to prepare the composite, polypropylene particles were dispersed in water-methanol mixture and $FeCl_3$ was added to be used for chemical oxidation. Addition of pyrrole started formation of polypyrrole particles in polypropylene suspension. The electrical and rheological properties of the composite were compared with polypropylene/polypyrrole composite prepared by melt mixing of pure polypropylene with chemically synthesized polypyrrole and with polypropylene/carbon black composites also prepared by melt mixing. Elemental analysis verified presence of polypyrrole in polypropylene matrix. The conductivity studies show that even a very small PPy amount present in composites results in a significant increase in conductivity. Processing conditions are observed to have a great effect on electrical conductivities of composites. The composite prepared by sintering PP particles covered with PPy shows about 7 orders of magnitude higher conductivity than the composite prepared by melt mixing of pure polypropylene with chemically synthesized polypyrrole whereas the conductivity of sintered PP/PPy composites is comparable to that of PP/Carbon black composite. The PP/CB and injection molded PP/PPy composites exhibit similar flow properties. However, for compresion molded PP/PPy composites a considerable increase of complex viscosity was observed [Pionteck. J, 1999].

10. Nanocomposites

Nanomaterials and nanocomposites have always existed in nature and have been used for centuries. However, it is only recently that characterization and control of structure at nanoscale have drawn intense interest for research and these materials start to represent new and exciting fields in material science. A nanocomposite is defined as a composite material where at least one of the dimensions of one of its constituents is on the nanometer size scale. In other words, nanocomposites can be considered as solid structures with nanometer-scale dimensional repeat distances between the different phases that constitute the structure. These materials typically consist of an inorganic (host) solid containing and an organic component or vice versa. They can consist of two or more inorganic/organic phases in some combinational form that at least one of the phases or features is in the nanosize.

In general, nanocomposite materials can exhibit different mechanical, electrical, optical, electrochemical, catalytic, and structural properties than those of each individual component. The multifunctional behavior for any specific property of the material is often more than the sum of the individual components [Mravcakova. M, 2006, Omoto. M, 1995, Ajayan. P. M, 2003 and Lee. E. S, 2004].

10.1 Polymer-based and polymer-filled nanocomposites

In recent years, the limits of optimizing composite properties of traditional micrometer-scale composite fillers have been reached due to the compromises of the obtained properties. Stiffness is traded for toughness, or toughness is obtained at the cost of optical clarity. In addition, regions of high or low volume fraction of filler often results in macroscopic defects which lead to breakdown or failure of the material. Recently, a new research area has provided the opportunity to overcome the limitations of traditional micrometer-scale polymer composites. This new investigation area is the nanoscale filled polymer composites where the filler is <100 nm in at least one dimension.

Implementation of the novel properties of nanocomposites strongly depends on processing methods that lead to controlled particle size distribution, dispersion, and interfacial interactions. Processing technologies for nanocomposites are different from those for composites with micrometer-scale fillers, and new developments in nanocomposite processing are among the reasons for their recent success.

Nanoscale fillers can be in many shapes and sizes, namely tube, plate-like or 3D particles. Fiber or tube fillers have a diameter <100 nm and an aspect ratio of at least 100. The aspect ratios can be as high as 106 (carbon nanotubes). Plate-like nanofillers are layered materials typically with a thickness on the order of 1 nm, but with an aspect ratio in the other two dimensions of at least 25. Three dimensional (3D) nanofillers are relatively equi-axed particles <100 nm in their largest dimension. This is a convenient way to discuss polymer nanocomposites, because the processing methods used and the properties achieved depend strongly on the geometry of the fillers.

10.2 Nanocomposites of polypyrrole

Eun Seong Lee and Jae Hyung Park prepared in situ formed procesable polypyrrole nanoparticle/amphiphilic elastomer composites which could have applications in biosenors, semiconductors, artificial muscles, polymeric batteries and electrostatic dissipation due to their process ability and considerable conductivities. The polymerization process of pyrrole was achieved by chemical oxidation of the pyrrole monomer by $FeCl_3.6H_2O$ in the presence of multiblock copolymer dissolved in methanol/water mixture. The multiblock copolymer was used as a stabilizer during polypyrrole synthesis and when cast after removing the dissolved polymers, served as a flexible and elastomeric matrix. The polymerization time, concentration of multiblock copolymer and the oxidant, reaction medium composition were optimized in terms of conductivity measurements and the highest conductivity was reported as 3.0 ± 0.2 Scm^{-1}. Mechanical properties such as tensile strength and elongation at break of the composites were found to increase with increasing amount of multiblock copolymer [Wu. T. M, 2008].

Tzong-Ming Wu and Shiang-Jie Yen have reported synthesis, characterization and properties of monodispersed magnetic coated multi-walled carbon nanotube/polypyrrole nanocomposites. Fe_3O_4 was used for coating multi-walled carbon nanotube (MWCNT). Fe_3O_4 coated c-MWCNT/PPy nanocomposites were synthesized via the in situ polymerization. The polymerization of pyrrole molecules was achieved on the surfaces of Fe_3O_4 coated c-MWCNT. The comparison of conductivities have shown that Fe_3O_4 coaed c-MWCNT/PPy nanocomposites have about 4 times higher conductivity that that of pure PPy matrix. Fe_3O_4 coated c- MWCNT/PPy nanocomposites were observed to exhibit ferromagnetic behaviour [Boukerma. K, 2006].

Kada Boukerma and Jean-Yves Piquemal prepared montmorillonite/polypyrrole nanocomposites and investigated their interfacial properties. The synthesis of MMT/PPy nanocomposites was achieved by in situ polymerization of pyrrole in the presence of MMT. Scanning electron microscopy results have shown that the surface morphology of the nanocomposites were more like the surface of untreated MMT. X-ray photoelectron spectroscopy (XPS) exhibited that the nanocomposites have MMT-rich surfaces which inicates intercalation of polypyrrole in the host galleries. The increase in interlamellar spacing was measured by transmission electron microscope. Invers gas chromatography measurements showed high surface energy of the nanocomposites [Mravcakova. M, 2006].

Miroslava Mravcakova and Kada Boukerma prepared montmorillonite/polypyrrole nanocomposites. The effects of organic modification of clay on the chemical and electrical properties were studied. The morphology investigations showed that the surface of MMT/PPy has a MMT-rich surface and relatively low conductivity (3.1×10^{-2} Scm^{-1}) indicating intercalation of PPy in the clay galleries. Whereas, the organically modified MMT/PPy nanocomposites have PPy-rich surface and higher conductivity indicating PPy formation on the surface of MMT. The dispersive contribution of surface energy of o-MMT was measured to be significantly low compared to that of MMT due to the stearly chains from the ammonium chlorides used for organic modification [Ranaweera. A. U, 2007].

A.U. Ranaweera and H.M.N Bandara prepared electronically conducting montmorillonite-Cu_2S and montmorillonite-Cu_2S-polypyrrole nanocomposites. MMT-Cu_2S nanocomposite was prepared by cation-exchange approach and its conductivity was measured as 3.03×10^{-4} Sm^{-1}. The polymerization of pyrrole was achieved between the layers of MMT-Cu_2S to obtain MMT-Cu_2S-PPy nanocomposite. The characterization was performed by XRD, FT-IR and impedance measurements. The electronic conductivity was reported as 2.65 Sm^{-1} [Dallas. P, 2007].

Panagiotis Dallas and Dimitrios Niarchos reported interfacial polymerization of pyrrole and in situ synthesis of polypyrrole/silver nanocomposites. The oxidizing agents used were Ag(I) or Fe(III). Depending on using different surfactants (SDS or DTAB) or not using any surfactant, the average diameter of polypyrrole structures was observed to be in the range of 200-300 nm. The electron microscopy images exhibited different morphologies of polypyrrole depending on using various surfactants or not using any as well as the size and shape of the silver nanocomposites. X-ray diffractometry showed amorphous structure of polymers. Further characterization was performed by thermogravinetric analysis and FT-IR spectroscopy [Carotenuto. G, 1996].

10.3 Nanoparticle/polymer composite processing

There are three general ways of dispersing nanofillers in polymers. The first is direct mixing of the polymer and the nanoparticles either as discrete phases or in solution. The second is in-situ polymerization in the presence of the nanoparticles, and the third is both in-situ formation of the nanoparticles and in-situ polymerization. Due to intimate mixing of the two phases, the latter can result in composites called hybrid nanocomposites [Lee. E. S, 2004].

10.4 Direct mixing

Direct mixing is a well-known and established polymer processing technique. When these traditional melt-mixing or elastomeric mixing methods are feasible, they are the fastest method for introducing new products to market. Although melt mixing has been successful in many cases, for some polymers, due to rapid viscosity increase with the addition of significant volume fractions of nanofiller, this processing method has limitations. There are many examples showing melt mixing method for composite production and exhibiting some limitations for the process [Lee. E. S, 2004].

10.5 Solution mixing

In solution mixing, in order to overcome the limitations of melt mixing method, both the polymer and the nanoparticles are dissolved or dispersed in solution. This method enables modification of the particle surface without drying, which reduces particle agglomeration. After dissolation the nanoparticle/polymer solution can be cast into a solid, or solvent evaporation or precipitation methods can be used for isolation of nanoparticle/polymer composite. Conventional techniques can be used for further processing [Lee. E. S, 2004].

10.6 In-situ polymerization

In in-situ polymerization, nanoscale particles are dispersed in the monomer or monomer solution, and the resulting mixture is polymerized by standard polymerization methods. This method provides the opportunity to graft the polymer onto the particle surface. Many different types of nanocomposites have been processed by in-situ polymerization. Some examples for in-situ polymerization are polypyrrole nanoparticle/amphiphilic elastomer composites; magnetite coated multi-walled carbon nanotube/polypyrrole nanocomposites and polypyrrole/ silver nanocomposites. The key to in-situ polymerization is appropriate dispersion of the filler in the monomer. This often requires modification of the particle surface because, although dispersion is easier in a liquid than in a viscous melt, the settling process is also more rapid.

11. Polypyrrole

Among the conjugated polymers, polypyrrole (PPy) is the most representative one for its easy polymerization and wide application in gas sensors, electrochromic devices and batteries. Polypyrrole can be produced in the form of powders, coatings, or films. It is intrinsically conductive, stable and can be quite easily produced also continuously. The preparation of polypyrrole by oxidation of pyrrole dates back to 1888 and by electrochemical polymerization to 1957. However, this organic p-system attracted general interest and was found to be electrically conductive in 1963. Polypyrrole has a high mechanical and chemical stability and can be produced continuously as flexible film (thickness 80 mm; trade name: Lutamer, BASF) by electrochemical techniques. Conductive polypyrrole films are obtained directly by anodic polymerization of pyrrole in aqueous or organic electrolytes.

Apart from electrochemical routes, polypyrrole can also be synthesized by simple chemical ways to obtain powders. Basically chemical oxidative polymerization methods can be used to synthesize bulk quantities of polypyrrole in a fast and easy way. Like other conducting polymers polypyrrole exhibit more limited environmental, thermal and chemical stability than conventional inert polymer due to the presence of dopant and its dynamic and electro active nature.

11.1 Synthesis of polypyrrole

Polypyrrole and many of its derivatives can be synthesized via simple chemical or electrochemical methods [120]. Photochemically initiated and enzyme-catalyzed polymerization routes have also been described but less developed. Different synthesis routes produce polypyrrole with different forms; chemical oxidations generally produce powders, while electrochemical synthesis leads to films deposited on the working electrode and enzymatic polymerization gives aqueous dispersions [Liu. Y. C, 2002, Tadros. T. H, 2005 and Wallace. G. G, 2003]. As mentioned above the electrochemical polymerization method is utilized extensively for production of electro active/conductive films. The film properties can be easily controlled by simply varying the electrolysis conditions such as electrode potential, current density, solvent, and electrolyte. It also enables control of thickness of the polymers. Electrochemical synthesis of polymers is a complex process and various factors such as the nature and concentration of monomer/electrolyte, cell conditions, the solvent, electrode, applied potential and temperature, pH affects the yield and the quality of the film.

Thus, optimization of all of the parameters in one experiment is difficult. In contrast, chemical polymerization does not require any special instruments; it is a rather simple and fast process. Chemical polymerization method involves oxidative polymerization of pyrrole monomer by chemical oxidants either in aqueous or non-aqueous solvents or oxidation by chemical vapor deposition in order to produce bulk polypyrrole as fine powders. Fe(III) chloride and water are found to be the best oxidant and solvent for chemical polymerization of pyrrole respectively regarding desirable conductivity characteristics.

Previous studies have shown that the optimum initial mole ratio of Fe(III)/Pyrrole for polymerization by aqueous Fe(III) chloride solution at 19 °C is 2.25 or 2.33. Also, several studies have revealed that factor such as solvent, reaction temperature, time, nature and concentration of oxidizing agent; affect the oxidation potential of the solution which affects the final conductivity of the product.

S.Goel and A. Gupta synthesized polypyrrole samples of different nanodimensions and morphologies by time dependent interfacial polymerization reaction. Pure chloroform was used as solvent for pyrrole and ammonium persulphate dissolved in HCl was used as the oxidizing solution. The polymerization occurred in the interface of organic and aqueous phases and polypyrrole was formed as thin layer on the interface. Morphology study of polypyrrole nanoparticles was done by scanning electron microscopy and transmission electron microscopy.

Yang Liu and Ying Chu synthesized polypyrrole nanoparticles through microemulsion polymerization. Alcohol-assisted microemulsion polymerization was performed in order to adjust the inner structure of polypyrrole nanoparticles for polymerization SDS was used as the surfactant, water was used as the solvent and aqueous solution of $NH_4S_2O_8$ was used as the oxidant. Characterization of polpyrrole was done by FT-IR and morphology study was performed by SEM and TEM. Hongxia Wang and Tong Lin synthesized polypyrrole nanoparticles by oxidation of pyrrole with ferric chloride solution during microemulsion polymerization process. Dodecyltrimethyl ammonium bromide (DTAB) was used as the surfactant. Particle characaterisation was performed by using FTIR, elemental analysis, UV-VIS spectra and SEM. Variation of particle size from about 50 to 100 and 100 to 200 nm with the change in surfactant concentration was reported. Xinyu Zhang and Sanjeev K. Manohar synthesized narrow pore-diameter polypyrrole nanotubes. The synthesis was performed by chemical oxidative polymerization of pyrrole using $FeCl_3$ oxidant and V_2O_5 nanofibers as the sacrificial template producing microns long electrically conducting polypyrrole nanotubes having 6 nm averages pore diameter. M.R. Karim and C.J. Lee synthesized polypyrrole by radiolysis polymerization method. Conducting PPy was synthesized by the in situ gamma radiation-induced chemical oxidative polymerization method. This method was reported to provide highly uniform polymer morphology. Jyongsik Jang and Joon H. Oh, synthesized polypyrrole nanoparticles via microemulsion polymerization with using various surfactants. Fe(III) chloride was used as the oxidant. The selective fabrication of amorphous polypyrrole nanoparticles as small as 2 nm in diameter using microemulsion polymerization at low temperature was reported.

12. Polypropylene

Polypropylene (PP) is a thermoplastic material that is produced by polymerization of propylene molecules into very long polymer molecule or chains (fig 5.). There are number of different ways to link the monomers together, but its most widely used form is made with catalysts that produce crystallizable polymer chains. The resulting product is a semi crystalline solid with good physical, mechanical, and thermal properties. Another form of PP produced in much lower volumes as a byproduct of semi crystalline PP production and having very poor mechanical and thermal properties, is a soft, tacky material used in adhesives, sealants, and caulk products. The above two products are often referred to as "isotactic" (crystallizable) PP (i-PP) and "atactic" (noncrystallizable) PP (-PP), respectively.

The average length of the polymer chains and the breadth of distribution of the polymer chain lengths determine the main properties of PP. In the solid state, the main properties of the PP reflect the type and amount of crystalline and amorphous regions formed from the polymer chains. Polypropylene has excellent and desirable physical, mechanical and thermal properties when used in room temperature applications. It is relatively stiff and has a high melting point, low density and relatively good resistance to impact.

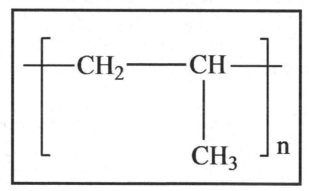

Fig. 5. Structural of Polypropylene.

Among conjugated polymers polypyrrole has attracted great interest due to its high conductivity, good thermal and environmental stability and ease of synthesis. However, it is an infusible, inprocessable Synthesis, characterization, and application of polyethylene glycol modified insulin polymer having relatively poor mechanical properties. On the other hand, polypropylene is a well-known insulating thermoplastic with outstanding mechanical properties. In this study, the synergistic assemblage of polypyrrole with polypropylene is investigated. Synthesis of polypyrrole nanoparticles via microemulsion polymerization, preparation of PP/PPy nanocomposites in order to provide some level of process ability to infusible and inprocessable PPy while inducing conductivity to insulating PP and preparation of PP/PPy nanocomposites with dispersant in order to improve the dispersion of PPy nanoparticles using identical procedures.

12.1 Preparation of PP/PPy nanocomposites

PP/PPy nanocomposites were prepared by melt mixing of pure PP with PPy using Brabender Plasti-Corder. The composition of nanocomposites varied between 1-20% PPy by weight. In order to provide a regular shape, the nanocomposites were pressed in a mould followed by fast cooling. The identical procedure is employed with addition of 2% by weight dispersant (SDS) during mixing process of pure PP with PPy.

12.2 Synthesis of polypyrrole nanoparticles

Synthesis of polypyrrole nanoparticles was achieved using micro-emulsion polymerization system by oxidation of pyrrole monomer with FeCl$_3$.6H$_2$O. As the oxidant was added, the color of the solution changed from colorless to deep greenish black which is an indication of oxidation of conducting polypyrrole. The reaction product polypyrrole was obtained in the form of black powder.

12.3 Scanning electron microscope analysis of polypyrrole nanoparticles

Scanning electron microscopy was performed in order to investigate the dimensions and the morphology of polypyrrole nanoparticles. The scanning electron micrographs of polypyrrole nanoparticles are presented in fig 6. The SEM micrographs of polypyrrole exhibited globular, nanometer-sized particles. The polypyrrole nanoparticles are observed

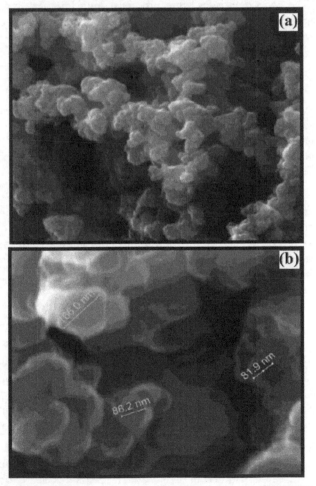

Fig. 6. SEM micrographs of PPy nanoparticles at magnifications of (a) 80000, (b) 300000.

to have a distribution of dimensions between 50 to 150 nm. The SEM results confirm that micro-emulsion polymerization system was successful in the synthesis of nano-sized polypyrrole particles. The SEM results prove that the micro-emulsion polymerization system provided similar dimensions of PPy nanoparticles with previous studies where 50 to 100 nm and 100 to 200 nm polypyrrole nanoparticles were reported.

12.4 Preparation of PP/PPy nanocomposites

The polypyrrole nanoparticles prepared by micro-emulsion polymerization system were mixed with polypropylene in order to provide some level of process ability to infusible and inprocessable polypyrrole while inducing conductivity to insulating polypropylene. In order to obtain PP/PPy nanocomposites, the polypyrrole nanoparticles were mixed with polypropylene by melt mixing technique followed by pressing to give a regular shape to nanocomposites. The nanocomposites were processed with injection molding and several black colored dog-bone shaped samples were obtained successfully. The composition of nanocomposites varied in the range of 1 to 20% by weight polypyrrole nanoparticles in polypropylene.

12.5 Characterization of PP/PPy nanocomposites

Mechanical properties of PP/PPy nanocomposites were investigated by tensile tests. The effect of loading different amounts of polypyrrole nanoparticles into thermoplastic polypropylene matrix and the changes in mechanical properties produced by incorporation of polypyrrole nanoparticles were examined. In order to understand the effect of using sodium dodecylsulphate as dispersant in PP/PPy nanocomposites, identical tests were performed also for the nanocomposites prepared with dispersant.

A stress-strain curve is known to provide information about both linear elastic properties and mechanical properties related to plastic deformation of a material. In order to specify a material as ductile or brittle, the response of the material to applied stress is investigated. The area under stress-strain curve corresponds to the energy required to break the material. As it is clearly seen in fig 7. pure PP is very ductile at a test rate of 5 cm/min and the area under the curve is very large indicating the great energy required to break the material. The Young's modulus, tensile strength and percentage strain at break values of pure polypropylene are 430 MPa, 27.8 MPa and %424 respectively.

The changes in mechanical properties that are produced by loading different amounts of polypyrrole nanoparticles can be well understood from stress-strain curves of PP/PPy nanocomposites which are illustrated in fig 8. through 3.8. As it is clearly observed in stress-strain curves of nanocomposites, addition of polypyrrole nanoparticles makes polypropylene matrix very brittle. In fact, addition of even the smallest amount of polypyrrole which is 1% causes a dramatic decrease in the energy required to break it.

The changes in mechanical properties that are produced by loading different amounts of polypyrrole nanoparticles can be well understood from stress-strain curves of PP/PPy nanocomposites which are illustrated in fig 8. through 3.8. As it is clearly observed in stress-strain curves of nanocomposites, addition of polypyrrole nanoparticles makes polypropylene matrix very brittle. In fact, addition of even the smallest amount of polypyrrole which is 1% causes a dramatic decrease in the energy required to break it.

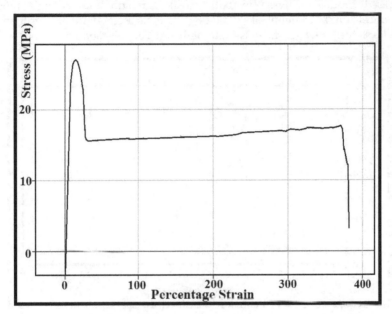

Fig. 7. Stress vs strain curve of pure PP.

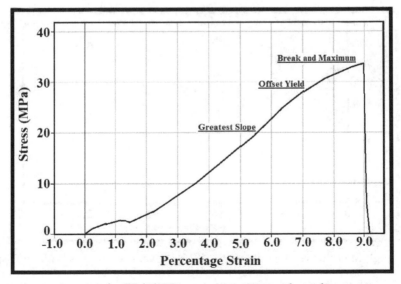

Fig. 8. Stress vs strain curve for PP/1%PPy nanocomposite without dispersant.

PPy nanoparticles in PP enhanced the strength and the stiffness of the nanocomposites. The greatest change for both properties was observed in PP/1%PPy nanocomposite. As it is seen in fig 9. and fig 10. incorporation of 1% PPy into PP matrix increased the tensile strength and Young's modulus of pure PP considerably. Increasing amount of PPy nanoparticles in PP matrix caused gradual increase in both tensile strength and Young's modulus of nanocomposites until addition of 10% PPy nanoparticles. Further addition of PPy nanoparticles slightly changes the tensile strength and Young's modulus.

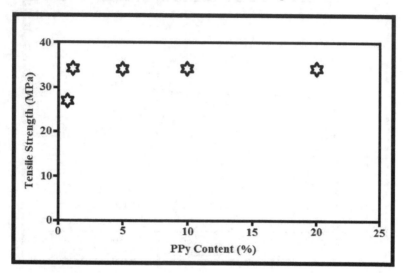

Fig. 9. Tensile strength vs PPy content for PP/PPy nanocomposites without dispersant.

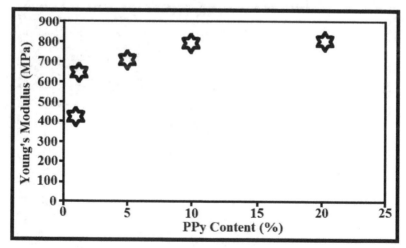

Fig. 10. Young's Modulus vs PPy Content For PP/PPy Nanocomposites Without Dispersant.

The tensile test results of PP/PPy nanocomposites show that incorporation of PPy nanoparticles in PP improves the strength and the stiffness while limiting the elongation of PP. In order to investigate the potential improvement in dispersion of PPy nanoparticles in PP, identical tensile tests were employed to nanocomposites prepared with dispersant. Due to the effect of dispersant, the interaction between PPy nanoparticles with PP matrix is expected to be improved. The Young's modulus, tensile strength and percentage strain at break values for nanocomposites prepared with dispersant are presented in Table. 1.

PPy Content (w%)	Young's Modulus (MPa)	Tensile Strength (MPa)	Percentage Strain at Break (%)
0	430±10	27.8±0.5	424±9.0
1	583±77	30.1±0.4	14.4±0.2
5	748±53	32.8±0.6	9.3±0.9
10	786±10	32.9±0.4	8.0±0.3
20	831±31	33.2±0.6	7.1±0.2

Table 1. Young's Modulus, Tensile Strength, Percentage Strain Values For PP, PP/PPy Nanocomposites With 2% Dispersant by Weight.

The change in percentage strain at break, Young's modulus and tensile strength with increasing polypyrrole content in nanocomposites prepared with dispersant are shown in fig 11. through fig 13. The gradual decrease in percentage strain at break values for increasing amounts of PPy is clearly seen in fig 11. Addition of 1% PPy caused a significant decrease in percentage strain since it prevents extension of PP matrix.

However, the decrease in percentage strain is relatively smaller due to binding effect of dispersant used. fig 12 and fig 13 exhibit the increase in tensile strength and Young's modulus of the nanocompoites with addition of PPy.

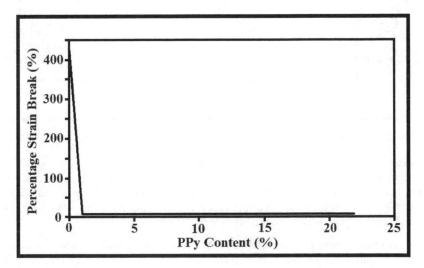

Fig. 11. Percentage strain at break vs PPy content for PP/PPy nanocomposites with dispersant.

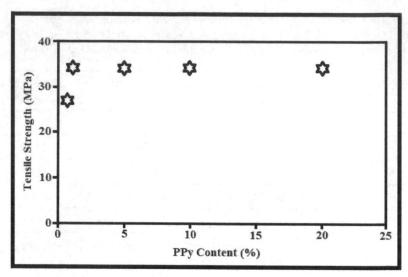

Fig. 12. Tensile strength vs PPy content for PP/PPy nanocomposites with dispersant.

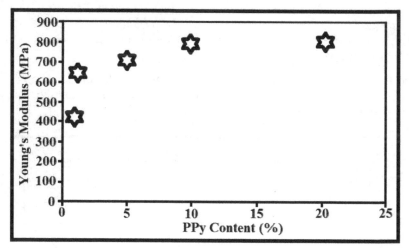

Fig. 13. Young's modulus vs PPy content for PP/PPy nanocomposites with dispersant.

The increase in both tensile strength and Young's modulus with increasing PPy content indicates the reinforcing action of PPy. In order to investigate the potential effect of dispersant in enhancement of dispersion of PPy nanoparticles, the change in tensile strength and percentage strain at break and Young's modulus for both nanocomposite sets prepared without and with dispersant are examined (Table. 1.).

The obtained results show that addition of PPy nanoparticles leads to a similar increase in tensile strength and Young's modulus of pure PP in both nanocomposite sets. Although the values are not identical, the values for nanocomposites involving dispersant did not exhibit significant difference compared to ones prepared without dispersant. However, the effect of dispersant is perceived in percentage strain at break values. The nanocomposites prepared using dispersant exhibited a regular decrease in percentage strain at break while a sudden decrease was observed for nanocomposites prepared without dispersant. The percentage strain at break values for PP/1%PPy nanocomposite prepared with and without dispersant are found to be 15.3 and 9.5 respectively. The higher decrease in nanocomposite prepared without dispersant can be explained by considering weaker interaction of polypyrrole with polypropylene. Same case is true for also nanocomposites with 5% polypyrrole content. Although, similar behavior was observed for nanocomposites with 10% and 20% polypyrrole content, the difference in values are not as considerable as the ones for nanocomposites with 1% and 5% PPy content.

13. Application of polypropylene derivatives in synthesis of nanomaterials

13.1 Metaloxide nanoparticles

Nanoparticles of MO (NP-CuO) were prepared by a novel sonochemistry route from metal acetate and sodium hydroxide in the presence of polypropylene derivatives such as polyethylene glycol, polypropylene glycol and polyvinyl alcohol. Variations in several parameters and their effects on the structural properties of nanoparticles (particle size and morphology) were investigated. 0.05 M solution of metal acetate in the presence of polyethylene glycol gave the best results. The characterizations were carried out by X-Ray diffraction, scanning electron microscopy, IR spectroscopy, thermal gravimetric analysis and differential thermal analysis.

Ultrasonic irradiation and stabilizer (PPG, PEG and PVA) can greatly enhance the conversion rate of precursor to nanometer- sized MO particles in the presence of polyethylene glycol (PEG), polypropylene glycol (PPG) and polyvinyl alcohol (PVA). Also the role of calcinations, time, and concentration of metal acetate solution and power of ultrasound wave on the size, morphology and chemical composition of nanoparticles was investigated. For the precursor, we used metal acetate dissolved in ethanol/water/(PEG or PPG or PVA). Then MO nanoparticles were directly obtained by addition of a solution of sodium hydroxide. The influence of several parameters on the size and morphology of MO particles was reported. The powders were characterized by powder X-Ray Diffraction (XRD), Scanning Electron Microscopy (SEM), IR spectroscopy, Thermal Gravimetric Analysis and Differential Thermal Analysis (TGA/DTA).

13.2 CuO nanoparticles

Different amounts of NaOH solution with a concentration of 0.1 M were added to the 0.1, 0.05, 0.025 M solutions of Cu acetate in ethanol/water. The obtained mixtures were sonicated for 30-60 min with different ultrasound powers. To investigate the role of surfactants on the size and morphology of nanoparticles, we used 0.5g of polypropylene derivatives in the reactions with optimized conditions. Table 1 shows the conditions of reactions in detail. A multiwave ultrasonic generator (Bandlin Sonopuls Gerate-Typ: UW 3200, Germany) equipped with a converter/transducer and titanium oscillator (horn), 12.5 mm in diameter, operating at 30 kHz with a maximum power output of 780 W, was used for the ultrasonic irradiation. The ultrasonic generator automatically adjusted the power level. The wave amplitude in each experiment was adjusted as needed. The X-ray powder diffraction (XRD) measurements were performed using a Philips diffractometer of X'pert Company with mono chromatized Cuk_α radiation. The crystallite sizes of the selected samples were estimated using the Scherrer method. TGA and DTA curves were recorded using a PL-STA 1500 device manufactured by Thermal Sciences. The samples were then characterized using a scanning electron microscope (SEM) (Philips XL 30) with gold coating.

Sample	$Cu(OAc)_2.2H_2O$	NaOH (0.1 M)	Aging time	Ultrasound power	Template
1	50 ml (0.1 M)	100 ml	1 hr	30W	PEG
2	50 ml (0.05 M)	100 ml	1 hr	30W	PEG
3	50 ml (0.025 M)	100 ml	1 hr	30W	PEG
4	50 ml (0.1 M)	100 ml	1 hr	45W	PEG
5	50 ml (0.05 M)	100 ml	1 hr	45W	PEG
6	50 ml (0.025 M)	100 ml	1 hr	45W	PEG
7	50 ml (0.1 M)	100 ml	30 min	30W	PEG
8	50 ml (0.05 M)	100 ml	30 min	30W	PEG
9	50 ml (0.025 M)	100 ml	30 min	30W	PEG
10	50 ml (0.1 M)	100 ml	30 min	45W	PEG
11	50 ml (0.05 M)	100 ml	30 min	45W	PEG
12	50 ml (0.025 M)	100 ml	30 min	45W	PEG
13	50 ml (0.1 M)	100 ml	1 hr	30W	PPG
14	50 ml (0.05 M)	100 ml	1 hr	30W	PPG
15	50 ml (0.025 M)	100 ml	1 hr	30W	PPG
16	50 ml (0.1 M)	100 ml	1 hr	45W	PPG
17	50 ml (0.05 M)	100 ml	1 hr	45W	PPG
18	50 ml (0.025 M)	100 ml	1 hr	45W	PPG
19	50 ml (0.1 M)	100 ml	30 min	30W	PPG
20	50 ml (0.05 M)	100 ml	30 min	30W	PPG
21	50 ml (0.025 M)	100 ml	30 min	30W	PPG
22	50 ml (0.1 M)	100 ml	30 min	45W	PPG
23	50 ml (0.05 M)	100 ml	30 min	45W	PPG
24	50 ml (0.025 M)	100 ml	30 min	45W	PPG
25	50 ml (0.1 M)	100 ml	1 hr	30W	PVA
26	50 ml (0.05 M)	100 ml	1 hr	30W	PVA
27	50 ml (0.025 M)	100 ml	1 hr	30W	PVA
28	50 ml (0.1 M)	100 ml	1 hr	45W	PVA
29	50 ml (0.05 M)	100 ml	1 hr	45W	PVA
30	50 ml (0.025 M)	100 ml	1 hr	45W	PVA
31	50 ml (0.1 M)	200 ml	30 min	30W	PVA
2	50 ml (0.05 M)	200 ml	30 min	30W	PVA
33	50 ml (0.025 M)	100 ml	30 min	30W	PVA
34	50 ml (0.1 M)	100 ml	30 min	45W	PVA
35	50 ml (0.05 M)	100 ml	30 min	45W	PVA
36	50 ml (0.025 M)	100 ml	30 min	45W	PVA

Table 2. Experimental Conditions For Preparation of CuO Nanoparticles.

Fig 14a. shows the XRD patterns of CuO nanoparticles. fig 1shows the XRD pattern of the direct sonochemicaly synthesized CuO nanoparticles and fig (14b, 14c and 14d) show the XRD patterns of this sample after calcinations for 2 hours at 500 °C. Sharp diffraction peaks shown in fig 14 indicates good crystallinity of CuO nanoparticles. No characteristic peak related to any impurity was observed. The broadening of the peaks indicates that the particles were of nanometer scale. The average size of the particles of the some samples (the best result from 36 reactions) was 80 nm, and the average size of these samples after calcinations at 500 °C for 2 hours were calculated as about 70 nm. These findings are in agreement with those observed from the SEM images.

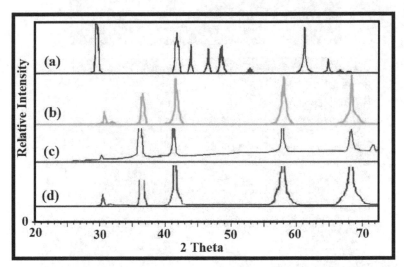

Fig. 14. X-Ray Powder Diffraction Pattern of CuO Nanoparticles (Sample No. 2) (a) Before Calcinations, (b) After Calcination (PEG Template), (c) After Calcination (PVA Template) and (d) After Calcinations (PPG Template).

The morphology, structure and size of the samples were investigated by Scanning Electron Microscopy (SEM). fig 15a indicates that the original morphology of the particles was approximately spherical with the diameter varying between 35 to 103 nm. The best morphology with smaller particles and good distribution was obtained for the sample number 2 before calcinations. fig 15b shows the SEM images of these samples after calcinations with different concentrations, it is clear from the fig 15b that the size of the particles has become small after calcinations. fig 15c shows the SEM images of the sample number 10. The role of PEG on the morphology of this sample is obvious. It has been reported that the presence of a capping molecule (such as polyvinyl alcohol, polyethylene glycol and polypropylene glycol) can alter the surface energy of crystallographic surfaces in order to promote the anisotropic growth of the nanoparticles. PEG is adsorbed on the crystal nuclei and helps the particles to grow separately. To investigate the size distribution of the nanoparticles, a particle size histogram was prepared for the sample No 2, (fig 16). Most of the particles possess sizes in the range from 30 to 103 nm. For further demonstration.

Fig. 15. The SEM images of CuO nanoparticles; (a) sample No. 2 before calcination, (b) samples No. 1(a), 3(b), 4(c), 5(d) after calcination, (c) sample No. 2 after calcinations.

Thermo-gravimetric analyses (TGA) were carried out to show that there was no difference between the curves of intermediate products and the one after calcinations. The results showed that there was no notable loss of weight in the TGA curves, proving the existence of CuO, which does not decompose at this temperature range. Further, the similarity of the TG curves of two samples shows the direct synthesis of CuO. fig 17a shows the TGA diagrams of the sample No. 2, and copper acetate and the compound before calcinations. fig 17b shows the DTA diagrams of the sample No. 2, copper acetate and the compound before calcinations.

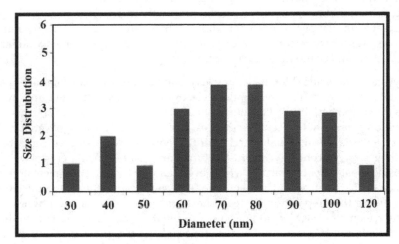

Fig. 16. Particle size histogram for the sample 2.

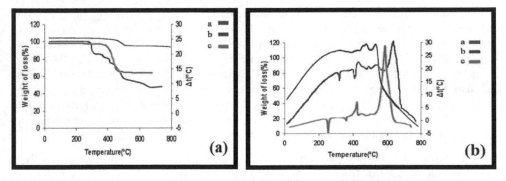

Fig. 17. The TG- DTA curves of CuO nanoparticles of the sample No. 2.

In order to investigate the role of sonication on the composition, size and morphology of the products, we carried out the reaction without sonication with the same conditions of the optimized sample. The XRD patterns of the obtained product corresponds to CuO but the SEM images show that the nanoparticles of the samples without using sonication have larger sizes as compared with the samples obtained via the sonochemical route.

It was indicated that the NP-CuO could decompose the organic pollutants by formation of exceed super oxides and/or hydroxyl radicals at the CuO interface.

13.3 Other M(II) oxide nanoparticles

Some nano-sized metal oxid (MO) such as ZnO, NiO and MnO were successfully synthesized by sonochemistry method in solutions at room temperature. The reactants used are $(M(Ac)_2.2H_2O)$ and sodium hydroxide (NaOH), and $H_2O/EtOH$ as a carrier in polyethylene glycol (PEG) template. Some of parameters such as effect concentration of NaOH solution, ultrasound power and sonicating time in growth and morphology of the nano-structures were investigated. The best morphology with smaller particles size and good distribution was obtained by using 0.025 M solution of NaOH and 45 w ultrasound powers in 1 h sonicating time. The particle size of the nano-sized metal oxid powders synthesized at room temperature is approximately between 40–80 nm. The resulting nano-sized powder was characterized by X-ray diffraction (XRD) measurements, Raman, BET, Solid state UV-vis and Scanning Electron Microscopy (SEM).

In recent years, metal oxide (MO) nanoparticles as a kind of functional material has attracted extensive interests due to its novel optical, electronic, magnetic, thermal and mechanical properties and potential application in catalyst, battery electrodes, gas sensors, electrochemical films, photo-electronic devices, and so on. In these applications, it is still needed for synthesizing high-quality and ultra-fine powders with required characteristics in terms of their size, morphology, microstructure, composition purity, crystallizability, etc. which are the most essential factors which eventually determine the microstructure and performance of the final products. Therefore, it is very important to control the powder properties during the preparation process. There are many chemical and physical methods to prepare nanometer MO, including the precipitation of metal acetate with NaOH. Recently, the interest preparation of nanometer MO has been growing. However, only few practical methods have been reported. The nanomaterials whose synthesis was reported are ZnS, Sb_2S_3, HgSe, SnS_2, CdS, CdSe, PbX (E = S, Se, Te, O), CuS, Ag_2Se and $CdCO_3$. But the sonochemical method for preparation of nanomaterials is very interesting, simple, cheap and safe. However, we have developed a simple sonochemical method to prepare NiO, MnO and ZnO nanostructures, wherein MO is synthesized by the reaction of $(CH_3CO_2)_2M.2H_2O$ and NaOH in an ultrasonic device. The MO nanoparticles have been characterized by X-ray powder diffraction (XRD), Raman spectroscopy, Solid state UV, BET and also the morphology and size of the nanostructures have been observed by scanning electron microscopy (SEM). We have performed these reactions in several conditions to find out the role of different factors such as the aging time of the reaction in the ultrasonic device and the concentration of the metal acetate on the morphology of nanostructures.

Typical procedure for preparation of MO nanoparticles: NaOH solution with a concentration of 0.1 M (100 ml) were added to the 0.05 and 0.025 M solutions of $M(CH_3COO)_2.2H_2O$ in ethanol/water. To investigate the role of surfactants on the size and morphology of nanoparticles, we used 0.5 gr of polyethylene glycol (PEG) in the reaction with optimized conditions. The mixtures were sonicated for 30-60 min, with different ultrasound powers followed by centrifuging with a centrifuge, and separation of the solid and liquid phases. The solid phase was washed for three times ethanol and water. Finally, the washed solid phase was calcinated at 500 ºC for 30 min. Table 1 shows the conditions of reactions in detail. A multiwave ultrasonic generator (Bandlin Sonopuls Gerate-Typ: UW 3200, Germany) equipped with a converter/transducer and titanium oscillator (horn), 12.5 mm in diameter, operating at 30 kHz with a maximum power output of 780W, was used for the ultrasonic irradiation. The ultrasonic generator automatically adjusted the power level. The wave amplitude in each experiment was adjusted as needed. X-ray powder diffraction (XRD) measurements were performed using a Philips diffractometer of X'pert Company with mono chromatized Cuk_α radiation. The samples were characterized with a scanning electron microscope (SEM) (Philips XL 30) with gold coating. Raman spectra were recorded on a Labram HR 800-Jobin Yvon Horbiba spectrometer. UVeVis spectra were measured with an HP 8453 diode array spectrophotometer. The specific surface area of samples was determined using the Brunauer–Emmet–Teller (BET) method in a volumetric adsorption apparatus (ASAP 2010 M, Micrometritics Instrument Corp).

Various conditions for preparation of MO nano-structures were summarized in Table. 3.

Series 1	$Mn(OAc)_2.2H_2O$	NaOH (0.1 M)	Aging time	Ultrasound power	Template
1	50 ml (0.05 M)	100 ml	1 hr	30W	PEG
2	50 ml (0.05 M)	100 ml	1 hr	45W	PEG
3	50 ml (0.025 M)	100 ml	1 hr	30W	PEG
4	50 ml (0.025 M)	100 ml	1 hr	45W	PEG
5	50 ml (0.05 M)	100 ml	30 min	30W	PEG
6	50 ml (0.05 M)	100 ml	30 min	45W	PEG
7	50 ml (0.025 M)	100 ml	30 min	30W	PEG
8	50 ml (0.025 M)	100 ml	30 min	45W	PEG
Series 2	$Ni(OAc)_2.2H_2O$	NaOH (0.1 M)	Aging time	Ultrasound power	Template
1	50 ml (0.05 M)	100 ml	1 hr	30W	PEG
2	50 ml (0.05 M)	100 ml	1 hr	45W	PEG
3	50 ml (0.025 M)	100 ml	1 hr	30W	PEG
4	50 ml (0.025 M)	100 ml	1 hr	45W	PEG
5	50 ml (0.05 M)	100 ml	30 min	30W	PEG
6	50 ml (0.05 M)	100 ml	30 min	45W	PEG
7	50 ml (0.025 M)	100 ml	30 min	30W	PEG
8	50 ml (0.025 M)	100 ml	30 min	45W	PEG
Series 3	$Zn(OAc)_2.2H_2O$	NaOH (0.1 M)	Aging time	Ultrasound power	Template
1	50 ml (0.05 M)	100 ml	1 hr	30W	PEG
2	50 ml (0.05 M)	100 ml	1 hr	45W	PEG
3	50 ml (0.025 M)	100 ml	1 hr	30W	PEG
4	50 ml (0.025 M)	100 ml	1 hr	45W	PEG
5	50 ml (0.05 M)	100 ml	30 min	30W	PEG
6	50 ml (0.05 M)	100 ml	30 min	45W	PEG
7	50 ml (0.025 M)	100 ml	30 min	30W	PEG
8	50 ml (0.025 M)	100 ml	30 min	45W	PEG

Table 3. Experimental conditions for the preparation of MO nanoparticles

The best morphology with smaller particles and good distribution was obtained for the sample number 4 in series 1, 2 and 3. fig 18(a,b and c) shows the XRD patterns of the direct sonochemicaly synthesized of the ZnO, NiO and MnO nanoparticles respectively. Sharp diffraction peaks shown in fig 18. indicate good crystallinity of MO nanoparticles. No characteristic peak related to any impurity was observed.

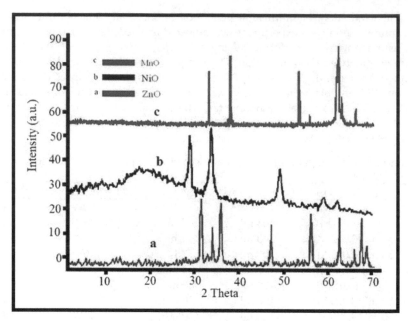

Fig. 18. X- Ray powder diffraction pattern of MO nanoparticles (a) ZnO, (b) NiO, (c) MnO.

The broadening of the peaks indicated that the particles were of nanometer scale. The morphology, structure and size of the samples are investigated by Scanning Electron Microscopy (SEM). fig 19 indicates that the original morphology of the ZnO, NiO and MnO particles are approximately spherical with the diameter varying between 40 to 80 nm. The role of PEG on the morphology of this sample is obvious. It has been reported that the presence of a capping molecule such as polyethylene glycol (PEG) can alter the surface energy of crystallographic surfaces, in order to promote the anisotropic growth of the nano particles. In this work PEG adsorbs on the crystal nuclei and it helps the particles to grow separately. To investigate the size distribution of the nano particles, a particle size histogram was prepared for MO nano particles, (fig 20).

Fig 21. demonstrates the UV-vis spectrum of the MO nanoparticles by ultrasonically dispersing in absolute ethanol. Strong absorption peak in the UV region can be observed. The absorption band gap Eg is usually achieved with the aid of the following equation:

$$(\alpha h\nu)^n = B(h\nu - E_g)$$

Concretely, $h\nu$ is the photo energy; a is the absorption coefficient; B is the constant related to the material; and n indicates either 2 or 1/2 for direct transition and indirect transition, respectively.

The inset of fig 21 gives us the typical $(ah\nu)^2 \sim h\nu$ curve for the MO samples calcinated at 500 °C. By the extrapolation of Eg. (1), we can get the present band gaps as 4.1, 3.7 and 3.9 Ev (layout for NiO, ZnO and MnO) indicating the small blue shifts upon size reductions for MO nanoparticles.

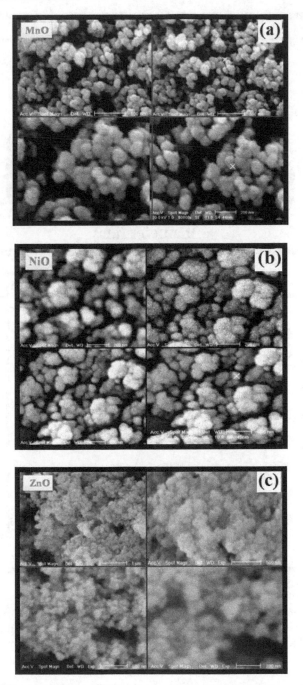

Fig. 19. Typical SEM micrographs of MO nano particles after calcinations: (a) ZnO, (b) NiO and (c) MnO.

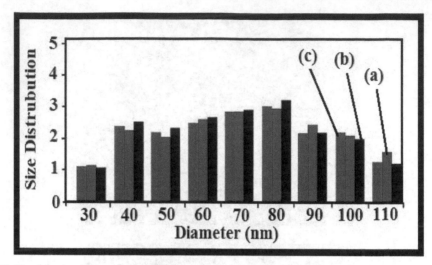

Fig. 20. Particle size Histogram of MnO, NiO and ZnO Nanopowders.

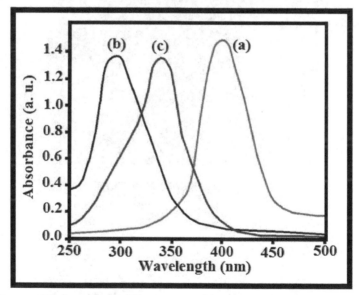

Fig. 21. Solid state UV absorption of MnO, NiO and ZnO nanoparticles

The surface area analysis was carried out on MO nanoparticles by BET method. Assuming the particles possess solid, spherical shape with smooth surface and same size, the surface area can be related to the average equivalent particle size by the equation:

$$D_{BET} = 6000/(\rho\ Sw)\ \text{(in nm)}$$

Where D_{BET} is the average diameter of a spherical particle; Sw represents the measured surface area of the powder in m^2/g; and ρ is the theoretical density in g/cm^3. fig 22 shows

BET plots of MO nanoparticles, the specific surface area of MO nanoparticles calculated using the multi-point BET-equation are 33, 40 and 53 m²/g (layout for NiO, ZnO and MnO), and the calculated average equivalent particle size is 36, 42 and 51 nm (layout for NiO, ZnO and MnO). We noticed that the particles size obtained from the BET and the SEM methods, agree very well with the result given by X-ray line broadening. The results of SEM observations and BET methods further confirmed and verified the relevant results obtained by XRD as mentioned above. Metal oxides are important catalyst in organic chemistry, gas sensors (such as CO_2, NO, SO_2, H_2O and CO) and battery cathode in the course of our research. The results of this investigation will be reported soon.

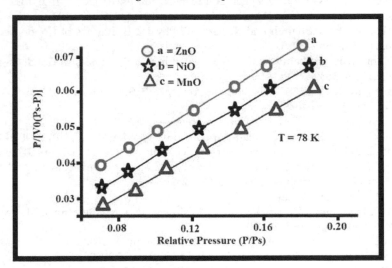

Fig. 22. BET plots of MnO, NiO and ZnO nanoparticles

Therefore Nano-sized of MnO, NiO and ZnO (MO) have been prepared by reaction between corresponding metal acetate and NaOH under ultrasound irradiation in solution at room temperature. Some of parameters such as effect concentration of NaOH solution, ultrasound power and sonicating time in growth and morphology of the nano-structures were investigated. The best morphology with smaller particles size and good distribution was obtained by using 0.025 M solution of metal acetate and 45 w ultrasound powers in 1 h sonicating time. Average particle sizes of the synthesized nano-sized MO powders were between 40-80 nm. Simple procedure, short reaction times, yields smaller particles and mild reaction conditions at room temperature are noteworthy advantages of this method. The specific surface area of MO nanoparticles calculated using the multi-point BET-equation are 33, 40 and 53 m²/g (layout for NiO, ZnO and MnO), and the calculated average equivalent particle size is 36, 42 and 51 nm (layout for NiO, ZnO and MnO). We noticed that the particles size obtained from the BET and the SEM methods, agree very well with the result given by X-ray line broadening. The results of SEM observations and BET methods further confirmed and verified the relevant results obtained by XRD as mentioned above. The infrared absorption band of the MO nanoparticles show blue-shifts compared with that of bulk MO [Alireza Aslani, 2008[a], 2009[b], 2010[c] and 2011[d]].

14. References

A. Duteil, R. Q U. M. Chaudret, C. Roucauj,. S. Bradley, Chem. Mater. 1993. 5: 341.

Advani. S. G, Processing and Properties of Nanocomposites, World Scientific, pp. 1, 2007.

Ajayan. P. M, Schadler, L.S., Braun, P.V., Nanocomposite Science and Technology, Wiley, VCH pp 10, 77-80, 111, 112, 2003.

Alireza Aslani and A. Morsali. Inorganica Chimica Acta. 362, (2009), 5012-5016.

Alireza Aslani and V. Oroojpour. Physica B, Physics of Condensed matter. 406, (2011), 144-149.

Alireza Aslani, A. Morsali and M. Zeller. Solid State Sciences, 10, (2008), 1591-1597.

Alireza Aslani, A. Morsali, V. T. Yilmaz and C. Kazak. 929, (2009), 187-192.

Alireza Aslani, A. R. B. Shamili and K. Kaviani. Physica B, Physics of Condensed matter, 405, (2010), 3972-3976.

Alireza Aslani, A. R. B. Shamili and S. Barzegar. Physica B, Physics of Condensed matter, 405 (2010), 3585-3589.

Alireza Aslani, M. R. Arefi, A. Babapoor, A. Amiri, K. B. Shuraki. Applied Surface Science, 257, (2011), 4885-4889.

Alireza Aslani, Physica B, Physics of Condensed matter. 406, (2011), 150-154.

Alireza Aslani, V. Oroojpour, M. Fallahi, Applied Surface Science, 257, (2011), 4056-4061.

Bhandarkar, S, and A. Bose, J. Colloid Interface Sci., Vol. 135, No. 2, 1990, pp. 541-550.

Bliznyuk. V. N, A. Campbell, and C. W. Frank, ACS Symposium Series 695, New York Oxford University Press, 1998, pp. 220-232.

Bonnell. D. A, New York: Wiley-VCH, 2001.

Boukerma. K, Piquemal. J. Y, Chehimi. M. M, Mravcakova. M, Omastova. M, Beaunier. P, Polymer, 47, pp 569-576, 2006.

Brust, M, et al., J. Chem. Soc. Chem. Comm., Vol. 7, 1994, pp. 801-802.

Carotenuto. G, Her.Y. S, Matijevic. E, Ind. Eng. Chem. Res., 35,2929, 1996.

Collier, C. P, et al., Science, Vol. 277, No. 5334, 1997, pp. 1978-1981.

Cortan, A. R, et al, J. Am. Chem. Soc., Vol. 112, No. 4, 1990, pp. 1327-1332.

Dallas. P, Niarchos. D, Vrbanic. D, Boukos. N, Pejovnik. S, Trapalis. C, Petridis. D, Polymer 48, pp 2007-2013, 2007.

Edelstein, A. S, and R. C. Cammarata, Bristol, PA: IoP Publishing, 1996.

Epstein. A. J, Platics Design Library, 1, 93, 1999.

F. Wakai, Y. K, S. S. N. M, K. I and K. N, Nature, 344, 6265, 421-423 (1990).

Freund. M. S, Deore B., Self-Doped Conducting Polymers, Wiley, pp.1,2, 10-12, 2006.

Gabriel. B. L, SEM, A User's Manual for Materials Science, American Society for Metals, 1985.

Greffet. J. J, and R. Carminati, Progr. Surf. Sci., Vol. 56, No. 3, 1997, pp. 133-237.

H. Hirai, Macromol. Chem. Suppl. 1985, 14, 55.

H. Hirai, Y. Nakao, N. Toshima, J. Macromol. Sci. Chem. 1979. 13, 727.

H. Maeda, J. Control. Release, 19, 315-324 (1992).

H. Suzuki, T. Ohno, J. Soc. Powder Technol, Jpn, 39, 877-884 (2002).

I. Matsui, J. Chem. Eng, Jpn, 38(8), 535-546 (2005).

Inzelt. G, Conducting Polymers A New Era in Electrochemistry, Springer, 1, 2008.

J. S. Bradley, in Clusters and Colloids, G. SCHMID (ed.) VCH, Weinheim, 1993, p. 459.

K. Ishikawa, J. Soc. Powder Technol, Jpn, 38, 731-740 (2001).

K. Ishikawa, K. Yoshikawa and N. Okada, Phys. Rev. B, 37, 5852-5855 (1988).

K. Kobayashi, J. Soc. Powder Technol, Jpn, 41, 473-478 (2004).

K. Megure, Y. Nakamura. Y. Hayashim, . Torizuka, K. Esumi, Bull. Chem. Soc.Jpn. 1988. 61. 347.

K. Niihara, J. Ceram. Soc. Jpn, 99 (10), 974-982 (1991).

K. Uchino, E. Sadanaga and T. Hirose, J. Am. Ceram. Soc, 72(8), 1555-1558 (1989).

Keyse. R. J, et al. Microscopy Handbook, Vol. 39, Oxford, England: BIOS Scientific Publishers, 1998.

Kitaigorodski. A. L, Organic Chemical Christallography, New York: Counsultants Bureau, 1961.

Kricheldorf. H. R, Nuyken, O., Swift, G., Handbook of Polymer Synthesis, Marcel Dekker., Ch. 12, pp 1,3,4, USA, 2005.

Lee. E. S, Park. J. H., Wallace. G. G., Bae. Y. H, Polymer International, 53:400-405, 2004.

Li, S, et al., IEEE Trans. on Magnetics, Vol. 37, No. 4, 2001, pp. 2350-2352.

Liu. Y. C, Materials Chemistry and Physics 77, pp 791-795, 2002.

Lover, T, et al., Chem. Mater. Vol. 9, No. 4, 1997, pp. 967-975.

Lubin. G, Handbook of Composites, Van Nostrand Reinhold Company Inc., USA, pp1, 2, 1982.

Lvov, Y. M, M. R. Byre and D. Bloor, Crystall. Reports, Vol. 39, No. 4, 1994, pp. 696-716.

Lvov, Y. M., et al, Phil. Mag. Lett., Vol. 59, No. 6, 1989, pp. 317-323.

Lvov. Y, and H. Mohwald, (eds.), New York: Basel/Marcel Dekker, 2000, pp. 125-167.

Lvov. Y. M, and L. A. Feigin, Kristallografiya, Vol. 32, No. 3, 1987, pp. 800-815.

M. Arakawa, Funsai (The Micrometrics), No 27, 54-64 (1983).

M. Arakawa, J. Soc. Powder Technol, Jpn, 42, 582-585 (2005).

M. E. Labib, R. Williams, Colloid Interface Sci. 1984. 97, 356.

M. Haruta, Catalysts, 36(6) 310-318 (1994).

M. Komiyama, H. Hirai, Bull. Chem. Soc. Jpn. 1983, 56, 2833.

M. Takashige, T. Nakamura, Jpn. J. Appl. Phys, 20, 43-46 (1981).

Mann, S., et al., J. Chem. Soc.-Dalton Trans., No. 4, 1983, pp. 771-774.

Mann, S., J. P. Hannington, and R. J. P. Williams, Nature, Vol. 324, No. 6097, 1986, pp. 565-567.

Mravcakova. M, Boukerma. K, Omastova. M, Chehimi. M. M, Materials Science and Engineering C, 26, pp 306-313, 2006.

Mravcakova. M, Omastova. M, Potschke. P, Pozsgy. A, Pukanszky. B., Pionteck. J, Adv. Technol., 17: 715-726, 2006.

N. Wada, Chem. Eng., 9, 17-21 (1984).

Nabok. A. V, et al, IEEE Trans. on Nanotechnology, Vol. 2, No. 1, 2003, pp. 44-49.

Nabok.A. V, et al., Thin Solid Films, Vol. 327-329, 1998, pp. 510-514.

Nakanishi, T. B. Ohtani, and K. Uosaki, J. Phys. Chem. B, Vol. 102, No. 9, 1998, pp. 1571-1577.

Netzer, L., and J. Sagiv, J. Am. Chem. Soc., Vol. 105, No. 3, 1983, pp. 674-676.

Omastova. M, Chodak. I, Pionteck. J, Potschke. P, Journal of Macromolecular Science, Part A, 35:7, 1117-1126, 1998.

Omoto. M, Yamamoto. T, Kise. H, Journal of Applied Polymer Science, Vol. 55, 283-287, 1995.

P. H. Hess, P. H. Parker, Appl. Polymer. Sci 1966. 10, 1915.

Petty. M. C, in Languir- Blodgett Films, G. G. Roberts, (ed.), New York: Plenum Press, 1990, pp. 133–221.

Petty. M. C, M. R. Bryce, and D. Bloor, (eds.), New York: Oxford University Press, 1995.

Pionteck. J, Omastova. M, Potschke. P, Simon. F, Chodak. I, Journal of Macromolecular Science, Part B, 38:5, 737-748, 1999.

Puntes, V. F, and K. M Krishnan, IEEE Trans. on Magnetics, Vol. 37, No. 4, 2001, pp. 2210–2212.

R. R. Karimi, A. R. B. Shamili, Alireza Aslani and K. Kaviani. Physica B, Physics of Condensed matter, 405, (2010), 3096–3100.

Ranaweera. A. U, Bandara. H. M. N, Rajapakse. R. M. G, Electrochimica Acta, 52, pp 7203-7209, 2007.

Rizza, R, et al, Chem. Mater., Vol. 9. No. 12, 1997, pp. 2969–2982.

Roberts, G. G, Adv. Phys., Vol. 34. No. 4, 1985, pp. 475–512.

Roberts, G. G, in Langmuir-Blodgett Films, G. G. Roberts, (ed.), New York: Plenum Press, 1990.

Roberts, G. G., et al., J. Verwey, (ed.), North Holland: Amsterdam, 1983, p. 141.

Rogach, A. L, et al., J. Phys. Chem. B, Vol. 103, No. 16, 1999, pp. 3065–3069.

Ross, J, and G. G. Roberts, Proceedings of 2nd International Meeting on Chemical Sensors, Bordeaux, France, 1986, p. 704.

S. Sato, N. Asai and M. Yonese, Colloid Polym. Sci, 274, 889-893 (1996).

Sanjay. K. Mazumdar, CRC Press LLC, USA, pp 4-6, 2002.

Sarathy, K. V., et al, J. Phys. Chem. B., Vol. 103, No. 3, 1999, pp. 399–401.

Schultz, D. L., et al, IEEE, 25th PVSC, Washington, D.C., May 13–17, 1996, pp. 929–932.

Stefanis. A, and A. A. G. Tomlinson, Enfield, NH: Trans Tech. Publications, 2001.

T. Sekino, Mater. Integr, 13(11) 50-54 (2000).

T. Yokoyama, Sokeizai, 3, 6-11 (2005).

Tadros. T. H, Applied Surfactants, Wiley-VCH Verlag GmbH and Co. KGaA, pp. 1-5, 2005.

Talapin, D. V, et al., Physica E, Vol. 14, No. 1–2, 2002, pp. 237–241.

Tredgold. R. H, Order in Thin Organic Films, Cambridge, England: Cambridge University Press, 1994.

Tsukruk. V. V, Progr. Polym. Sci., Vol. 22, No. 2, 1997, pp. 247–311.

Ulman. A, (ed.), Boston, MA: Butterworth- Heinemann, 1995.

Ulman. A, From Langmuir-Blodgett to Self- Assembly, Boston, MA: Academic Press, 1991.

V. H. Thiele, J. Kowallik. J. Colloid Sci. 1965. 20, 679.

Vogel, W, et al, Vol. 16, No. 4, 2000, pp. 2032–2037.

Wallace. G. G, Spinks, G. M, Kane-Maguire. L. A. P, Teasdale. P. R, CRC Press LCC, USA, pp.51, 2003.

Wegner. G, Molecular Crystals and Liquid Crystals Science, Vol. 234, 1993, pp. 283–316.

Wu. T. M, Yen. S. J, Chen. E. C, Chiang. R. K, Journal of Polymer Science: Part B: Polymer Physics, Vol. 46, 727-733, 2008.

Y. Kurokawa, Y. Hosoya, Surface, 34(2) 100-106 (1996).

Yaacob, I. I, S. Bhandarkar, and A. Bose, J Mater. Research, Vol. 8, No. 3, 1993, pp. 573–577.

Yarwood. J, Analyt. Proc., Vol. 30, 1993, pp. 13–18.

3

Polypropylene Nanocomposite Reinforced with Rice Straw Fibril and Fibril Aggregates

Yan Wu[1], Dingguo Zhou[1], Siqun Wang[2],
Yang Zhang[1] and Zhihui Wu[1]
[1]Nanjing Forestry University
[2]University of Tennessee
[1]China
[2]USA

1. Introduction

High thermoplastic content composites are those in which the thermoplastic component exists in a continuous matrix and the lignocellulosic component serves as reinforcing filler. The great majority of reinforced thermoplastic composites available commercially use inorganic materials as their reinforcing fillers, e.g., glass, clays, and minerals. These materials are heavy, abrasive to processing equipment, and non-renewable. In recent years, lignocellulosic fillers used to reinforce thermoplastics, especially polypropylene (PP), have expanded due to their strength, low density, relatively high aspect ratio and environmental benefits. Lignocellulosic materials such as wood fiber, wood flour, cellulose fiber, flax, and hemp have been given more attention by polymer manufactures (Bataille et al. 1989; Karnani et al. 1997). Futhermore, the micro/nanofibrils isolated from natural fibers have much higher mechanical properties as reported by Sakurada et al. (1962) that the cellulose crystal regions are a bundle of stretched cellulose chain molecules with Young`s modulus of 150 GPa and strength in the order of 10 GPa. Thus the micro/nanofibrils has the potential as the reinforcing materials to create innovative nanocomposites (Herrick et al. 1983; Stenstad et al. 2008; Turbak et al. 1983). Nevertheless, such fibers are used only to a limited extent in industrial practice, which may be explained by difficulties in achieving acceptable dispersion levels (Helbert 1996).

Two main methods, the chemical method, mainly by strong acid hydrolysis and mechanical method including a high intensity ultrasonication (Cheng, 2007a, 2007b; Wang & Cheng 2009), a high-pressure homogenizer treatment (Dufresne et al. 1997; Herrick et al. 1983; Nakagaito and Yano 2005; Stenstad et al. 2008; Turbak et al. 1983), a high pressure grinder treatment and a microfluidizer (Zimmermann et al. 2004), have been used to generate cellulose products. The product came from chemical method was described by cellulosic whisker or cellulose nanocrystal. However, the product isolated from mechanical method was defined by cellulose microfibril or microfibrillated cellulose.

Rice straw represents a potentially valuable source of fiber and has the potential to alleviate the shortage of wood fiber and petroleum resources. It is easy to obtain and its fiber is more flexible than wood fiber, which can highly reduced wear of the processing machinery, together with its abundance and low price, giving it more utilization value. Therefore, it can be used as a direct substitute for wood composites, and also can be used to make plastics composites (Wu et al. 2009). The fibril and fibril aggregates were generated from rice straw pulp cellulose fiber by a high intensity ultrasonication (HIUS) treatment and were used to reinforce polypropylene (PP) by a novel compounding machine, a minilab extruder with a materials cycle system.

The objective of this work was to use rice straw pulp cellulose fiber to prepare environmental-friendly rice straw fibril and fibril aggregates (RSF) and evaluate the fibril and fibril aggregates as a novel reinforcing material to compound polypropylene (PP)/ RSF nanocomposite. The scanning electron microscopy (SEM), wide angle X-ray diffraction (WAXD), laser diameter instrument (LDI) were used to evaluate the characteristics of RSF. The RSF/PP nanocomposite was prepared by novel extrusion process. The interface compatibility and tensile properties of nanocomposite were investigated by FTIR and tensile test, respectively.

2. Materials and methods

The isotactic polypropylene (iPP) was supplied by FiberVisions, Georgia, in the form of homopolymer pellets with a melt flow index of 35g/10 min (230 °C, 160 g) and a density of 0.91 g cm⁻³. The reinforcing filler was rice straw fibril and fibril aggregates (RSF) obtained from rice straw pulp fiber (Taonan paper and pulp company, Jilin province, North-east of China) that was cut to pass a screen (room temperature and relative humidity of 30%) with holes of 1 mm in diameter by a Willey mill before treatment. The maleated polypropylene (MAPP) was used as compatibilizing agent and Epolene G-3003 P has an acid number of 6 and a molecular mass of 125 722.

2.1 Fibril isolation

The milled rice straw pulp fiber was soaked in distilled water for more than 24 h, and then treated by high intensity ultrasonication (Sonics & Materials. INC, CT, 20kHz, Model 1500 W) for 30 min with 80% power level. After ultrasonication treatment, the obtained RSF aqua compound was kept in frozen.

2.2 Freeze drying

In order to avoid the aggregation of isolated RSF, the frozen RSF aqua compound was freeze-dry in Food Science, the University of Tennessee, Knoxville, TN, USA. The conditions of freeze drier (Virtis Genesis 12 EL) were -20 °C to + 20 °C over 4 days in 5 °C increments.

2.3 Compounding

The freeze-dry RSF was milled by a food processor and kept into a dessicator. The RSF/ PP nanocomposite was made by blending PP pellets with 1~6% MAPP (ratio of PP weight, wt%) and 2~11% RSF (ratio of total weight, wt%). All materials were then fed into a Haake Mini-Lab twin-screw extruder (Thermo Electron Corp., Hamburg, Germany). The blends were processed for 10 min, 20 min, 30 min at 50 rpm and 180 °C, 190 °C and 200 °C using a counter-rotating screw configuration, respectively. The RSF/ PP nanocomposite was then extruded through a 2.5 mm cylindrical die. The extruder strands were granulated and hot pressing at a temperature of 175 °C and a pressure of 5 MPa for 10 min. The obtained sheets with nominal thickness of 270 μm were conditioned at 23 ± 2 °C and 50 ± 5% relative humidity for not less than 40 h prior to tensile test in accordance with Procedure A of Practice D 618.

Sample	PP	MAPP	RSF
PP	100	0	0
PP2M	98	2	0
PP2M2R	96	2	2
PP2M5R	93	2	5
PP2M8R	90	2	8
PP2M11R	87	2	11
PP1M	99	1	0
PP2M	98	2	0
PP3M	97	3	0
PP4M	96	4	0
PP5M	95	5	0
PP6M	94	6	0
PP1M5R	94	1	5
PP2M5R	93	2	5
PP3M5R	92	3	5
PP4M5R	91	4	5
PP5M5R	90	5	5
PP6M5R	89	6	5

Table 1. Proportion of RSF/PP nanocomposite (by weight, %)

2.4 Fibril diameter test

The laser diameter instrument (Winner 2005, Qingdao, China) was used to investigate the diameter distribution after HIUS treatment. The sample concentration for testing was 1.526% and three levels were tested for the sample.

2.5 Cellulose crystallinity

The wide angle X-ray diffraction (WAXD) was used to study the crystallinities of treated and untreated rice straw fiber and fibril aggregates. The Segal method was used to calculate the crystallinities of the samples (Cheng et al. 2007b; Thygesen et al. 2005). The equipment (Material Data. Inc. DX-2000) was a pinhole type camera that recorded the patterns on Fuji image plates. The operating voltage was 40 kV, current was 30 mA, and the exposed period was 3000s using CuKa radiation with a wavelength of 0.15418 nm. The crystallinity is defined as the ratio of the amount of crystalline cellulose to the total amount of sample material including crystalline and amorphous parts. The Turley method was used to calculate the crystallinity (CrI) of the samples (Thygesen et al. 2005).

2.6 Mechanical characteristic

The tensile measurements were conducted on an Instron testing machine (Model 5567) with a length of 20 mm between the top and bottom clamps, a crosshead speed of 1 mm/min, and a load cell of 30 kN (Cheng et al. 2007a). The specimens were cut to drum shapes with width of 5 mm for the narrow portion and total length of 40 mm. Eight specimens were tested for each extruder condition according to the ASTM D 882 standard test method for tensile properties of thin plastic sheeting (ASTM D882-02). The overall significant differences of the influences on the tensile modulus and strength of polypropylene reinforced with RSF under different extruder conditions were conducted using a Statistic Analysis System (SAS) JMP version 6.0.2 software (SAS Institute, Cary, NC, USA).

2.7 FTIR testing

The functions of samples were tested by FTIR (Nicolet 380) accompanying with ATR. The sample scanning times was 64, the ratio of differentiate was 8.000, sampling plus was 2.0, the speed of moving lens was 0.6329, diaphragm was 100.00, wave range was 4000~400cm-1.

2.8 Morphology characteristic

Polarized light microscopy (PLM, Olympus-BX51) and a digital image analysis software package (ImageJ) were used to observe the distributions of the RSF in the RSF/PP nanocomposite. The fractured surfaces of nanocomposite after tensile test were investigated by a scanning electron microscopy (SEM, LEO 1525). The freeze-dry samples were observed by SEM. The voltages were 5-10 kV and various magnification levels were used to obtain images.

3. Results and discussion

3.1 Morphology characteristic

Figure 1 was the distribution of RSF diameters treated by HIUS. As seen in the figure, the distribution of diameters of RSF ranged from 0.1 μm to 80 μm by HIUS treatment. The percentage was 6.3% of RSF which diameters were less than 500 nm; almost 90 percents of RSF distributed between 7.0 μm and 80 μm; the average diameter was 41 μm.

3.2 Crystallinity of fibers and fibrils

According to (Thygesen et al. 2005), the crystallinity of treated rice straw cellulose fiber was 72.9%, which was higher than untreated rice straw cellulose fiber of 71.3%. The reason may be that some of the amorphous cellulose were degraded and removed during the ultrasonication treatment. High crystalline fibers and fibril aggregates could be more effective in achieving higher reinforcement for composite materials (Eichhorn & Young 2001).

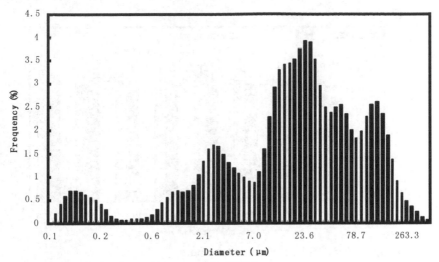

Fig. 1. Distribution of RSF diameters treated by HIUS

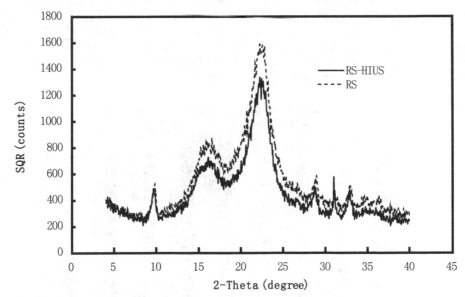

Fig. 2. CrI of untreated sample and sample treated by HPH

3.3 Effect of compounding conditions on RSF/PP nanocomposite tensile properties

The tensile strength of RSF/PP nanocomposite is shown in figure 3. The tensile strength of 5% rice straw fibril reinforcing PP nanocomposite was lower than PP/MAPP polymer. For the RSF/PP nanocomposite, the tensile strength increased with increasing cycle time from 10 min to 30 min at 180 °C. And the maximum value of tensile strength was 31.2 Mpa that appeared at the conditions of 190 °C, 20 min.

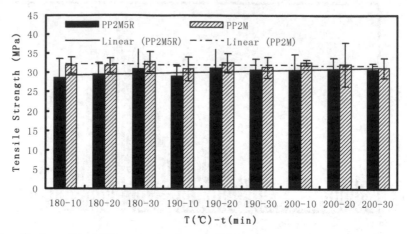

Fig. 3. Tensile strength of different extruder conditions

Figure 4 is the elastic modulus of PP/MAPP polymer and RSF/PP nanocomposite. The elastic modulus increased after added the 5% rice straw fibrils into the PP/MAPP polymer. For the PP/MAPP polymer, the elastic modulus increased significantly ($R^2 = 0.53$) with increasing compounding temperature and extruder cycle time. However, for the RSF/PP nanocomposite, the elastic modulus decreased with increasing compounding temperature and extruder cycle time, but this trend was not distinct.

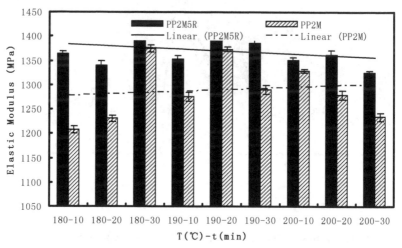

Fig. 4. Elastic modulus of different extruder conditions

The elongation at break was higher in PP/MAPP polymer than RSF/PP nanocomposite as shown in figure 5. With increasing compounding temperature and extruder cycle time the elongation at break decreased significantly ($R^2=0.71$) in PP/MAPP polymer, however, there had no big difference in RSF/PP nanocomposite.

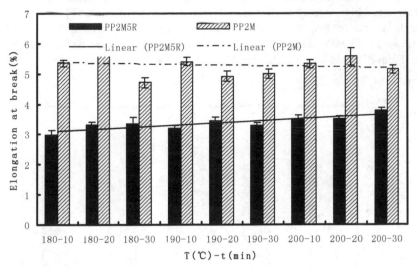

Fig. 5. Elongation of different extruder conditions

3.4 Effect of different RSF loadings on RSF/PP nanocomposite tensile properties

The tensile strength of RSF/PP nanocomposite with different fibril loadings is shown in figure 6. As the reference, the tensile strength of PP/MAPP polymer was also tested. The tensile strength at 5% RSF loading was up to the maximum value, 31.7 MPa, which was a little higher than the value of PP/MAPP polymer, 30.8 MPa. With increasing the fibril loadings the tensile strength decreased, but not distinct ($R^2 = 0.23$).

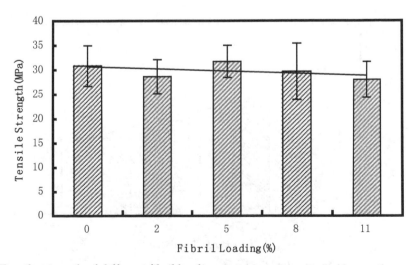

Fig. 6. Tensile strength of different fibril loadings

Figure 7 shows the elastic modulus of RSF/PP nanocomposite with different fibril loadings. The values were higher in RSF/PP nanocomposite than in PP/MAPP polymer. The fibril loadings from 2% to 8%, the elastic modulus increased significantly (R^2 = 0.70). The maximum was 1621 MPa at the 8% RSF, which was 17% higher than the value of PP/MAPP polymer. From 8% to 11% of RSF, the elastic modulus decreased.

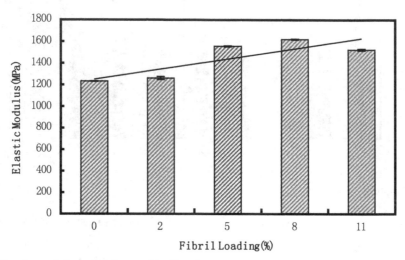

Fig. 7. Elastic modulus of different fibril loadings

The elongation at break showed a significant decreasing trend (R^2 = 0.89) with increasing the fibril loading (figure 8).

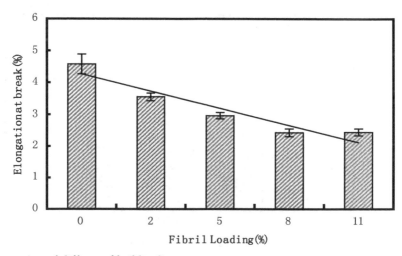

Fig. 8. Elongation of different fibril loadings

3.5 Effect of MAPP content on RSF/PP nanocomposite tensile properties

The tensile strength of PP/MAPP polymer decreased with increasing MAPP content from 1% to 6%, but not significant (figure 9). However, it showed a slightly increasing trend in RSF/PP nanocomposite as the MAPP content from 2% to 5%, then decreased when MAPP up to 6%.

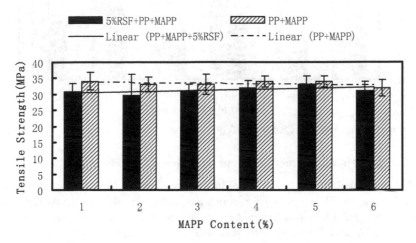

Fig. 9. Tensile strength of different MAPP contents

The elastic modulus and elongation at break appeared decreasing trend with increasing MAPP content both in PP/MAPP polymer and RSF/PP nanocomposite, as shown in figures 10 and 11. But the trends were not distinct according to the linear analysis.

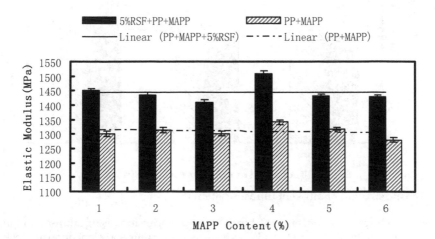

Fig. 10. Elastic modulus of different MAPP contents

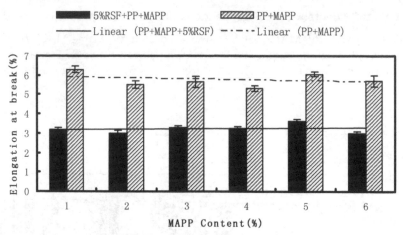

Fig. 11. Elongation at break of different MAPP contents

3.6 Effect of different ultrasonication treat condition on tensile properties

Figure 12 is the tensile strength of RSF/PP nanocomposite with different ultrasonication treat time RSF as the filler. For the ultrasonication treatment, the rice straw cellulose fiber content was 0.5% and 1%, respectively. As seen in this figure, the tensile strength increased distinctly ($R^2=0.70$ and $R^2=0.96$) with increasing ultrasonication treat time. The tensile strength of 0.5% rice straw cellulose fiber content was lower than 1% rice straw cellulose fiber content at different ultrasonication treat time.

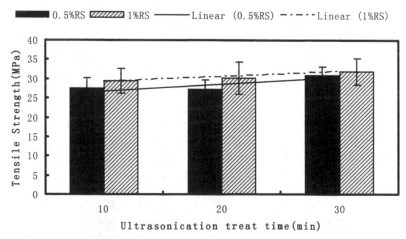

Fig. 12. Tensile strength of different ultrasonication treat time

The elastic modulus and elongation at break increased with increasing ultrasonication treat time as shown in figures 13 and 14. And also the elastic modulus and elongation at break were higher with 1% rice straw cellulose fiber content treated by ultrasonication as filler than 0.5%.

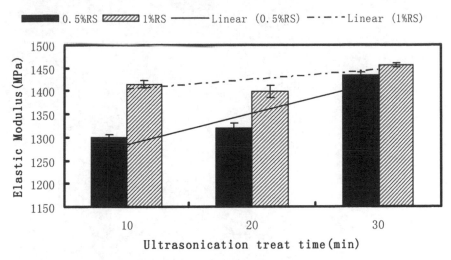

Fig. 13. Elastic modulus of different ultrasonication treat time

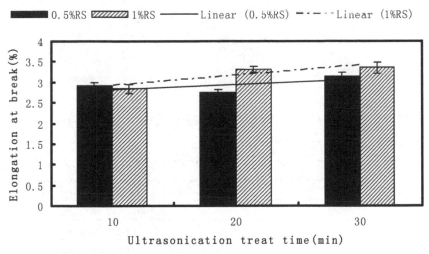

Fig. 14. Elongation at break of different ultrasonication treat time

3.7 FTIR analysis of RSF/PP nanocomposite

Figure 15 showed the FTIR spectra of PP, RSF/PP, RSF from up to down.

When adding MAPP and RSF into PP matrix, the FTIR spectra showed prodigious changes. CH_3 deformation of asymmetry stretching and CH_2 symmetry stretching moved to higher wavenumbers. Moreover, the frequency of CH_3 symmetry deformation decreased (1374CM-1 to 1372 CM-1). The existence of C-O-C stretching at 1224 CM-1, 1074 CM-1 and 1028 CM-1 indicated that PP and RSF were consistent by adding MAPP. Table 2 listed the vibration peaks and assignments of peaks of RSF/PP nanocomposite.

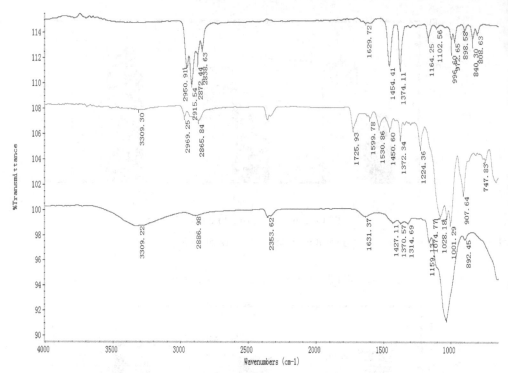

Fig. 15. FTIR curves of PP, RSF/PP nanocomposite and RSF

Wavenumber (cm-1)	Assignments
3309	-OH stretching
2969	CH3 deformation of asymmetry stretching
2865	CH2 symmetry stretching
1725	C=O stretching
1450	CH3 asymmetry deformation
1450	CH2 shearing
1372	CH3 symmetry deformation
1224, 1074, 1028	C-O-C stretching

Table 2. FTIR vibration peaks and assignments of peaks of RSF/PP nanocomposite

3.8 Morphology characteristics of fibers and fibrils

After ultrasonication treatment, the rice straw cellulose fiber changed into small size fibrils. The SEM pictures showed the appearance of untreated (figure 16a) and treated (figure 16b) rice straw cellulose fiber.

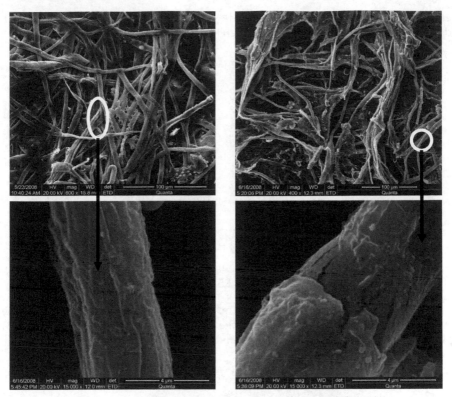

Fig. 16. SEM photos of untreated sample (a, left) and treated sample (b, right)

It can be seen from figure 13b inset figure that the fibrils peeled off from the fibers and the ultrasonication treatment made the surface of fibers producing fractures. The fibrils and microfibrils can form a kind of micro/nano-order-unit web-like network structure (Nakagaito & Yano, 2005), which can greatly expanded in the surface area that characteristic the RSF and increase the compounding abilities between fibrils/mircrofibrils and polymer. After the ultrasonication treatment, fibrils, fibrils aggregates and cellulose fibers together existed, and also the fibrils had a widely width (or diameter) ranges from tens of nanometer to micrometers as shown in figure 1. The fibrils with diameter less than 500 nm were peeled from the fibers, however, the fibril aggregates with diameter around 25 μm took the most place.

3.9 Fracture cross section morphology of RSF/PP nanocomposite

The fracture cross section morphology of RSF/PP nanocomposite after tensile test was observed by scanning electron microscopy (SEM). As shown in figure 17(a), the rice straw fibrils embedded into the polymer and exhibited better interaction with polymer. The fibrils were interconnected with the polymer and didn`t pull out from the polymer during the tensile test. However, the bigger size fiber still can be seen in the nanocomposite and it appeared the gas between the fiber and the polymer (figure 17b), which proved that the bigger size cellulose fiber and the polymer can not bond together easily or the compatibility was not as good as the fibril filler. And this may be the reason that the tensile strength in RSF/PP nanocomposite was lower than PP/MAPP polymer. To improve the degree and homogenization of fibril will be a good way to increase the compatibility between RSF and polymer.

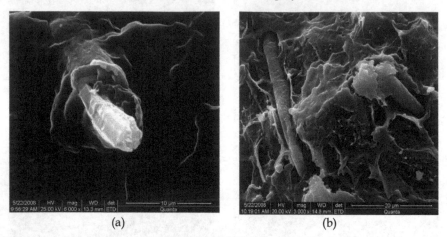

(a)   (b)

Fig. 17. Fracture cross section SEM photos of RSF/PP nanocomposite after tensile test

It was seen in figure 18 of PLM picture, the distribution of RSF in the RSF/PP nanocomposite was good, which indicated the minilab extruder with a cycle system was a suitable machine to compound the polymer and cellulose fibrils.

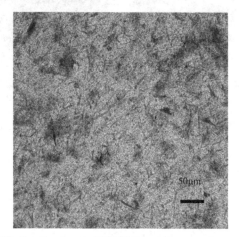

Fig. 18. PLM photo of RSF distribution in RSF/PP nanocomposite

4. Conclusions

The distribution of diameters of RSF ranged from 0.1 μm to 80 μm by HIUS treatment. The percentage was 6.3% of RSF which diameters were less than 500 nm; almost 90 percents of RSF distributed between 7.0 μm and 80 μm; the average diameter was 41 μm. The relative crystallinities of untreated rice straw cellulose fibers and rice straw cellulose fibers treated by HIUS were 71.3% and 72.9%, respectively.

The elastic modulus increased of RSF/PP nanocomposite comparing with PP/MAPP polymer. However, the tensile strength and elongation at break of 5% rice straw fibril reinforcing PP nanocomposite was lower than PP/MAPP polymer at different extruder compounding conditions. The tensile strength at 5% RSF loading was up to the maximum value, 31.7 MPa, which was a little higher than the value of PP/MAPP polymer, 30.8 MPa. The maximum was 1621 MPa at the 8% RSF, which was 17% higher than the value of PP/MAPP polymer. The elongation at break showed a significant decreasing trend with increasing the fibril loading. There was no distinct influence of tensile strength and elongation of nanocomposite and PP/MAPP polymer with increasing MAPP content. The maximum of elastic modulus was 1509 MPa at the 4% of MAPP. The tensile properties increased distinctly with increasing ultrasonication treat time. The tensile strength of 0.5% rice straw cellulose fiber content was lower than 1% rice straw cellulose fiber content at different ultrasonication treat time.

After adding MAPP and RSF into PP matrix, the FTIR spectra had big changes. The absorption peak of ester bonds (C-O-C) appeared at 1224 CM-1, 1074 CM-1 and 1028 CM-1, which proved that there had been a good compatibility between PP matrix and RSF. The SEM images showed: the fibrils were embedded into the PP/MAPP matrix, during the tensile test, which were not pulled out from the matrix.

5. Acknowledgment

The work is supported by Natural Science Foundation of China (31100417) and (31070492).

6. References

ASTM D (D 882 - 02) Standard test method for tensile properties of thin plastic sheeting.

Bataille, P.; Ricard, L. & Sapieha, S. (1989). Effects of cellulose fibers in polypropylene composites. *Polymer Composites*, Vol. 10, pp. 103, ISSN 1548-0569

Cheng, Q.; Wang, S.; Rials, T. & Lee, S. (2007a). Physical and mechanical properties of polyvinyl alcohol and polypropylene composite materials reinforced with fibril aggregates isolated from regenerated cellulose fibers. *Cellulose*, Vol. 14, pp. 593-602, ISSN 0969-0239

Cheng, Q.; Wang, S.; Zhou, D.; Zhang, Y. & Rials, T. (2007b). Lyocell-derived cellulose fibril and its biodegradable nanocomposite. *Journal of Nanjing Forestry University*, Vol. 31, No. 4, pp. 21-26, ISSN 1000-2006

Dufresne, A.; Cavaille, J.Y. & Vignon, M.R. (1997). Mechanical behavior of sheets prepared from sugar beet cellulose microfibrils. *Journal of Applied Polymer Science*, Vol. 64, No. 6, pp. 633-639, ISSN 1097-4628

Eichhorn, S.J. & Young, R.J. (2001). The Young's modulus of a microcrystalline cellulose. *Cellulose*, Vol. 8, No. 3, pp. 197-207, ISSN 0969-0239

Herrick, F.W.; Casebier, R.L.; Hamilton, J.K. & Sandberg, K.R. (1983). Microfibrillated cellulose: morphology and accessibility. *Journal of Applied Polymer Science*, Vol. 37, pp. 797-813, ISSN 1097-4628

Helbert, W. J.; Cavaillé, Y.; Dufresne, A. (1996). Thermoplastic nanocomposites filled with wheat straw cellulose whiskers. Part I: processing and mechanical behavior. *Polymer Composites*, Vol. 17, pp. 604-611, ISSN 1548-0569

Karnani, R.; Mohan, K. & Ramini, N. (1997). Biofiber-reinforced polypropylene composites. *Polymer Engineering & Science*, Vol. 37, pp. 476, ISSN 1548-2634

Nakagaito, A.N. & Yano, H. (2005). Novel high-strength biocomposites based on micriofibrillated cellulose having nano-order-unit web-like network structure. Journal of Applied Physics, Vol. 80, pp. 155-159, ISSN 0021-8979

Sakurada, I.; Nukushina, Y. & Ito, T. (1962). Experimental determination of elastic modulus of crystalline regions in oriented polymer. *Journal of Polymer Science Part B: Polymer Physics*, Vol, 57, No. 165, pp. 651-660, ISSN 1099-0488

Stenstad, P.; Andresen, M. & Tanem, B. (2008). Chemical surface modifications of microfibrillated cellulose. *Cellulose*, Vol. 15, pp. 35-45, ISSN 0969-0239

Thygesen, A.; Oddershede, J.; Lilholt, H.; Thosmsen, A.B. & Stahl, K. (2005). On the determination of crystallinity and cellulose content in plant fibres. *Cellulose*, Vol. 12, No. 6, pp. 563-576, ISSN 0969-0239

Turbak, A.F.; Snyder, F.W. & Sandberg, K.R. (1983). Microfibrilated cellulose, a new cellulos product: properties, uses, and commercial potential. *Journal of Applied Polymer Science*, Vol. 37, pp. 815-827, ISSN 1097-4628

Wang, S. & Cheng, Q. (2009). A novel method to isolate fibrils from cellulose fibers by high intensity ultrasonication. Part I. Process optimization. *Journal of Applied Polymer Science*, Vol. 113, pp. 1270-1275, ISSN 1097-4628

Wu, Y.; Zhou, D.G.; Wang, S.Q.; Zhang, Y. (2009) Polypropylene composites reinforced with rice straw micro/nano fibrils isolated by high intensity ultrasonication. *BioResources*, Vol. 4, pp. 1487-1497, ISSN 1930-2126

Zimmermann, T.; Pohler, E. & Geiger, T. (2004). Cellulose fibrils for polymer reinforcement. *Advanced Engineering Materials*, Vol. 6, No. 9, pp. 754-761, ISSN 1527-2648

4

Rheological Properties of Surface Treated Glass Fiber Reinforced Polypropylenes in Molten State

Yosuke Nishitani[1], Chiharu Ishii[2] and Takeshi Kitano[3]
[1]Kogakuin University,
[2]Hosei University,
[3]Tomas Bata University in Zlin
[1,2]Japan
[3]Czech Republic

1. Introduction

Fiber reinforced thermoplastics (FRTP) are widely used to industrial applications such as automobiles and electric devices in recent years (Thomason & Vlug, 1996; Kitano et al., 2000; Nishitani et al., 1999, 2001; Hausnerova et al., 2006). Since the machine and the electric devices became small and lightweight, the good balance of the physical properties and processability of FRTP is desired strongly. Polyolefines represent a group of the most common used polymers as matrix for FRTP, because they are inexpensive, easily processed and recycled (Nishitani et al., 1998a, 1998b, 2007). On the other hand, glass fiber (GF) is the most used as reinforcement for FRTP so as to modify the mechanical properties such as stiffness, strength, toughness, heat resistance and so on, and the other properties. Hence, glass fiber reinforced polypropylene composites (GF/PP) are of particular interest in these fields. However, it is difficult to form a strong bond between filler (fiber) and polymer due to poor wettability of the filler (fiber) especially in nonpolar polymer (Shenoy, 1999). The interfacial bond can be enhanced and the mechanical properties of the composites will be improved by suitable surface treatment. Most of fillers (fibers) are pretreated before they are used as secondary phases in composite materials. As typical surface treatment for polymer matrix composites , we often use the coupling agent as follows: silanes, azidsilanes, titantes and organopolysiloxianes, and so on (Shenoy, 1999). In particular, the most effective methods for controlling of the interface and interphase adhesion between GF and PP in GF/PP composites in industrial fields are considered to use silane coupling agents (Mader, 1996; Nishitani et al. 1998b). Many authors have investigated the effect of surface treatment on the various physical and chemical properties of GF/PP composites (Yue & Quek, 1994; Thomason & Schoolenberg, 1994; Mader, 1996, 2001; Lee & Jang, 1997; Kikuchi et al., 1997; Van Den Oever & Peijs, 1998; Hamada et al., 2000). However, their properties will be able to still be improved more by the suitable control of the interphase adhesion between GF and PP.

In order to achieve higher performance in FRTP such as GF/PP, there is a key issue that the rheological properties in molten state is very critical for these materials to understand proccesability, internal microstructures, their change and structure property relationship (Hausnerova, 2006, 2008a, 2008b; Nishitani, 2010a, 2010b, 2010c). In particular, although FRTPs undergo various flows during processing by flow molding such as injection, extrusion and compression, generally, the effect of surface treatment on processing properties has not been studied enough (Boaira & Chaffey, 1977; Han et al., 1981; Bigg 1982; Luo et al., 1983; Sani et al., 1985; Khan & Prud'Homme, 1987; Nishitani et al. 1998b, 2007; Shenoy 1999). Therefore, we need a proper rheological study on the effect of surface treatment taking into account the various factors such as type of fiber, its size and size distribution, and degree of agglomeration. However, since these factors are interrelated, the determination of the effect of surface treatment on the interphase adhesion is thought to be necessarily a complicated task. Han et al. showed that there were viscosity decrease and the first normal stress difference increase for $CaCO_3/PP$ composites regardless of the type of coupling agent used (Han et al., 1981). In contrast, in the case of glass bead filled PP the effect of surface treatment was not so distinct, the viscosity increased after treatment by aminosilane coupling agent, and decreased when octylsilane and titanate coupling agents were used (Khan & Prud'Homme, 1987). Similar complex rheological behavior appeared as the results of surface treatment of $CaSiO_3$ filled PA6 (Luo et al., 1983) and ferrite filled PP (Saini et al., 1985).

Recently, we studied the effect of silane coupling agents on the rheological properties of short and long glass fiber reinforced polypropylene composites in molten state (Nishitani et al., 1998b, 2007). Surface treatment by silane coupling agents increased the storage modulus and reduced the peak in the loss tangent. Furthermore, decrease in the dynamic viscosity was affected by the concentration of coupling agents. Nevertheless, according to our survey of the previous results, the effect of surface treatment on the rheological properties is still not well known. It is therefore necessary to investigate systematically for further understanding of it. The objective of this chapter is to report the results on the effect of surface treatment by silane coupling agent on the rheological properties of short and long glass fiber reinforced polypropylene composites in the molten state which were obtained mainly in our previous studies.

2. Rheological properties of surface treated short glass fiber reinforced polypropylenes

Short fiber reinforced thermoplastic composites have been employed extensively in the plastic industry because of the excellent combination of their mechanical properties, chemical resistance, moderate cost, and recyclability performance. Short glass fiber reinforced polypropylene composites (GF/PP) are of particular interest (Nishitani et al., 2007). Despite great effort expended on research into GF/PP composites, their properties can still be improved by control of the interphase adhesion between GF and PP. One of the methods of improving the adhesion between polymer and fiber is the use of surface treatment agents (Shenoy, 1999). Treatment agents, when properly chosen, bring considerable improvement, particularly in mechanical and thermal properties. A typical surface treatment for GFs is a coupling agent. Another method involves the addition of PP modified by maleic anhydride (mPP) (Sasagi & Ide, 1980). Mechanical properties are improved by the increase in interfacial adhesion. The lower the degree of grafting of maleic anhydride to the modified PP and the higher its molecular weight is, the more effective the improvement is. Recently, a combination of these methods was investigated (Peltonen et al., 1995; Nishitani et al., 1998c, 2007), but the mechanism is still not fully understood. In particular, there have been few efforts to investigate the rheological properties of GF/PP composites by these combinations. The aim of this section is to report the results of the effect of the fiber surface treatment with aminosilane coupling agent on the rheological properties of short glass fiber reinforced polypropylene composites (GF/PP) and of GF/PP composites supplemented by maleic anhydride modified PP (GF/mPP/PP). Rheological functions in the molten state were measured under steady state shear (viscous and elastic properties) and oscillatory flow (viscoelastic properties) regimes.

2.1 Materials and methods

The effect of surface treatment on the rheological properties of short glass fiber reinforced polypropylene (GF/PP) in molten state was investigated in this section. The materials used in this section were GF/PP and the same materials added with maleic anhydride modified PP (GF/mPP/PP). The polymers used were polypropylene (PP, Sumitomo Novelen, Sumitomo Chemical Co., Japan) and maleic anhydride modified PP (mPP, Sumitomo Chemical Co.) Glass fiber (GF, Micro Glass Fiber, Nippon Glass Fiber Co., Japan) was filled with PP. Surface treatment by aminosilane coupling agent (ASC, γ-aminopropyl-trietoxysilane, A1100, Nippon Unika Co., Japan) with different concentrations was performed on glass fibers. The GFs were pre-treated with ASC in the following way. ASC was mixed with 20 times its volume of water, then the hydrolyses ASC was further diluted in water to concentrations of 0.2, 0.5 and 1.0 wt.%, and the aqueous solutions were applied to GFs. The GF content was fixed with 20wt.% and 40wt.%. The compositions and average fiber lengths of the samples are shown in Table 1.

| Composition (wt.%) | PP | 20GF | | | | | | | | 40GF | | | |
		GF/PP				GF/mPP/PP				GF/PP		GF/mPP/PP	
PP	100	80	80	80	80	75	75	75	75	60	60	55	55
mPP	-	-	-	-	-	5	5	5	5	-	-	5	5
GF	-	20	20	20	20	20	20	20	20	40	40	40	40
ASC	-	-	0.2	0.5	1	-	0.2	0.5	1	-	1	-	1
Fiber length (mm)	Original	1.038	-	-	1.177	1.020	-	-	1.134	-	-	-	-
	Remixed	0.648	-	-	0.731	0.620	-	-	0.674	-	-	-	-

Table 1. Composition and average fiber length of GF/PP and GF/mPP/PP composites.

The compounding of PP with GF was carried out at 200°C in a specially developed elastic extruder (Kataoka et al, 1976; Zang et al, 1995, Nishitani et al 1998c, 2007). The weighed amounts of the pellets used for making the 3mm thick compression molded samples were placed into the mold cavity. They were compressed for 3min under 5MPa at 200°C. Finally, the test specimens for the rheological measurements were cut from the molded sheets. And second time-mixed composites were also prepared in order to investigate the influence of this process on the effect of surface treatment.

The steady-state shear flow properties in the low shear rate region and the dynamic functions were measured using a rotational viscometer (cone-plate type, RGM 151-S, Nippon Rheology Kiki Co., Ltd., Japan). The cone radius R was 21.5mm, the gap between the central area of the cone and plate H was kept at 175μm, and the cone angle θ was 4°. The measurements were carried out at 200°C. Steady state shear properties (shear viscosity η, and the first normal stress difference N_1) as well as dynamic functions (storage and loss moduli G', G'', respectively, dynamic viscosity η' and complex viscosity $|\eta^*|$) were determined as functions of the shear rate γ and angular frequency ω from 10^{-2} to 10 1/s (rad/s), respectively. The strain amplitude was chosen to be 25% (oscillatory angle $\pm 1°$) which is considered to be a large value in order to compare the results of the oscillatory and steady shear regimes.

2.2 Shear rate dependences

The shear viscosity η and the first normal stress difference N_1 are plotted against share rate $\dot{\gamma}$ for both the untreated and 1wt.% ASC treated GF/PP composites in Fig.1. In the lower shear region, between 0.05 and 0.5 1/s, the character of the viscosity curves of all the GF filled systems were similar to that of pure PP – i.e. they were nearly independent of the shear rate. It is thought that increase in shear viscosity was caused by enhanced resistance to shear flow arising from structure composed by reinforcing fibers in the system, and also from the increased fiber-polymer matrix interactions supported by surface treatment. In the shear rate region higher than 0.5 1/s, where the PP behaves in a pseudoplastic manner, the fibers tend to orient in the flow direction, and the effect of surface treatment by ASC diminishes diminishes(Fig.1(a)). The first normal stress difference N_1 data shown in Fig. 1(b) indicates an increase in elastic property as a result of the surface treatment by ASC. In general, the first normal stress difference N_1 increases with an increase in fiber length, with structure formation by the fibers, and with the elastic properties of fibers themselves. The degree of dispersion of fibers in a polymer matrix is also important factor influencing on the elastic properties of fiber reinforced composites in the molten state. When the fibers are surface treated, the structure formed can be changed easily. Furthermore, the apparent fiber modulus of a treated system will be higher than that of an untreated one because of the softened phase, which is able to change the structure in the region of the polymer- fiber interface. Although the first normal stress differences of untreated GF/PP composites in the high shear rate region reach the values close to those of the pure polymer matrix, those of the treated materials are higher.

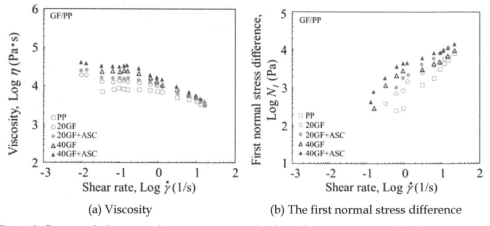

(a) Viscosity   (b) The first normal stress difference

Fig. 1. Influence of silane coupling agent on steady shear flow properties of GF/PP composites: (a) viscosity as a function of shear rate, (b) the first normal stress difference as a function of shear rate

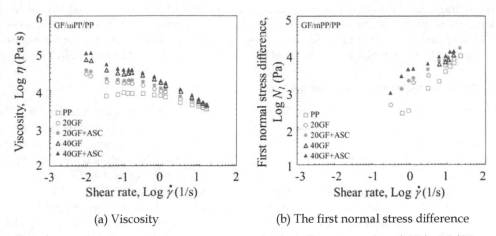

(a) Viscosity   (b) The first normal stress difference

Fig. 2. Influence of silane coupling agent on steady shear flow properties of GF/mPP/PP composites: (a) viscosity as a function of shear rate, (b) the first normal stress difference as a function of shear rate

In the low shear rate region, the first normal stress difference increases with the fiber content, which may be due to the hydrodynamic effect associated with fiber orientation in the flow direction. The effect of surface treatment was identical to that in the high shear rate region. The shear viscosity and the first normal stress difference as a function of shear rate for maleic anhydride modified PP (5wt.%) added composites (GF/mPP/PP compositese) are plotted in Fig. 2. Although the overall level of the both functions η and N_1 was higher than that in GF/PP composites (Fig. 1), the effect of surface treatment by 1wt.% ASC on η and N_1 of GF/PP composites is similar to that of GF/mPP/PP.

The results demonstrated in Figs. 1 and 2 are somewhat different from those obtained in studies of particulate filled polymer systems, for example, mica filled PP (Boaira & Chaffey, 1977) or calcium carbonate filled PP (Han et al., 1981), where the increased values of shear viscosity and the first normal stress difference caused by the filler addition could be reduced to a greater or lesser extent – by surface treatment. Such discrepancies in the results can be explained by the different roles played by the coupling agent during processing as follows. First, it forms chemical bonds between the inorganic filler and polymer matrix, and assists physical adhesion by Van der Walls's forces. Second, it may behave as a lubricant, which decreases the friction between the filler and polymer matrix. Third, it changes the interfacial energy of the filler, and simultaneously supports better filler dispersion in the polymer matrix, and reduces agglomeration by wetting. Finally, a surface treatment agent can behave as an additive to make deformation of fiber assembly easier and to make the viscoelastic properties of a matrix polymer lower. It is thought that the mechanisms mentioned above do not occur separately, and therefore it is difficult to distinguish a particular type.

Regarding GF/PP and GF/mPP/PP composites investigated in this section, the adhesion between fiber and polymer matrix was improved. In addition, because of the improved protection against the breakage of fibers, the fiber length of treated materials can be kept longer throughout the compounding and processing than that of untreated ones. Finally, the addition of maleic anhydride modified PP (mPP) increases the rheological properties (viscosity and the first normal difference), because mPP probably not only improves the surface treatment, but also forms chemical bonds between fiber and polymer matrix.

2.3 Angular frequency dependences

The storage and loss moduli G' and G'' as a function of angular frequency ω for GF/PP composites are shown in Fig. 3. The storage modulus G' of the GF/PP composites treated by 1 wt.% ASC was scarcely higher than that of the untreated ones, and it was almost independent of fiber content (20 or 40 wt.%), although generally G' should increase with fiber loading level (Fig.3(a)). Also, the loss modulus G'' of the treated GF/PP composites (Fig. 3(b)) was only slightly higher than that of the untreated composites. Unlike G', G'' increased with increasing the fiber content. Although G'' of the untreated composites (20 and 40 wt.% GF) increased over the whole angular frequency range covered by the experiments, this trend was valid only for angular frequencies higher than 0.1 rad/s in the case of the treated materials. In the frequency region lower than 0.1 rad/s, the surface treatment by ASC initiated the behaviour in which G'' was independent of ω, indicating a second rubbery plateau, i.e. the long-scale relaxation time. This behaviour implies the presence of an apparent yield stress, which will be attributed to the strong fiber structure formation imparted by the surface treatment. The properties of GF/mPP/PP composites, demonstrated in Fig. 4, showed the same trend of G' against frequency dependency as for GF/PP composites in Fig. 3(a). However, with 20wt.% GF and no ASC treatment, there was no evidence of plateau behaviour in the loss modulus.

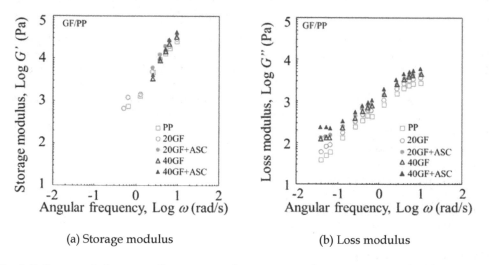

(a) Storage modulus   (b) Loss modulus

Fig. 3. Influence of silane coupling agent on dynamic viscoelastic properties of GF/PP composites: (a) storage modulus as a function of angular frequency, (b) loss modulus as a function of angular frequency

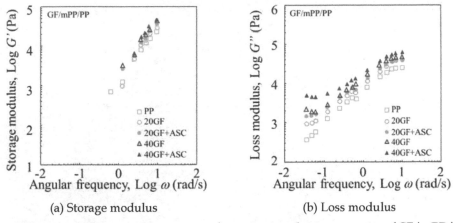

(a) Storage modulus   (b) Loss modulus

Fig. 4. Influence of silane coupling agent on dynamic viscoelastic properties of GF/mPP/PP composites: (a) storage modulus as a function of angular frequency, (b) loss modulus as a function of angular frequency

2.4 Comparison of steady shear and dynamic flow data

Cox and Merz (Cox & Merz, 1958) observed a close similarity between the steady state shear and complex viscosities determined at corresponding values of the shear rate and angular frequency, respectively. This empirical rule can be written as follows:

$$\eta(\dot{\gamma}) = |\eta^*(\omega)| = \sqrt{\eta'^2(\omega) + \eta''^2(\omega)} \tag{1}$$

Where η, $|\eta^*|$ are the shear and complex viscosities, and η', η'' are the real and imaginary parts of the complex viscosity $|\eta^*|$, respectively. The validity of this relation for polymer melts and concentrated solutions has been amply verified. The steady state shear and complex viscosities both approached their limiting (zero viscosity) values as their arguments go to zero, and they decreased in a similar way with increasing γ or ω, although at high shear rates (angular frequency) $|\eta^*|$ may fall more rapidly than η. As proposed by Dealy (Dealy & Wissbrun, 1990), the Cox-Merz rule should be generally valid for flexible molecules; however, it is not suitable for almost any fiber filled polymer systems (Kitano et al., 1984a, 1984b; Li et al., 1997; Nishitani et al., 2007), since the viscosity increment caused by the fber addition would be different under the two types of flow.

In Fig. 5, double-logarithmic plots of η vs. γ and $|\eta^*|$ vs. ω are shown for both GF/PP and GF/mPP/PP composites. The η and $|\eta^*|$ curves coincided relatively well in their plateau (Newtonian) regions at shear rates (angular frequencies) from 0.1 to 1 1/s (ras/s) for untreated samples. Nevertheless, the $|\eta^*|$ values of the GF/PP treated by 1wt.% ASC, shown in Fig. 5(a), were clearly higher than the η level. In the case of the treated GF/mPP /PP, in Fig. 5 (b), η vs. γ and $|\eta^*|$ vs. ω curves showed the same tendency as the untreated systems. Such behaviour supports the idea of the ASC surface treatment having different roles in GF/PP and GF/mPP/PP composites.

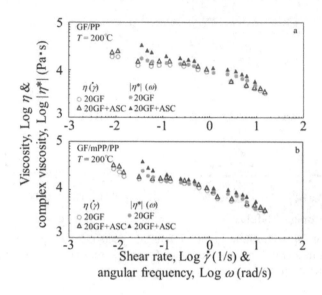

Fig. 5. Shear viscosity versus shear rate and complex viscosity versus angular frequency: (a) GF/PP composites, (b) GF/mPP/PP composites

2.5 Effect of coupling agent concentration

In order to clarify the influence of the concentration of the aminosilane coupling agent (ASC) on the rheological properties of GF/PP and GF/mPP/PP composites, both steady shear (η, N_1) and dynamic ($|\eta^*|$, G') functions are plotted against the ASC concentration C in Figs. 6 and 7. The content of fibers was fixed at 20wt.%., and the relative values representing the ratio of treated GF/PP and GF/mPP/PP to untreated GF/PP and GF/mPP/PP, respectively, are shown at two shear rates (angular frequencies) of 0.1 and 9.2 1/s (rad/s). The relationship between the relative viscosity η_r obtained from the steady shear flow measurements, and the ASC concentration C (wt.%) is demonstrated in Fig. 6(a). Generally, η_r increased slightly with increasing C, and for higher shear rate (9.2 1/s) it was almost independent of C for both systems. In order to discuss about the elastic response, the relation between the relative normal stress difference N_{1r}, and C is plotted in Fig. 6(b). It was found that whereas the N_{1r} of GF/mPP/PP composites showed a maximum at 0.5 wt.% ASC, that of GF/PP composites increased monotonously with increasing C. To conclude, the N_{1r} variation with amount of aminosilane coupling agent was much larger than that of η_r. Figs. 7(a) and 7(b) show the effect of the ASC concentration C on the viscoelastic properties. While the relative complex viscosity $|\eta^*|_r$ of the GF/PP system, in Fig. 7(a), increased with increasing C, it was almost independent of the amount of ASC for the GF/mPP/PP composite. The dependency of the relative storage modulus G'_r on the ASC concentration C of both composites shown in Fig. 7(b) was similar to that of $|\eta^*|_r$ vs. C curve for the GF/PP composites. However, the G'_r vs. C curve for GF/mPP/PP composites showed a peak at 0.2 wt.% ASC. In general, G'_r reached smaller values than N_{1r} shown in Fig. 6(b). As one explanation of this phenomenon, different deformation modes under the particular flow can be considered.

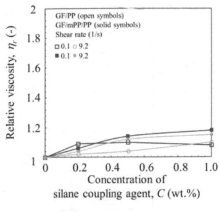
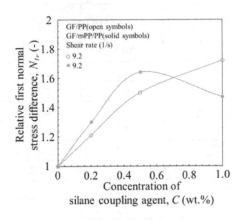

(a) Relative viscosity

(b) Relative first normal stress difference

Fig. 6. Relationship between relative values of steady shear flow functions and concentration of silane coupling agnet for GF/PP and GF/mPP/PP composites: (a) relative viscosity, (b) relative first normal stress difference

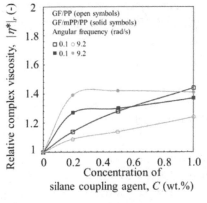
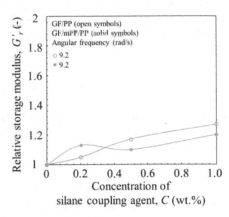

(a) Relative complex viscosity

(b) Relative storage modulus

Fig. 7. Relationship between relative values of oscillatory flow functions and concentration of silane coupling agnet for GF/PP and GF/mPP/PP composites: (a) relative complex viscosity, (b) relative storage modulus

Finally, the effect of surface treatment with regard to the fiber content is described and summarized in Table 2. The increase in the ASC concentration seems to make the processing of composites more difficult due to the increase of viscous and elastic properties of the composites. However, mechanical properties such as tensile, bending and impact strength are highly enhanced by ASC treatment (Nishitani et al. 1998b, 1998d). Consequently, the optimum concentration of surface treatment agent has to be determined individually, according to the performance required for a particular product.

Relative values	$\dot{\gamma}\,(1/s)$ or $\omega\,(\text{rad}/s)$	1 wt.% ASC			
		GF/PP		GF/mPP/PP	
		20 wt.% GF	40 wt.% GF	20 wt.% GF	40 wt.% GF
η/η_0	0.1	1.06	1.35	1.18	1.16
	9.2	1.10	1.16	1.17	1.02
$\|\eta^*\|/\|\eta^*\|_0$	0.1	1.45	1.46	1.38	1.91
	9.2	1.25	1.18	1.42	1.13
N_1/N_{10}	0.1	-	1.82	-	-
	9.2	1.72	1.68	1.47	1.42
G'/G'_0	0.1	-	-	-	-
	9.2	1.27	1.24	1.20	1.17

Table 2. Relative values of rheological properties for GF/PP and GF/mPP/PP composites

2.6 Yield stress

The presence or absence of a yield stress is of great importance in the molding of filler filled or fiber reinforced polymers, and also for the physical stability of many industrial products. Casson proposed an equation describing the steady state shear flow properties of the suspensions of solid particles in Newtonian liquids (Casson, 1959), so as to easily evaluate yield stresses:

$$\tau^{1/2} = \tau_0^{1/2} + k_1 \dot{\gamma}^{1/2} \tag{2}$$

Where τ is the shear stress, τ_0 represents its yield value, $\dot{\gamma}$ stands for shear rate, and k_1 is a constant. If the plots of $\tau^{1/2}$ against $\dot{\gamma}^{1/2}$ give straight lines, the yield stress can be obtained by extrapolation to $\gamma=0$.

In this section, we discuss about the existence of the yield stresses of the GF/PP and GF/mPP/PP composites in the molten state and their relationship among the yield stress, content of dispersed fillers (fibers), and the structure of fiber assembly (Figs. 8 and 9). Fig.8 shows Casson's plots for GF/PP and the GF/mPP/PP composites. The line for pure PP is straight ones passing through the origin, suggesting the absence of a yield point. Concerning filled PP, the yield stresses, σ_y is determined only in the very low shear rate region ($\dot{\gamma}^{1/2}$ less than 0.4 $1/s^{1/2}$), where the flow data can be fitted by the straight lines. The σ_y of GF/PP composites, which is estimated from Fig. 8(a), seems to increase slightly with increasing GF content, and also by the ASC surface treatment of GF. The rate of σ_y increase for the GF/mPP/PP, Fig. 8(b), is larger than that for the GF/PP. This means that the addition of modified PP strongly influences on the yield stress appearance.

Such a trend can be seen clearly in Fig. 9, where the yield stresses of both composites are plotted as a function of the weight fraction of glass fiber ϕ_w in semi-logarithmic co-ordinates. The surface treatment of GF by silane coupling agent, and further the addition of modified PP increases the yield stress of the composites in the same manner as the increment of fiber content. This behaviour is attributed to the strong structure formed by the fiber assembly, which are coupled by ASC, and strengthened by modified PP. Finally, it should be noted here that a yield stress is usually responsible for long scale relaxation times (the second rubbery plateau behaviour).

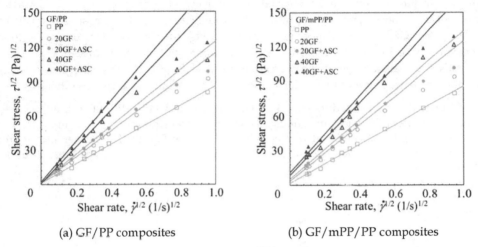

(a) GF/PP composites   (b) GF/mPP/PP composites

Fig. 8. Casson's plots: (a) GF/PP composites, (b) GF/mPP/PP composites

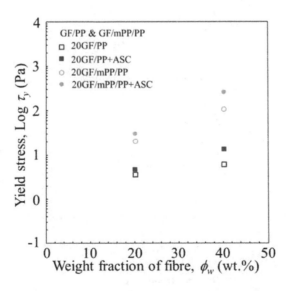

Fig. 9. Relationship between yield value of shear stress and weight fraction of fiber for GF/PP and GF/mPP/PP composites

2.7 Effet of re-mixing process on rheological properties

Re-mixing (second compounding) of short fiber reinforced polymers is generally carried out to avoid the non-uniform dispersion of fillers or fibers in a polymer matrix. Composites processed by injection molding undergo the second mixing in an extruder or injection molding machine. In this section, we investigated the influence of re-mixing on the ASC surface treatment. The plots of η vs. γ for the mixed GF/PP and GF/mPP/PP composites and remixed GF/PP-R and GF/mPP/PP-R ones are depicted in Fig. 10. The viscosities of the re-mixed composites were lower, and there was no clear difference between untreated and treated materials. The reason might be the decrease in the fiber length after re-mixing (see Table 1), and the simultaneous diminution of the influence of fiber length on the viscous properties.

In addition, the viscosity difference between GF/PP and GF/mPP/PP composites was reduced after second mixing. In contrast, the complex viscosity of re-mixed composites was different for treated and untreated materials, as shown in Fig. 11. The former showed the same tendency as the shear viscosity, but the latter values were not much affected by re-mixing.

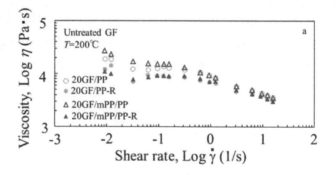

(a) Untreated composites

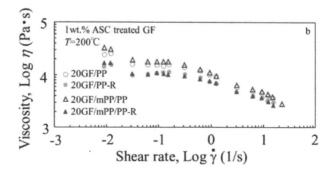

(b) ASC treated composites

Fig. 10. Effect of re-mixing on viscosity as a function of shear rate curves: (a) untreated composites, (b) ASC treated composites

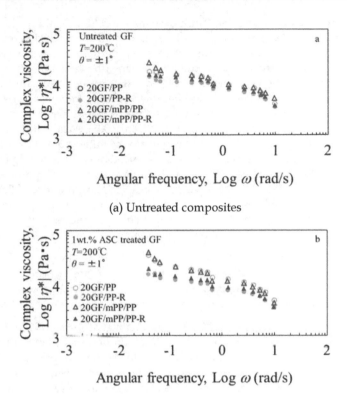

(a) Untreated composites

(b) ASC treated composites

Fig. 11. Effect of re-mixing on complex viscosity as a function of angular frequency curves: (a) untreated composites, (b) ASC treated composites

3. Rheological properties of surface treated long glass fiber reinforced polypropylenes

Fiber reinforced thermoplastic composites are usually fabricated by a flow molding process. For high performance fiber reinforced composites it is important to analyse the flow /deformation behaviour of fiber filled systems during the polymer processing, and to investigate the flow mechanisms and any changes in the internal structure of these systems. The flow properties of short fiber filled systems, whose reinforcement fiber length is up to several 100 micrometers, have been studied extensively (Czarnecki & White, 1980; Laun, 1984; Mutel & Kamal, 1986; Ausias et al, 1992; Basu et al, 1992; Greene & Wikes, 1995; Kim & Song, 1997). For these systems, non-uniformity of orientation or dispersion of fibers arises, depending on flow conditions, and it is well known that this anisotropic or heterogeneous structure will cause anisotropic physical properties (Kitano et al, 1981). On the other hand, researches into continuous or long fiber filled systems have not been widespread, because the measuring or analysing methods required are not established yet, and the flow behaviour becomes more complicated in this case (Davis & Mcalea, 1990; Groves & Stocks, 1991; Groves et al, 1992; Davis & Manson 1993; Greene & Wikes, 1995; Nishitani et al 1998a, 1998b, 2001).

The measuring methods for short fiber filled systems are seldom valid for long fiber filled ones. Steady shear flow measurements, usually used for short fiber filled systems, cannot be applied to long fiber filled ones, since the heterogeneity of the latter system is higher than that of the former, and the structure composed by the long fiber assembly changes gradually under large shear deformations. Then, the stable flow state cannot be obtained. Therefore, it is common for the flow properties of long fiber filled systems to be evaluated by dynamic viscoelastic properties, measured by means of oscillatory flow experiments under small strain/deformation amplitudes (Nishitani et al 1998b, 2001). When measuring the flow properties of fiber filled systems it is desirable to retain the initial state of fiber orientation or dispersion. The change in the initial state is easily occurred during the setting up of a sample when a cone-plate typed rheometer is used. Therefore, it is more appropriate to use a parallel-plate typed rheometer for the viscoelastic properties of long fiber filled systems. In addition, it is desirable to know the flow properties of continuous or long fiber filled composites in order to clarify unsolved problems such as the impregnation of matrix resin and the uniformity of the fiber distribution in composites.

Furthermore, various surface treatments are used in order to achieve the high performance. The methods of surface treatment for long fiber filled systems are the same ones as for short fiber filled systems, however the effect of the surface treatment on the rheological properties of long fiber filled composites is not understood enough yet.

In our previous works (Kitano et al, 1994, 2000; Nishitani et al, 1998a, 1998b, 1999, 2001) we investigated the fabrication methods and physical properties of long fiber reinforced polypropylenes, which were compression molded polypropylene fibers mixed homogeneously with long reinforcement fibers such as glass fiber, carbon fiber, aramid fiber, polyvinyl alchol fiber and polyamide 6 fiber by an apparatus called a "fiber separating and flying machine". The fabrication method employed here is a discontinuous and dry process similar to the stampable sheet molding method. This is superior to other manufacturing methods from the point of view of thermal efficiency, isotropic physical properties (because of the completely separated and homogeneously dispersed reinforcement fibers), and applicability to hybrid composites.

The objective of this section is to report the effect of surface treatment on the rheological properties, which is the dynamic viscoelastic properties in the molten state, investigated experimentally, for long glass fiber reinforced polypropylene composites (GFL/PP) prepared by the mentioned above. The present section discusses the dynamic viscoelastic properties in terms of various factors: angular frequency, concentration of silane coupling agent, various kinds of silane coupling agents, volume fraction of fiber, temperature and strain amplitude.

3.1 Materials and methods

The materials used in this section were surface treated long glass fiber reinforced polypropylene melts. Polypropylene fiber (PP, Showa aroma, Showa Denko K. K., Japan) was used as the matrix. Glass fiber (GF, Micro Glass Roving, Nihon Glass Fiber Co., Ltd., Japan, fiber diameter d=13μm) was used as reinforcement fibers. Surface treatment by different kinds of silane coupling agents with different concentrations was performed on glass fibers. Four types of silane coupling agents: aminosilane (ASC, A-1100, Nippon Unicar

Co. Ltd., Japan), diaminosilane (DAS, SH-6020P, Dow Corning Toray Co., Ltd., Japan), acrylsilane (ACS, A174, Nippn Unicar Co., Ltd.) and epoxysilane (ESC, A187, Nippon Unicar Co., Ltd.) were used as surface treatment agent. These details: code, component name and concentration are listed in Table 3.

Coupling agent	Code	Component Name	Concentration C (wt.%)	Volume fraction of fiber V_f (vol.%)
Aminosilane	ASC	γ-aminopropyltrietoxysilane	0.2, 0.5, 1.0	5, 10, 20, 30
Diaminosilane	DAS	γ-(2-minoethyl)aminopropyltrimetoxysilane	0.5	10, 20
Acrylsilane	ACS	γ-methacryloxypropyltrimetoxysilane	1.0	10, 20
Epoxysilane	ESC	γ-glycidxypropyltrimetoxysilane	1.0	10, 20

Table 3. Name of coupling agent, its component and code, its concentration and volume fraction of GF

All of the composite materials were compression molded from mixed mats prepared by the apparatus called a "fiber separating and flying machine" (Nishitani et al, 1999, 2001). This apparatus is schematically illustrated in Fig. 12. Continuous GF with volume fraction of 5, 10, 20 and 30 vol.% were fed into the "fiber separating and flying machine" apparatus simultaneously with PP fibers which are previously cut in 100 to 150mm. The collected mixed mats prepared by this apparatus were used as a base material for compression molding. Then the mixed mats were cut into 150 x 150 mm pieces, the weighed amounts of the mats for molded samples with 3mm thickness were piled into the mold cavity. These mats were kept in an air circulation oven at 120°C for 3 hours in order to remove the absorbed water in the fibers. Then the dried mats were compression-molded by a hot press in 3mm thick sheets for 3minutes under 5MPa at 200°C. Test specimens for viscoelastic properties measurements were cut from the sheets. The average length of glass fibers was 21.6mm, their orientation was generally three dimensional, and the degree of orientation gradually changed with the fiber content.

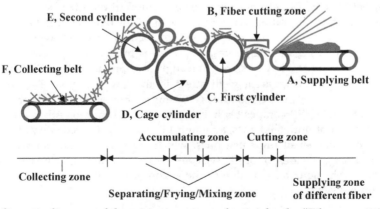

Fig. 12. Schematic diagram of the mixture mats production by the "Fiber separating and flying machine" apparatus

For long fiber reinforced composites it is difficult to estimate their rheological properties by using a cone-plate typed or capillary typed rheometer under steady shear flow. Therefore, specially designed rheometer is used for continuous fiber reinforced composites (Nishitani et al, 1998b, 1998c, 2001). In our experiments, a parallel-plate typed rheometer (151-S, Nippon Rheology Kiki Co., Ltd., Japan) was used to measure the dynamic viscoelastic properties of the samples. The specimens were compression-molded sheets cut into disks of 27 to 28 mm diameter. The gap between the two plates of 12.5mm radius was fixed at 3mm. Under such gap condition, a test specimen was slightly compressed in the molten state. The angular frequency varied from 10^{-2} to 10^{2} rad/s, and the forced oscillatory angle of lower plate was used at 0.5, 1.0 and 2.0 degree, corresponding to the strain amplitude of 4.4%, 8.7% and 17.5% in shear unit, respectively. The measurements were carried out for all the samples at 180, 200 and 220°C. Linear viscoelastic properties such as storage modulus G', loss modulus G'', dynamic viscosity η' and loss tangent tan δ were determined.

3.2 Angular frequency dependence

First, the effect of the surface treatment on the rheological properties of long glass fiber reinforced polypropylene composites (GFL/PP) is discussed. The rheological properties are evaluated by the dynamic viscoelastic properties in the molten state, which are strongly dependent on the internal micro structure formation of the polymer composites. We shall discuss the angular frequency dependence, which is the basic variable in dynamic viscoelastic properties. All the results discussed in this section were measured under an oscillatory angle of 0.5° (strain amplitude 4.4%) and the range of angular frequency was from 10^{-2} to 10^{2} rad/s. Fig. 13(a) shows the effect of the concentration of aminosilane coupling agent (ASC) on the relationship between the storage modulus G' and angular frequency ω of GFL/PP composites. G' of untreated GFL/PP increases with increasing ω. G' of ASC treated GFL/PP increases with increasing the concentration of ASC (from 0.1 to 1.0 wt.%) and shows the typical G' curve similar to highly filled systems although the content of GF in the composites is 10vol.%. The slopes of G' against ω become small in low ω region, indicating the "second rubbery plateau" i.e. the long-scale relaxation time (Ferry, 1980; Nishitani et al, 2001, 2007, 2010a, 2010b). This tendency is highly enhanced by the increase of the ASC concentration. This behaviour may be attributed to the fiber network formation, which is due to high aspect ratio, and the interfacial interaction between GF and PP. Fig. 13(b) shows the effect of the concentration of ASC on the relationship between the dynamic viscosity η' and ω of GFL/PP composites. η' decreases monotonously with increasing ω, and the slopes of η' against ω have a slope of an angle of approximately -45°. This behaviour shows the presence of an apparent yield stress in the low ω region. The effect of the concentration of ASC on η' or loss modulus G'' seems to show more complex behaviour. η' of the ASC treated GFL/PP composites having the concentration of ASC of less than 0.5 wt.% is less than that of untreated ones, although 1.0wt.% ASC treated GFL/PP composites shows higher η' than that of untreated ones, which means the treatment with low ASC concentration rather decreases η' or G'' (viscous properties) although increase G' (elastic property). These behaviour may be due to the action of the coupling agent as a wetting agent, lubricant or plasticizer.

In Fig. 14(a), the storage modulus G' is plotted against the angular frequency ω for various silane coupling agents treated GFL/PP composites. G' of all treated GFL/PP is higher than

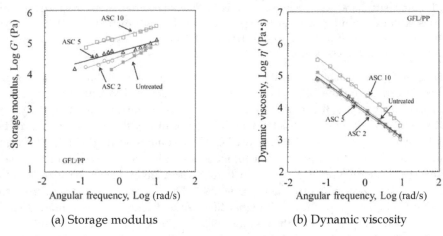

(a) Storage modulus   (b) Dynamic viscosity

Fig. 13. Influence of concentration of aminosilane coupling agent on dynamic viscoelastic properties for GFL/PP composites: (a) storage modulus versus angular frequency, (b) dynamic viscosity versus angular frequency

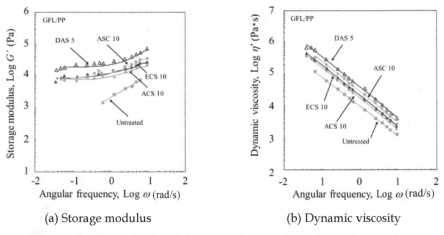

(a) Storage modulus   (b) Dynamic viscosity

Fig. 14. Influence of different kinds of silane coupling agent on dynamic viscoelastic properties for GFL/PP composites: (a) storage modulus versus angular frequency, (b) dynamic viscosity versus angular frequency

that of untreated GFL/PP composites in a wide range ω region and shows the typical G' curve similar to that of highly filled systems, indicating the "second rubbery plateau", however the values of G' change with the types of silane coupling agents and increase in the following order: untreated < aclylsilane (ASC) < epoxysilane (ESC) < aminosilane (ASC) < diaminosilane (DAS). Fig. 14(b) shows the dynamic viscosity η' of various silane coupling agent treated GFL/PP composites as a function of ω. η' decreases monotonously with increasing ω, and the slopes of η' against ω have a slope of an angle -45°. As with G', η' of various silane coupling agent treated GFL/PP increases in comparison with η' of untreated

GFL/PP composites, and this increase of η' is in the same order as G'. In particular, both properties such as G' and η' appear conspicuously to be high for the GFL/PP composites with diaminosiane coupling agent (DAS). This trend showed the same tendency also by the mechanical properties such as impact strength and so on (Nishitani et al., 1998a)

Loss tangent tan δ, which is defined as the ratio of the loss modulus G'' and storage modulus G', can clarify the correlation between the elastic and viscous properties of the materials. Fig. 15 shows the relationship between tan δ and angular frequency ω for untreated and ASC 10 (1.0wt.%) treated GFL/PP composites with various volume fractions of fiber from 5 to 30 vol.%. Tan δ of untreated GFL/PP composites (Fig. 15 (a)) with volume fraction of fiber V_f less than 10vol.% (GFL10) increases with V_f and tan δ - ω curves have the peak. In high V_f region with V_f higher than 10vol.%, tan δ becomes smaller with increasing fiber content, and independent of ω in 30 vol.% (GFL30). It may be thought from these results that GF in the composites with higher fiber content contributes to a more dominant role for the elastic properties than the viscous ones. Tan δ of GFL30 is independent of ω, and these composites behave like solid materials because of the high rigidity of the structure composed with rigid fibers. On the other hand, tan δ of ASC 10 treated GFL/PP composites are shown in Fig. 15 (b) as a function of ω. Tan δ decreases gradually with increasing V_f. For composites with V_f higher than 20 vol.%, tan δ is independent of ω, behaving like solid materials. Although tan δ - ω curves of untreated GFL/PP composites have the peak values up to 20 vol.% (GFL20), those of ASC 10 treated GFL/PP composites have them up to 10 vol.% (GFL10). This fact indicates the decrease in viscous properties with high fiber content. By the surface treatment with ASC, this tendency is sifted to low fiber content region compared with untreated GFL/PP composites. These behaviour may be attributed to the strong interfacial interaction between GF and PP.

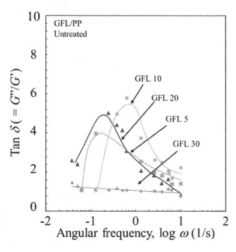
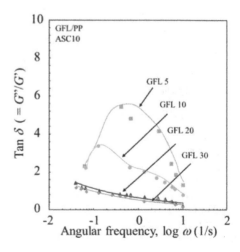

(a) Untreated GFL/PP composites   (b) ASC 10 treated GFL/PP composites

Fig. 15. Influence of volume fraction of fiber on loss tangent for GFL/PP composites: (a) Untreated GFL/PP composites, (b) ASC 10 treated GFL/PP composites

In general, the influence of surface treatment by coupling agents on the dynamic viscoelastic properties of the polymer composites is complex. This is because the surface treatment by silane coupling agents play two of multiple roles simultaneously, as a coupling agent which improves the adhesion between fiber (filler) and polymer, a lubricant which reduce the friction, a plasticizer which helps to make the fiber and the polymer softer, a wetting agent which reduces agglomeration and a an additive to make deformation of fiber assembly easier and to lower the viscoelastic properties of a matrix polymer. It is thought that the mechanisms mentioned above do not occur separately, and therefore it is very difficult to distinguish a particular type. From the results in this section, it can be concluded that the surface treatment by the various silane coupling agents improves the viscoelastic properties. Although the viscous properties are dominant in low concentration of silane coupling agent, which remarkably shows the role acting by plasticizer and internal lubricant, the elastic properties increase with the concentration of silane coupling agent. Thus, it is found that the role of the coupling agent, which forms chemical bonds between GF and PP and physical adhesion, becomes lager gradually with the increase of the concentration. Moreover, with increasing fiber content, the effect of fiber itself on the viscoelastic properties of GFL/PP composites becomes higher than the effect of surface treatment on them, and then the elastic properties increase gradually.

3.3 Effect of coupling agent concentration

In general, rheological properties such as storage modulus G' and dynamic viscosity η' are considered to be sensitive indicators for the quantitative analysis of morphological change in the composite materials. To more clarify the influence of silane coupling agent on the rheological properties of GFL/PP composites, both the storage modulus G' and the dynamic viscosity η' are plotted against the aminosilane coupling agent (ASC) concentration C as a parameter of angular frequency (ω=0.127, 1.257 and 9.488 rad/s) in Figs. 16 and 17. In Fig. 16 (a), G' of GFL5 (GF content is 5vol.%) increases gradually with the increase of the ASC concentration C, except for data at 0.127 rad/s in low C region. G' of GFL10 (Fig. 16(b)) increases monotonously with increasing C as with G' of GFL5, and the increasing ratio of GFL10 is higher than that of GFL5. G' of GFL20 (Fig. 16(a)) increases rapidly with the surface treatment and has a constant value with C higher than 0.5 wt.%. However, in high fiber content region (GFL30, Fig. 16 (b)), G' shows the complex behaviour and has a minimum value. Moreover, the difference by the angular frequency becomes smaller with increasing the fiber content. On the other hand, η' against C curves, in Fig. 17, shows the different dependence on fiber content. η' of GFL5 and GFL20 shows the same tendency as G', and η' of GFL10 and GFL30 has a minimum value. It is found from these results that the storage modulus G', although showing the complex behaviour in high fiber content increases with the ASC concentration C according to fiber content. The dynamic viscosity η' shows the two kinds of tendencies, which increase with increasing C and have a minimum value. Therefore, the effect of ASC concentration on the viscoelastic properties shows the existence of an optimum concentration for systems at each volume fraction of fiber. In addition, these tendencies are similar in a wide range of angular frequency.

3.4 Effect of volume fraction of fiber

In order to clarify the effect of volume fraction of fiber on the viscoelastic properties of surface treated GFL/PP composites in this section, the relationship between the storage modulus G' and the dynamic viscosity η' and volume fraction of fiber V_f at a typical angular frequency ω (ω =1.257 rad/s) are shown in Fig. 18 as a parameter of ASC concentration. The dependence of G' on V_f changes remarkably with the ASC concentration. G' of untreated GFL/PP composites increases with increasing V_f. On the other hand, G' of treated systems increases monotonously with V_f until it reaches the value of 2 to 5 x 10^6 Pa, and finally levels off at high V_f. This means that the elastic properties of treated GFL/PP composites are more dominant at relatively low contents than that of untreated systems. As with G', η' reaches the value of 2 to 5 x 10^6 Pa. In addition, the degrees of the increase of G' and η' become higher according to the ASC concentration order.

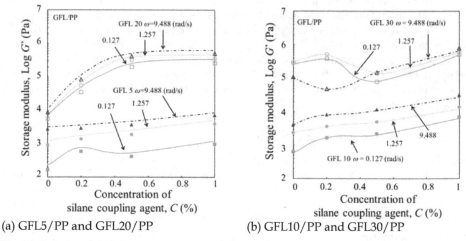

(a) GFL5/PP and GFL20/PP   (b) GFL10/PP and GFL30/PP

Fig. 16. Relationship between storage modulus and concentration of aminosilane coupling agent for GFL/PP composites: (a) GFL5/PP and GFL20/PP, (b) GFL10/PP and GFL30/PP

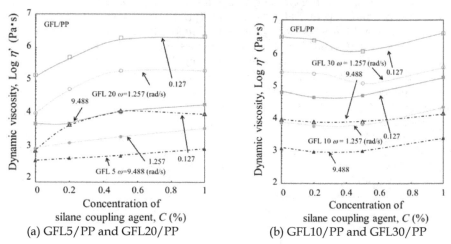

(a) GFL5/PP and GFL20/PP   (b) GFL10/PP and GFL30/PP

Fig. 17. Relationship between dynamic viscosity and concentration of aminosilane coupling agent for GFL/PP composites: (a) GFL5/PP and GFL20/PP, (b) GFL10/PP and GFL30/PP

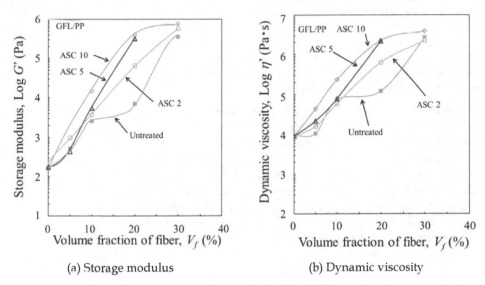

(a) Storage modulus (b) Dynamic viscosity

Fig. 18. Relationship between dynamic viscoelastic properties and volume fraction of fiber for GFL/PP composites at the angular frequency of 1.257 rad/s: (a) storage modulus, (b) dynamic viscosity

3.5 Temperature dependences

The influence of temperature on the viscoelastic properties of surface treated GFL/PP composites will be discussed here. Fig. 19 shows the storage modulus G' (Fig. 19(a)) and the dynamic viscosity η' (Fig. 19(b)) as a function of angular frequency ω as a parameter of temperature T for untreated and surface treated GFL10/PP composites. G' of untreated GFL/PP composites decreases with increasing T. However, G' of ASC10 treated GFL/PP composites corresponds approximately to the measurements at different temperatures. As with G', η' of treated systems at different temperatures are in accord with the one in a wide range of ω. It is found from the results of the measurements at different temperature that the surface treatment decreases the temperature dependence of viscoelastic behaviour, especially of elastic ones, although the untreated systems show the clear temperature dependences. Thus, the difference of processability with different temperatures is minimal by performing the surface treatment by silane coupling agent. These temperature dependences were almost same for the systems with different fiber content although not shown here. Accordingly, it is found from the results that the viscoelastic properties originating in the fiber network formation are dominant in the measuring ω region, the temperature dependences of them are relatively small and also these tendencies become stronger by the surface treatment of long GF.

3.6 Strain amplitude dependences

According to the report of Mutel and Kamal (Mutel & Kamal, 1986), the viscoelastic properties such as the storage modulus G' and the loss modulus G'' of polymer in the molten state show the strain amplitude dependence if Lissajous figure (torque – shear rate loop) does not draw the perfect (harmonic) ellipse. A strain dependent but harmonic stress region was observed for filled melts in oscillatory shear as well as an non-harmonic stress region at low frequencies and large strain amplitudes. The strain dependent harmonic stress region is thought to be the result of the much longer time scales for the relaxation of fluid compared to the period of oscillations. In our experimental data in this section, the Lissajous figures almost draw the perfect ellipses.

To more clarify the effect of strain amplitude dependence on the viscoelastic properties of surface treated GFL/PP composites, the influence of the oscillatory angle Θ or strain amplitude on the storage modulus G' and the dynamic viscosity η' of untreated and treated GFL/PP composites at different angular frequencies ω is shown in Figs. 20, 21, 22 and 23. The strain amplitude was chosen to be 4.4, 8.7 and 17.5 % (which is corresponding to oscillatory angle of 0.5, 1.0 and 2.0°, respectively). Here, all the data were calculated by assuming the linearity was maintained. Although the effect of surface treatment on the strain amplitude dependences does not remarkably appear, G' decreases with increasing the strain amplitude and also with increasing V_f. And the degree of decrease of G' decreases with increasing the angular frequency. Thus, the smaller the angular frequency is, the higher the strain amplitude dependence for G' is in the same strain amplitude. The strain amplitude dependence of dynamic viscosity η' has the same tendency as that of G' although its dependence of η' is smaller than that of G'.

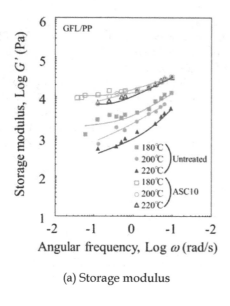

(a) Storage modulus

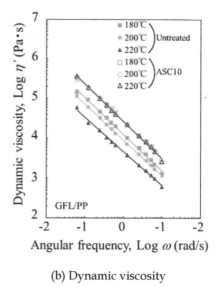

(b) Dynamic viscosity

Fig. 19. Influence of temperature on dynamic viscoelastic properties for GFL/PP composites: (a) storage modulus, (b) dynamic viscosity

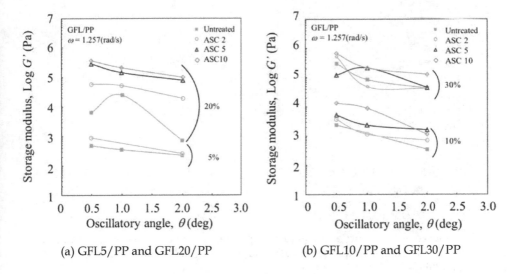

Fig. 20. Influence of oscillatory angle (strain amplitude) on storage modulus for GFL/PP composites at the angular frequency of 1.257 rad/s; (a) GFL5/PP and GFL20, (b) GFL10/PP and GFL30/PP

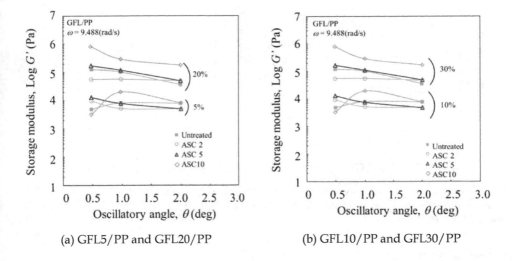

Fig. 21. Influence of oscillatory angle (strain amplitude) on storage modulus for GFL/PP composites at the angular frequency of 9.488 rad/s; (a) GFL5/PP and GFL20, (b) GFL10/PP and GFL30/PP

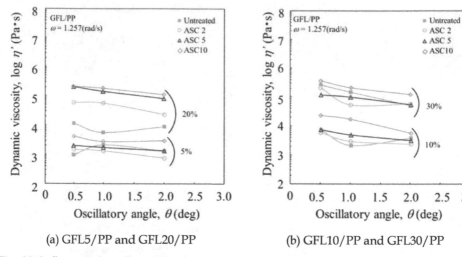

(a) GFL5/PP and GFL20/PP   (b) GFL10/PP and GFL30/PP

Fig. 22. Influence of oscillatory angle (strain amplitude) on dynamic viscosity for GFL/PP composites at the angular frequency of 1.257 rad/s; (a) GFL5/PP and GFL20, (b) GFL10/PP and GFL30/PP

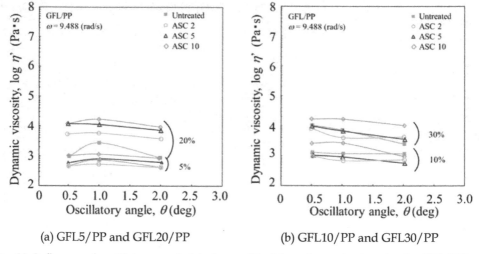

(a) GFL5/PP and GFL20/PP   (b) GFL10/PP and GFL30/PP

Fig. 23. Influence of oscillatory angle (strain amplitude) on dynamic viscosity for GFL/PP composites at the angular frequency of 9.488 rad/s; (a) GFL5/PP and GFL20, (b) GFL10/PP and GFL30/PP

4. Conclusion

This chapter discusses the rheological properties of surface treated short and long glass fiber reinforced polypropylene composites in molten state. Glass fibers were surface-treated by silane coupling agent.

The first part deals with the effect of surface treatment on the rheological properties of 20 and 40 wt.% short glass fiber reinforced polypropylenes (GF/PP) and of the same composites containing maleic anhydride modified polypropylene (GF/mPP/PP). These GF/PP and GF/mPP/PP composites were extruded by an extruder specially designed for mixing. Steady state shear and dynamic viscoelastic properties in the molten state were measured using a cone plate typed rheometer. Surface treatment increased the rheological properties of the molten composites. This trend was more pronounced for composites containing modified PP (mPP). In the case of the GF/PP composites the increase was monotonous for all measured functions, while in the case of the GF/mPP/PP ones, the parameters representing elasticity showed peaks at 0.5wt.% aminosilane coupling agent(ASC, γ-aminopropyltrimethoxysilane). The Cox-Merz empirical law was found to be applicable to untreated GF/PP and GF/mPP/PP composites at low shear rates (or angular frequencies), where the materials showed the Newtonian plateau. As for ASC surface treated composites, GF/mPP/PP followed the same trend. On the other hand, GF/PP exhibited a higher complex viscosity than the steady state shear one over the whole range of angular frequencies (shear rates) covered by the experiments Surface treatment enhanced apparent yield stress, as did with increasing the glass fiber content. The effect of the latter was more pronounced in the case of the GF/mPP/PP series. The viscosities of the re-mixed composites became lower than those of the original ones, and the difference between untreated and ASC treated systems was detectable only in the case of the complex viscosity of the GF/mPP/PP composites.

The second part deals with the influence of surface treatment on the rheological behaviour of long glass fiber reinforced polypropylene melts (GFL/PP). Surface treatment by different kinds of silane coupling agents (aminosilane, diaminosilane, epoxysilane and acrylsilane) with different concentrations was performed on glass fibers. Glass fiber and polypropylene fiber mixed mats were prepared by a mixing machine which is called "Fiber separating/ flying machine" and then compression molded. Rheological properties on a rotational parallel plate rheometer were evaluated. Dynamic viscoelastic properties are discussed in terms of various factors: angular frequency, concentration of silane coupling agent, various kinds of silane coupling agents, volume fraction of fiber, temperature and strain amplitude. Surface treatment was shown to increase storage modulus and also reduce the peak of loss tangent, which means that the adhesion between the polymer matrix and the fibers is improved. Dynamic viscosity, however, was changed by the concentration of silane coupling agent. This is because the coupling agent acts to couple the fiber to the polymer matrix, as a wetting agent or as an internal lubricant. Diaminosilane coupling agent was the most effective of the various silane systems for enhancement of rheological behaviour. The dependence of the viscoelastic properties on the concentration of silane coupling agent showed the existence of an optimum concentration for systems at each volume fraction of fiber from the point of view of the processing of these composites. From the results of the measurement at different temperatures and strain amplitude, it was found that surface treatments decrease the temperature dependence of rheological behaviour, and that the dependence of viscoelastic properties on strain amplitude was minimal.

5. Acknowledgment

The authors would thank Nippon Glass Fiber Co. Ltd. for donating the glass fiber for this study. This research is a part of results of research in GF/PP composites research group of workshop on interfacial scientific of composite material. We would like to thank the BERC and the Ogasawara Foundation for the Promotion of Science & Engineering for founding this study. A part of this study was financially supported by the grant from Kogakuin University. This article was also created with support of Operational Programme Research and Development for Innovations co-funded by the European Regional Development Fund (ERDF) and national budget of Czech Republic, within the framework of project Centre of Polymer Systems (reg. number: CZ.1.05/2.1.00/03.0111).

6. References

Ausias, G.; Agassant, J.F.; Vincent, M.; Lafleur, P.G.; Lavoie, P.A. & Carreau, P.J. (1992). Rheology of Short Glass Fiber Reinforced Polypropylene. *Journal of Rheology*, Vol.36, No.4, pp. 525-542 ISSN 0148-6055

Basu, D.; Banerjee, A.N. & Misra, A. (1992). Comparative Rheological Studies on Jute-Fibre- and Glass-Fibre-Filled Polypropylene Composite Melts. *Journal of Applied Polymer Science*, Vol.46, No.11, pp. 1999-2009 ISSN 0021-8995

Bigg, D.M. (1982). Rheological Analysis of Highly Loaded Polymeric Composites Filled with Non-Agglomerating Spherical Filler Particles. *Polymer Engineering and Science*, Vol.22, No.8, pp. 512-518 ISSN 0032-3888

Boaira, M. S. & Chaffey, C. E. (1997). Effect of Coupling Agents on the Mechanical and Rheological Properties of Mica Reinforced Polypropylene. *Polymer Engineering and Science*, Vol.17, No.10, pp. 715-718 ISSN 0032-3888

Cantwell, W.J.; Tato, W.; Kausch, H.H. & Jacquemet, R. (1992). Influence of a Fiber-matrix Coupling Agent on the Properties of a Glass Fiber/Polypropylene GMT. *Journal of Thermoplastic Composite Materials*, Vol.5, No.4, pp. 304-317 ISSN 0892-7057

Casson, N. (1959) *Rheology of Disperse Systems*, pp. 84-104. Pergamon Press, London

Cox, W.P. & Merz, E.H. (1958). Correlation of Dynamic and Steady Flow Viscosities. *Journal of Polymer Science*, Vol.28, pp.619-622 ISSN 0021-8995

Czarnecki, L. & White, J.L. (1980). Shear Flow Rheological Rroperties, Fiber Damage, and Mastication Characteristics of Aramid-, Glass-, and Cellulose-Fiber-Reinforced Polystyrene Melts. *Journal of Applied Polymer Science*, Vol.25, No.6, pp.1217-1244 ISSN 0021-8995

Davies, P. (1993). Rheological Properties of Stampable Tthermoelastic Composites. *Journal of Thermoplastic Composite Materials*, Vol.6, No.3, pp. 239-254 ISSN 0892-7057

Davis, S.M. & McAlea, K.P. (1990). Stamping Rheology of Glass Mat Reinforced Thermoplastic Composites. *Polymer Composites*, Vol.11, pp. 368-378 ISSN 0272-8397

Dealy, J.M. & Wissbrun, K.F. (1990). *Melt Rheology and Its Role in Plastics Processing*. Springer ISBN 978-0792358862

Fejes-Kozma, Zs. & Karger-Kocsis, J. (1994). Fracture Mechanical Characterization of a Glass Fiber Mat-reinforced Polypropylene by Instrumented Impact Bending. *Journal of Reinforced Plastics and Composites*, Vol.13, No.9, pp. 822-834 ISSN 0731-6844

Ferry, J. D. (1980). *Viscoelastic Properties of Polymers, 3rd Edition*, Wiley Press, ISBN 978-0471048947, New York

Greene, J.P.& Wilkes, J.O. (1995). Steady-State and Dynamic Properties of Concentrated Fiber-Filled Thermoplastics. *Polymer Engineering and Science*, Vol.35, No.21, pp. 1670-1681 ISSN 0032-3888

Groves, D.J. & Stocks, D.M. (1991). Rheology of Thermoplastic-Carbon Fibre Composite in the Elastic and Viscoelastic States. *Composites Manufacturing*, Vol.2, pp.179-184

Groves, D.J.; Bellamy, A.M. & Stocks, D.M. (1992). Anisotropic Rheology of Continuous Fibre Thermoplastic Composites. *Composites*, Vol.23, No.2, pp. 75-80 ISSN 0010-4361

Hamada, H. ; Fujihara, K. & Harada, A. (2000). Influence of Sizing Conditions on Bending Properties of Continuous Glass Fiber Reinforced Polypropylene Composites. *Composites Part A: Applied Science and Manufacturing*, Vol.31, No.9, pp. 979-990 ISSN 1359-835X

Han, C.D. ; Van den Weghe, T. ; Shete, P. & Haw, J.R. (1981). Effects of Coupling Agents on the Rheological roperties, Processability, and Mechanical Properties of Filled Polypropylene. *Polymer Engineering and Science*, Vol.21, pp. 196-204 ISSN 0032-3888

Hausnerova, B. ; Honkova, N. ; Lengalova, A. ; Kitano, T. & Saha, P. (2006). Rheology and Fiber Degradation during Shear Flow of Carbon-Fiber-Reinforced Polypropylenes. Polymer Science, Ser.A., Vol.48, No.9, pp. 1628-1639 ISSN 0965-545X

Hausnerova, B. ; Honkova, N. ; Lengalova, A. ; Kitano, T. & Saha, P. (2008). Rheological Behavior of Fiber-Filled Polymer Melts at Low Shear Rate, Part 1 Modeling of Rheological Properties. *Polimery*, Vol.53, pp. 507-512 ISSN 0032-2725

Hausnerova, B. ; Honkova, N. ; Lengalova, A. ; Kitano, T. & Saha, P. (2008). Rheological Behavior of Fiber-Filled Polymer Melts at Low Shear Rate, Part 21 Experimental Investigation. *Polimery*, Vol.53, pp. 649-656 ISSN 0032-2725

Hong-Lie, L. ; Han, C. D. & Jovan, M. (1983). Effects of Coupling Agents on the Rheological Behavior and Physical/Mechanical Properties of Filled Nylon 6. *Journal of Applied Polymer Science*, Vol.28, No.11, pp. 3387-3398 ISSN 0021-8995

Kataoka, T. ; Kitano, T. Onishi, S. & Nakama K. (1976) Mixing Effect of Filler and Polymer by an Elastic Extruder. *Rheol. Acta*, Vol.15, pp. 268-270 ISSN 0035-4511

Khan, S.A. & Prud'Homme, R.K. (1987). Melt Rheology of Filled Thermoplastics. *Reviews in Chemical Engineering*, Vol.4, No.3-4, pp. 205-270 ISSN 0167-8299

Kikuchi, S. ; Fujita, Y. ; Sano, K. ; Inoguchi, H. ; Hiragushi, M.& Hamada, H. (1997). The Effect of GF/PP Matrix Interfacial Properties on the Weldline Strength in Short Glass Fiber Reinforced Polypropylene. *Composite Interfaces*, Vol.4, No.6, pp. 367-378 ISSN 0927-6440

Kim, J.K. & Song, J.H. (1997). Rheological Properties and Fiber Orientations of Short Fiber-Reinforced Plastics. *Journal of Rheology*, Vol.41, No.5, pp.1061-1085 ISSN 0148-6055

Kitano, T.; Kataoka, T. & Shirota, T. (1981). An Empirical Equation of the Relative Viscosity of Polymer Melts Filled with Various Inorganic Fillers. *Rheologica Acta*, Vol.20, No.2, pp.207-209 ISSN 0035-4511

Kitano, T. ; Kataoka, T. & Nagatsuka, Y. (1984). Dynamic Flow Properties of Vinylon Fibre and Glass Fiber Reinforced Polyethylene Melts. *Rheologica Acta*, Vol.23, No.4, pp.408-416 ISSN 0035-4511

Kitano, T.; Nagatsuka, Y.; Lee, M.; Kimijima, K & Oyanagi, Y. (1994). A Method for the Production of Randomly Oriented Fiber Reinforced Thermoplastic Composites and Their Mechanical Properties. *Seikei-Kakou (the Journal of Japanese Society of Polymer Processing, in Japanese)*, Vol.6, No.12, pp. 904-915 ISSN 0915-4027

Kitano, T. ; Haghani, E. ; Tanegashima, T. & Saha P. (2000). Mechanical Properties of Glass Fiber/Organic Fiber Mixed Mat Reinforced Thermoplastic Composites. *Polymer Composite*, Vol. 21, No.4, pp. 493-505 ISSN 0272-8397

Laun, H.M. (1984). Orientation Effects and Rheology of Short Glass Fiber-Reinforced Thermoplastics. *Colloid & Polymer Science*, Vol.262, No.4, pp.257-269 ISSN 0303-402X

Lee, N.-J. & Jang, J. (1997). The Use of a Mixed Coupling Agent System to Improve the Performance of Polypropylene-Based Composites Reinforced with Short-Glass-Fibre Mat. *Composites Science and Technology*, Vol.57, No.12, pp. 1559-1569 ISSN 0266-3538

Li, S. ; Järvelä, P.K. & Järvelä, P.A. (1997). A Comparison Between Apparent Viscosity and Dynamic Complex Viscosity for Polypropylene/Maleated Polypropylene Blends. *Polymer Engineering and Science*, Vol.37, No.1, pp. 18-23 ISSN 0032-3888

Mäder, E. ; Jacobasch, H.-J. ; Grundke, K. & Gietzelt, T. (1996). Influence of an Optimized Interphase on the Properties of Polypropylene/Glass Fibre Composites. *Composites Part A: Applied Science and Manufacturing*, Vol.27A No.9, pp. 907-912 ISSN 1359-835X

Mäder, E. ; Moos, E. & Karger-Kocsis, J. (2001). Role of Film Formers in Glass Fibre Reinforced Polypropylene - New Insights and Relation to Mechanical Properties. *Composites Part A: Applied Science and Manufacturing*, Vol.32, No.5, pp. 631-639 ISSN 1359-835X

Mutel, A.T. & Kamal, M.R. (1986). Characterization of the Rheological Behavior of Fiber-Filled Polypropylene Melts under Steady and Oscillatory Shear using Cone-and-Plate and Rotational Parallel Plate Rheometry. Polymer Composites, Vol.7, No.5, pp. 283-294 ISSN0272-8397

Nishitani, Y. ; Sekiguchi, I. ; Nakamura, K. ; Nagatsuka Y. & Kitano, T. (1998). Fabrication of Glass Fiber Reinforced Polypropylenes and their Physical Properties : 1. Influence of Surface Treatment of the Mechanical Properties. *Seikei-Kakou (the Journal of Japanese Society of Polymer Processing, in Japanese)*, Vol.10, No.2, pp. 129-138 ISSN 0915-4027

Nishitani, Y. ; Sekiguchi, I. ; Nakamura, K. ; Nagatsuka Y. & Kitano, T. (1998). Fabrication of Glass Fiber Reinforced Polypropylenes and their Physical Properties : 2. Influence of Surface Treatment of the Rheological Properties in Molten State. *Seikei-Kakou (the Journal of Japanese Society of Polymer Processing, in Japanese)*, Vol.10, No.2, pp. 139-148 ISSN 0915-4027

Nishitani, Y. ; Kitano, T. ; Nagatsuka, Y. ; Nakamura, K. & Sekiguchi, I. (1998). Influence of Surface Treatment on the Mechanical Properties of Short Glass Fiber Reinforced Polypropylenes and their Viscoelastic Properties in Molten State.. *Reserch Report of Kogakuin University (in Japanese)*, Vol.84, pp. 11-20 ISSN 0368-5098

Nishitani, Y. ; Sekiguchi, I. ; Nakamura, K. ; Tai, N. ; Nagatsuka Y. & Kitano, T. (1998). Influence of Fiber Length on the Mechanical Properties for Glass Fiber Reinforced Polypropylenes. *Kyouka Plastics (the Journal of Japan Reinforced Platics Society, in Japanese)*, Vol.44, No.5, pp. 197-203 ISSN 0452-9685

Nishitani, Y. ; Sekiguchi, I. ; Yoshimitsu, Y. ; Tahira, K. ; Saha, P. ; Nagatsuka Y. & Kitano, T. (1999). Long Glass Fibre Reinforced Polypropylenes : Fabrication and

Mechanical Properties. *Polymers & Polymer Composites*, Vol.7, No.3, pp. 205-215 ISSN 0967-3911

Nishitani, Y. ; Sekiguchi, I. ; Hausnerova, B. ; Nagatsuka Y. & Kitano, T. (2001). Dynamic Viscoelastic Properties of Long Organic Fibre Reinforced Polypropylene in Molten State. *Polymers & Polymer Composites*, Vol.9, No.3, pp. 199-211 ISSN 0967-3911

Nishitani, Y. ; Sekiguchi, I. ; Hausnerova, B. ; Zdrazilova, N. & Kitano, T. (2007). Rheological Properties of Aminosilane Surface Treated Short Glass Fibre Reinforced Polypropylenes. Part 1 : Steady Shear and Oscillatory Flow Properties in Molten State. *Polymers & Polymer Composites*, Vol.15, pp. 111-119 ISSN 0967-3911

Nishitani, Y. ; Ohashi, K. ; Sekiguchi, I. ; Ishii, C. & Kitano, T. (2010). Influence of Additon of Styrene-Ethylene/Butylene-Styrene Copolymer on Rheological, Mechanical and Tribological Properties of Polyamide Nanocomposites. *Polymer Composites*, Vol.31, No.1, pp. 68-76 ISSN 0272-8397

Nishitani, Y. ; Yamada, Y. ; Sekiguchi, I. ; Ishii, C. & Kitano, T. (2010). Effects of Addition of Functionalized SEBS on Rheological, Mechanical, and Tribological Properties of Polyamide 6 Nanocomposites. *Polymer Engineering and Science*, Vol.50, No.1, pp. 100-112 ISSN 0032-3888

Nishitani, Y. ; Sekiguchi, I. & Kitano, T. (2010). Rheological Properties of Various Carbon Fibers Filled PBT Composites. *Proceedings of the Polymer Processing Society 26th Annual Meeting –PPS-26-*, R01-134, Banff, Canada, July 4-8, 2010

Peltonen, P. ; Pääkkönen, E.J. ; Järvelä, P.K. & Törmälä, P. (1995). The Influence of Adhesion Promoters on the Properties of Injection Moulded Long-Glass-Fibre Polypropylene. *Plastics, Rubber and Composites Processing and Applications*, Vol.23, pp. 111-126 ISSN 0959-8111

Saini, D.R. ; Shenoy, A.V. & Nadkarni, V.M. (1985). Effect of Surface Treatments on Rheological, Mechanical and Magnetic Properties of Ferrite-Filled Polymeric Systems. *Polymer Engineering and Science*, Vol.25 , No.13, pp. 807-811 ISSN 0032-3888

Sasagi, I. & Ide, F. (1981). Effect of Grafting of Unsaturated Carboxylic Acid on Glassfiber-Reinforced Polypropylene. *Koubunshi Ronbunshu (the Journal of the society of Polymer Science, Japan, in Japanese)*, Vol.38, No.2, pp. 67-74 ISSN 0386-2186

Shenoy, A. V. (1999). *Rheology of Filled Polymer Systems*, Kluwer Academic Publishers, ISBN 0-412-83100-7, Dordrecht, The Netherlands

Thomason, J.L. & Schoolenberg, G.E. (1994). An Investigation of Glass Fibre/Polypropylene Interface Strength and its Effect on Composite Properties. *Composites*, Vol.25, No.3, pp. 197-203 ISSN 0010-4361

Thomason, J.L. & Vlug, M.A. (1996). Influence of Fibre Length and Concentration on the Properties of Glass Fibre-reinforced Polypropylene: 1. Tensile and Flexural Modulus. *Composites Part A: Applied Science and Manufacturing*, Vol.27, No.6, pp. 477-484 ISSN 1359-835X

Van Den Oever, M. & Peijs, T. (1998). Continuous-Glass-Fibre-Reinforced Polypropylene Composites II. Influence of Maleic-Anhydride Modified Polypropylene on Fatigue Behaviour. *Composites Part A: Applied Science and Manufacturing*, Vol.29, No.3, pp. 227-239 ISSN 1359-835X

Wu, H.F.; Dwight, D.W. & Huff, N.T., (1997) Effects of Silane Coupling Agents on the Interphase and Performance of Glass-fiber-reinforced Polymer Composites, *Composites Science and Technology*, Vol.57, No.8, pp. 975-983 ISSN 0266-3538

Yue, C.Y. & Quek, M.Y. (1994). The Interfacial Properties of Fibrous Composites - Part III Effect of the Thickness of the Silane Coupling Agent. *Journal of Materials Science*, Vol.29, No.9, pp. 2487-2490 ISSN 0022-2461

Zang, Z. ; Kitano, T. & Hatakeyama, T. (1995). Crystallization Behavior of Carbon Fiber Reinforced Polyamides: 1. Dynamic and Isothermal Crystallization. *International Polymer Processing*, Vol.10, No.2, pp. 165-171 ISSN 0930-777X

5

Composites Made of Polypropylene Nonwoven Fabric with Plasmas Layers

Maciej Jaroszewski, Janina Pospieszna, Jan Ziaja and Mariusz Ozimek
Wrocław University of Technology, Institute of Electrical Engineering Fundamentals
Poland

1. Introduction

Engineering of materials used for shielding from electromagnetic fields is currently one of the most extensively developing field of applications of composite materials (Bula et al., 2006; Jaroszewski & Ziaja, 2010; Koprowska et al., 2004, 2008; Sarto et al. 2003, 2005; Wei et al., 2006; Ziaja et al., 2008, 2009, 2010). The choice of suitable materials for the shields and their appropriate arrangement have an essential meaning. Development of lightweight and resistant to environmental exposure shielding materials is possible by using substrates of polypropylene and plasma technology (Ziaja&Jaroszewski, 2011).

The shields for suppression of electric field were made in the form of composites of polypropylene unwoven fabrics with deposited plasma layers. Additional advantage of the application of the method is the possibility of plasma cleaning of a fabric surface and modifying its surface properties. The unique properties of pulse plasma make possible to obtain metallic and dielectric coatings on polypropylene fabrics, which are not achievable by standard methods. The coatings are characterized by a good adhesion to the substrates.

The surface of the samples was examined in two ways: by metallurgical microscope Nikon MA200 and scanning microscope Quanta 200 in the low vacuum mode. To identify the structure of the obtained layers the X-Ray radiography was used. Additionally properties of the composites was studied using impedance spectroscopy. The method of impedance spectroscopy allows one to connect the measured frequency characteristics with the physical structure of tested material and the alternations in the structure. This method has been used by the authors to determine the properties of plasma layers deposited on a polypropylene nonwoven fabric (Jaroszewski et al., 2010a; Pospieszna et al., 2010; Pospieszna et al., 2010b).

2. Polypropylene nonwoven fabric with plasma layers in EM technique

Polypropylene materials (PP), because of their electric properties (such as surface resistivity ρ_s, volume resistivity ρ_v, dielectric loss factor $tg\delta$, permittivity ε), mechanical properties and resistance to noxious agents (resistance to acids, bases, salts and organic solvents) are used in various industries. Polypropylene materials characterise, also, with the lowest specific density among widely used polymers. Those properties predispose polypropylene to be used as a substrate for composite protective screens shielding people and electric or electronic devices against noxious activity of electromagnetic (EM) fields. Composite shields are fabricated through metallizing film surfaces or PP nonwoven fabric. Metallic layer thickness does not exceed several micrometres. One or several types of metal as well as conductive metal oxides can make up a metallic layer. Due to their lightness and mechanical strength, those composites are an alternative to classic EM field shielding materials.

Due to characteristic surface properties, PP is a very difficult material to metallize. Ideal for that purpose is the magnetron sputtering method described in papers (Ziaja&Jaroszewski, 2011; Ziaja et al., 2010; Ziaja et al., 2008). Obtained coefficients of shielding effectiveness (SE) of popular composites based on PP/Me matrixes exceed up to 60dB (Me=Zn SE exceeds 60 dB, Me=Cu SE approx. 35 dB, Ti approx. 30 dB).

SE does not depend solely on type of material and its surface and volume resistivity, but also on fabrication (crystal structure) of applied layers. Crucial for SE value is not only the number and resistivity of conductive bridges forming on nonwoven fabrics' surface, but also the specific surface area of applied layers. The more expanded the surface the higher the SE value. Therefore, in order to evaluate those composites' suitability for electromagnetic field shielding screens – apart from electric properties – surface morphology of applied layers has to be known.

3. Morphology of PP plasma composites

As means to determine crystal structure of layers, x-ray diffraction method was used. Surface morphologies and chemical composition was determined by a scanning microscope equipped with an x-ray microprobe.

Phase composition of samples was analysed by means of x-ray examination carried out on DRON-2 diffractometer producing Fe filtered Co radiation of $\lambda = 1.7902$ Å wavelength. Scanning was carried out according to the wide angle x-ray scattering method at $\Delta2\Theta = 0.05°$ spread with scattering angle $2\Theta = (40 \div 65°)$. Diffraction patterns were analysed by the Xrayan programme through comparison of interplanar d-spacing of reflection intensity I to PDF files data.

Morphology analysis of PP/Me composites' surface was carried out using the VEGA/SBH scanning electron microscope manufactured by TESCAN. The microscope is intended for scanning conductive samples in a high-vacuum chamber. At 30kV voltage the maximum resolution is 3.0 nm. Magnification ranges between 6-1000 000 times at specimen current from 1pA to 2uA. The electron optical column is composed of four lenses enabling smooth system configuration and scanning at an optimum resolution. The microscope is additionally equipped with an EDS system attachment [scattered electrons energy analysis] INCA PENTAFTx3 of 133 eV resolution. One of system parts is an analyser enabling point to line analysis of samples' chemical composition.

Structural and electric properties of PP/Me composites depend on magnetron sputtering process parameters (current density at sputtered electrode, pressure and composition of working gases, distance between nonwoven fabric and the target, deposition rate). By changing those parameters, chemical composition and structure of deposited layers can be manipulated, thus their electrical parameters can be modified. Usually, layer thickness is adjusted by changing gun power or deposition rate. X-ray radiography examination of layers deposited at the same rate, but at increasing power emitted on sputtered electrode are characterised by increased intensity of characteristic lines. Examples of PP/Zn composites' x-ray spectra were presented in fig. 1. One can note, that at low gun power obtained Zn layers are half-amorphous, indicative of which is lack of Zn-characteristic reflections (fig. 1a). EM field shielding effectiveness is low and does not exceed 10 dB. By increasing the power, increased are not only the layers but their crystallisation as well, causing conductive bridges to form (fig. 1b, 1c, 1d). Surface conductivity decreases and coefficient of shielding effectiveness ranges between 10-30dB. Further power increase induces more conductive bridges of expanded specific surface area to form.

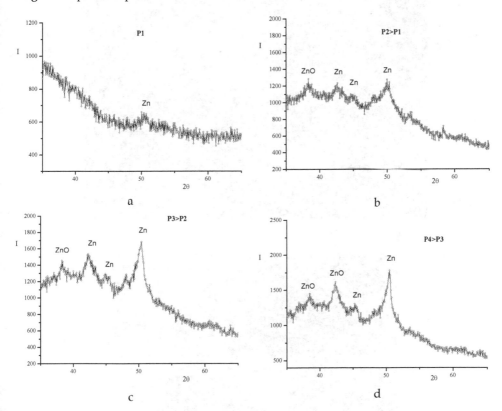

Fig. 1. X-ray radiography spectra of zinc layers deposited on polypropylene nonwoven fabric at different emitted power (P1, P2, P3, P4)

Not only surface fibres are coated with metal (fig 2a), but also the nonwoven fabrics' internal fibres (fig. 2b) and areas between fibres (fig. 2c). Similar crystallization are observable for zinc layers deposited on polypropylene film (fig. 3). A layer obtained in that manner characterises with large specific surface area, which disperses electromagnetic field and increases shielding coefficient SE at the same time. Cross-section of such composite is presented in Fig. 4. It comes to one's attention that metallic layers on fibre surface are solid and uniform. Best composites characterise with SE~ 60 dB, which not only stems from low surface resistivity, but also from expanded surface of metallic layers (fig. 5).

Similar results are observable for other metallic layers, e.g. Al, Ti. Higher layer crystallisation and tighter texture are also notable. Reference book (Ziaja&Jaroszewski, 2011) discusses the method of crystallising Ti layers.

a

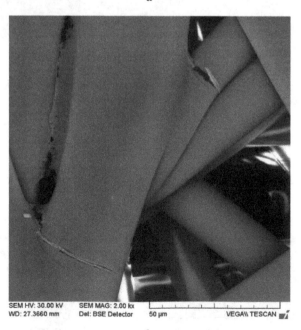

b

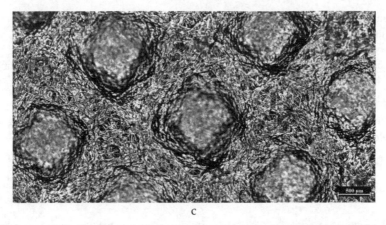

c

Fig. 2. Surface morphology of Zn layers deposited on polypropylene nonwoven fabric PP Film

Layers deposited on PP films display different behaviour. Layers obtained at highest gun powers remain amorphous. At identical layer deposition parameters the materials characterises with higher SE than the nonwoven fabric. It stems from continuous structure of layers deposited on film as opposed to conductive mesh formed on the surface of nonwoven fabrics. As in case of Ni, Fe or Al layers, SE of film is 10 dB higher.

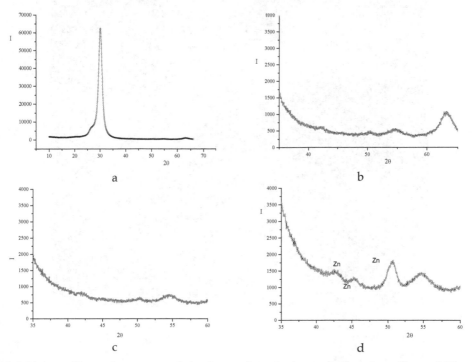

Fig. 3. X-ray radiography spectra of zinc layers deposited on polypropylene film at different emitted current densities

Similar relations in samples' morphology, were noticed by authors in composites with carbon plasma layers deposited on PP nonwoven fabric. They are still, however, characterised by lower SE compared to composites with metallic layers. Likewise promising results of using carbon layers were presented (Wang et al., 2011), where C layers were deposited by silk-screen printing. Presented layers were in form of short and long nanotubes, whose length was critical for SE.

Based on the above-mentioned, nonwoven fabrics with layers are an eligible alternative for classic shielding mats.

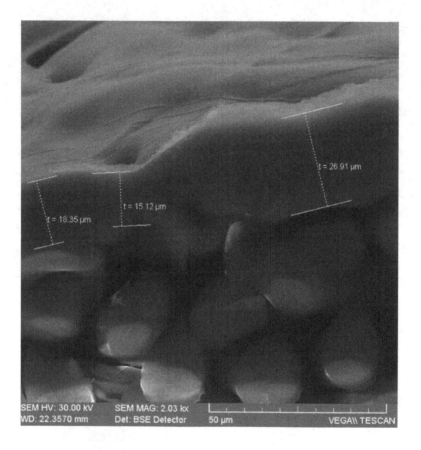

Fig. 4. Cross-section of zinc layers deposited on polypropylene nonwoven fabric

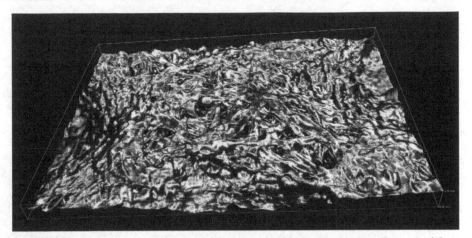

Fig. 5. Expanded surface of metallic layers deposited on polypropylene nonwoven fabric

4. Determination of composites dielectric properties by impedance spectroscopy

Impedance spectroscopy is a modern and exceptionally effective tool for analysing different materials, including complex composites systems. The method draws on measuring linear electric answer of examined material to excitation in form of low amplitude, sinusoidally alternating voltage $u(t) = Um \sin(\omega t + \psi_u)$. The voltage applied to the electrode system, between which examined specimen is placed, induces sinusoidal current $i(t) = Im \sin(\omega t + \psi_i)$ to flow through the sample with effective current I, at phase displacement by angle $\varphi = \psi_u - \psi_i$. Analysing that answer in wide spectrum of frequencies yields useful information on conductivity and polarisation phenomena taking place in the examined material.

In the field of frequencies, spectral transmittance $\underline{H}(\omega)$ is usually used for linear circuits to describe electrical answers. It is defined as a relation of complex responses $\underline{Y}$ and excitation $\underline{X}$ of investigated circuit:

$$H(\omega) = Y/X = Y\,ej\psi y / X\,ej\psi x = Y/X\,e\,j(\psi y - \psi x) = Y/X\,e\,j\,\varphi \qquad (1)$$

where $\psi_y - \psi_x = \varphi$.
Its module

$$|\underline{H}(\omega)| = Y/X \qquad (2)$$

and argument

$$\varphi = \varphi(\omega) = \arg \underline{H}(\omega) \qquad (3)$$

are known as amplitude and phase characteristic of spectral transmittance.

In impedance spectroscopy, spectral transmittance $\underline{H}(\omega)$ usually has a form of either complex impedance $\underline{Z}(\omega)$ or complex admittance $\underline{Y}(\omega)$.

For a given two-terminal network, complex impedance is defined as relation of complex voltage $\underline{U}$ values and that voltage-induced current $\underline{I}$:

$$\underline{Z}(\omega) = \frac{\underline{U}(\omega)}{\underline{I}(\omega)} = |Z(\omega)| e^{j\varphi(\omega)} \tag{4}$$

$$\underline{Z}(\omega) = \mathrm{Re}\,\underline{Z} + \mathrm{Im}\,\underline{Z} = R + jX \tag{5}$$

where $\mathrm{Re}(\underline{Z})$ and $\mathrm{Im}\,(\underline{Z})$ are real and imaginary parts of complex impedance (resistance R and reactance X).

Expression describing complex admittance becomes:

$$\underline{Y}(\omega) = \frac{1}{\underline{Z}(\omega)} = \frac{\underline{I}(\omega)}{\underline{U}(\omega)} = |\underline{Y}(\omega)| e^{-j\varphi(\omega)} \tag{6}$$

$$\underline{Y}(\omega) = \mathrm{Re}\,\underline{Y} + \mathrm{Im}\,\underline{Y} = G + jB \tag{7}$$

where $\mathrm{Re}(\underline{Y})$ and $\mathrm{Im}(\underline{Y})$ are real and imaginary part of complex admittance (conductance G and susceptance B).

In practice, the examined sample can be assigned with an equivalent electrical model in form of either parallel or serial connection of resistor and capacitor (Fig. 6), which can be specified by relevant real and imaginary parts of investigated transmittance (admittance and impedance). Assumed model enables to separate active and passive current parts, in case of parallel equivalent of the sample (Fig. 6a) and voltage – in case of serial model (Fig. 6b).

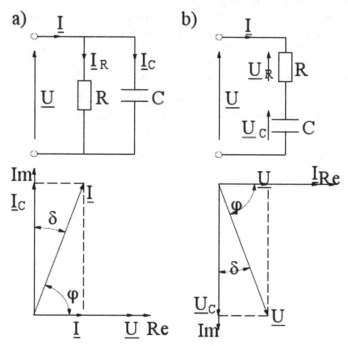

Fig. 6. Capacitor equivalent circuit with real dielectric phase vectors: a) parallel circuit, b) serial circuit.

In equivalent electrical model in form of parallel connection of resistance R resistor and ideal capacitor of C capacity, the active current part $I_R = U / R$ (U voltage phase) and passive $I_C = j \omega C U$ (voltage lags the current by $\pi/2$ phase) represent leakage current and capacitor charging current respectively. Complex admittance measured for that circuit can be represented by:

$$\underline{Y}(\omega) = \underline{I}(\omega) / \underline{U}(\omega) = (\underline{I_R} + \underline{I_C}) / \underline{U} = 1 / R_p + j \omega C_p =$$
$$= G_p + j \omega C_p = j \omega (C_p + G_p / j \omega) = j \omega (C' - j C'') = j \omega \underline{C} \qquad (8)$$

where

$$\underline{C}(\omega) = (C' - j C'') = \underline{Y}(\omega) / j \omega \qquad (9)$$

are defined as complex capacity and C' and C" are its real and imaginary parts respectively.

Complex material parameters can be determined using impedance measurement of simple test structures of given electrode geometry and dimensions of specimen (distance d and electrode surface S) and equivalent complex admittance circuit. Such material parameter is complex conductivity σ (ω), which is obtained by multiplying two-terminal network's complex admittance expression by the d/S parameter:

$$d / S (G_{p.} + j \omega \underline{C_p}) = \sigma' + j \omega \varepsilon_0 \varepsilon_r = \sigma'(\omega) + j \sigma''(\omega) = \underline{\sigma}(\omega) \qquad (10)$$

where:

σ' – material conductivity,
ε_0 – vacuum permittivity,
ε_r – relative permittivity of the material,
Its real and imaginary parts are σ' and σ'' respectively.

Introduction of complex capacity C requires introduction of the notion of complex relative permittivity ε, to which it is proportionate by definition:

$$\underline{C}(\omega) = C_o \underline{\varepsilon} = C_o (\varepsilon' - j \varepsilon'') \qquad (11)$$

where: C_o – geometric capacitance of the electrode system, real part ε' and imaginary part ε'' of complex permittivity ε determine storage capacity and energy diffusive power respectively, however, energy losses are connected both to polarisation and insulation leakage current.

The measure of lag between dielectric polarisation and changes in electric field is the dielectric loss factor defined as:

$$tg\delta = \frac{I_R}{I_C} = \frac{1}{\omega RC} = \frac{\varepsilon''}{\varepsilon'} \qquad (12)$$

where the angle δ is the cofactor of phase displacement φ to 90°.

Obtained through measurements quantities: impedance Z(ω) and admittance Y(ω) can be adopted as basic values enabling further processing of results of measurements in order to give an adequate account of given material properties which are of interest to us.

When research results are subject to interpretation through impedance spectroscopy, one has to bear in mind, that measured quantities are a measure of analysed system's properties, which is composed of electrodes and located between them material. Hence, the measured quantities give a picture of the entire circuit in sinsuidally alternating electric field, including lead resistance and inductance, stray capacitance and phenomena related to electrode polarisation.

Dielectric response of the composite non-woven PP / plasma layer is a function of both the physical structure of substrate and applied layer. So far, studies carried out by impedance spectroscopy allowed us to:

- identify the degree of porosity of the substrate on which the plasma layer was applied(Jaroszewski et al., 2010),
- determine the effect of surface resistivity of the composite relaxation processes(Pospieszna et al., 2010, Pospieszna & Jaroszewski, 2010),
- identify the strong dependence of dielectric composite properties upon a number of formed metal/metal-oxide layers (Ziaja&Jaroszewski, 2011).

5. Conclusions

Polypropylene in form of nonwoven fabric is promising material for EM shield composites. However, the use of this material is dependent on the possibility to cover it with another material exhibit conductive properties. This is possible only by using magnetron techniques.

The processes of formation of conductive bridges at the nonwoven fabric surface and on the fibres inside are critical to the screening factor. These processes can be examined both by analysing the surface morphology and dielectric properties.

6. Acknowledgment

This publication was prepared with the key project – POIG no. 01.03.01-00-006/08 co-financed from the founds of European Regional Development Found within the framework of the Operational Programme Innovative Economy.

7. References

Bula K., Koprowska J., Janukiewicz J. (2006). Application of Cathode Sputtering for Obtaining Ultra-thin Metallic Coatings on Textile Products, Fibres & Textiles in EE, Vol. 14, No. 5 (59) (2006) pp.75 – 79

Jaroszewski M., Ziaja J. (2010). Zinck-unvowen fabric composite obtained by magnetron sputtering, Proceedings of Twelfth International Conference on Plasma Surface Engineering; September 13 - 17, 2010, PSE 2010, Garmisch-Partenkirchen, Germany, PSE 2010

Jaroszewski M., Pospieszna J., Ziaja J. (2010). Dielectric properties of polypropylene fabrics with carbon plasma coatings for applications in the technique of electromagnetic field shielding, J. Non-Cryst. Solids, Volume 356, Issues 11-17, 2010, 625-628

Koprowska J., Ziaja J., Janukiewicz J. (2008). Plasma Metallization Textiles as Shields for Electromagnetic Fields, EMC Europe 2008, Hamburg, Germany, September 8-12, 2008, pp. 493-496

Koprowska J., Pietranik M., Stawski W. (2004). New Type of Textiles with Shielding Properties, Fibres &Textiles in Eastern Europe, vol. 12, (2004), n.3 (47), 39-42

Pospieszna J., Jaroszewski M., Bretuj W., Tchórzewski M. (2010a). Influence of surface and volume electrical resistivity on dielectric properties of carbon-polypropylene fabric composite obtained by plasma deposition, Electrotech. Rev. 2010, R. 86, nr 5, pp. 275-278

Pospieszna J., Jaroszewski M., Szafran G. (2010b). Influence of substratum on dielectric properties of plasma carbon films, Electrotech. Rev. 2010, R. 86, nr 11b/2010, pp. 308-310

Sarto F., Sarto M.S., Larciprete M.C., Sibilia C. (2003). Transparent films for electromagnetic shielding of plastics, Rev. Adv. Mater. Sci., (2003), n.5, 329-336

Sarto M. S., Li Voti R., Sarto F., Larciprete M. C. (2005). Nanolayered Lightweight Flexible Shields with Multidirectional Optical Transparency, IEEE Trans. on EMC, vol. 47, No 3, (2005) pp.602- 611

Wang L.B., See K.Y., Zhang J.W., Salam B., Lu A.C.W. (2011). Ultrathin and flexible screen-printed metasurfaces for EMI shielding applications, IEEE Transactions on Electromagnetic Compatibility, Vol. 53, No. 3, August 2011, pp. 700-704

Wei Q. F., Xu W. Z., Ye H., Huang F. L. (2006). Surface Functionalization of Polymer Fibres by Sputtering Coating, J. Industrial Textiles , Vol. 35 No. 4 (2006) pp. 287-294

Ziaja J., Jaroszewski M. (2011); EMI Shielding using Composite Materials with Plasma Layers, Electromagnetic Waves, Vitaliy Zhurbenko (Ed.), ISBN: 978-953-307-304-0, InTech, Available from: http://www.intechopen.com/articles/show/title/emi-shielding-using-composite-materials-with-plasma-layers

Ziaja J., Ozimek M., Janukiewicz J. (2010). Application of thin films prepared by impulse magnetron sputtering for shielding of electromagnetic fields, Electrotech. Rev. 2010, R. 86, nr 5, pp. 222-224

Ziaja J., Ozimek M., Koprowska J. (2009). Metallic and oxide Zn and Ti layers on textile as shields for electromagnetic fields, EMC Europe 2009 Workshop, Athens, Greece, 11-12 June 2009, pp. 30-33

Ziaja J., Koprowska J., Janukiewicz J. (2008). The use of plasma metallization in the manufacture of textile screens for protection against electromagnetic fields, Fibres & Textiles in Eastern Europe. 2008, vol. 16, nr 5, pp. 64-66

Ziaja J., Koprowska J., Janukiewicz J., (2008a). Using of plasma metallization for fabrication of fabric screens against electromagnetic field, FIBRES & TEXTILES in Eastern Europe 5, s. 70-72

Ziaja J., Koprowska J., Żyłka P. (2008b). Influence of nonwoven structures on surface resistivity of plasma titanium films. Proceedings of 6th International Conference ELMECO-6 : electromagnetic devices and processes in environment protection joint with 9th Seminar "Applications of Superconductors" AoS-9, Nałęczów, Poland, June 24-27, 2008. s. 95-96

6

The Effects of Adding Nano-Calcium Carbonate Particles on the Mechanical and Shrinkage Characteristics and Molding Process Consistency of PP/nano-CaCO3 Nanocomposites

Karim Shelesh-Nezhad, Hamed Orang and Mahdi Motallebi
University of Tabriz
Iran

1. Introduction

The performance characteristics of plastic injection molded parts are depended upon the compositions of processing raw materials, parts and molds design specifications as well as the processing conditions. By appropriate incorporation of additives into a polymer matrix, it is possible to achieve the desired characteristics in the molded parts. Polypropylene (PP) is a semi-crystalline thermoplastic and is widely used for general applications. PP possesses the advantages of processing ease, very resistant to moisture absorption, and good chemical resistance to solvents. However, its applications as an engineering thermoplastic are limited due to its high shrinkage rate and relatively poor impact resistance at room or low temperatures (Lam et al., 2009).

Recent developments in fillers and reinforcements technology have made it possible to enhance the properties and applications of PP. In formulating different compositions of materials, it is essential to bring into account the cost effectiveness, adding value, ease of processing and wide range of applications. Types, shapes, concentrations and dimensional conditions of fillers and reinforcements may directly affect the processing ease, production consistency, molding cycle time, parts dimensional conditions, mechanical, thermal as well as tribological properties. Platelike and layered particles may lead to poor impact strength, and the impact resistance is in reverse proportion to particle size (DeBoest, 1988; Mohd Ishak et al., 2008). These fillers may act either as sites of stress concentration or micro cracks initiator. Micron-sized spherical fillers such as CaCO3 have marginal influence on the impact resistance (DeBoest, 1988). While nano-sized CaCO3 may act as a nucleating agent (Avella et al., 2006) and impact modifier (K. Yang & Q. Yang, 2006) in a polymer matrix. The presence of nano-CaCO3 may possibly facilitate the mobilization of macromolecular chains and improve the ability of matrix polymer to adapt to deformation and hence to increase the ductility and impact strength of composites. The nanoparticles may also initiate micro-void formations which locally deform the matrix surrounding the particles and initiate mass plastic deformation and, in consequence, increase the toughness and impact energy (Kemal et al., 2009). Stiffness or Young's modulus can be readily improved by adding either micro- or nano-particles since rigid inorganic particles generally have a much higher stiffness than polymer matrices. However, strength strongly depends on the stress transfer between the particles and the matrix. For well-bonded particles, the applied stress can be effectively transferred to the particles from the matrix; this clearly improves the strength (Fu et al., 2008).

Due to its non-polar chemical structure, PP interacts poorly with the typically polar fillers such as CaCO3, and optimum dispersion is normally difficult to achieve. Compatibilisers are frequently used to improve the interfacial adhesion between CaCO3 and PP, in order to gain the envisaged enhancement in mechanical properties (Fuad et al., 2010). Bi-functional molecules such as maleic-anhydride grafted PP (PP-g-MAH) are commonly used as compatibilisers for PP and CaCO3 (Roberts & Constable, 2003).

The crystallographic morphology of PP matrix can be noticeably altered by the presence of nano-CaCO3 because of its nucleating effect (Lin et al., 2011; Zhang et al., 2004). The coated CaCO3 nanoparticles affect the crystallization of PP in two ways: by serving as heterogeneous nucleation sites and also by reducing the spherulitic growth rate due to block the diffusion of polymer chains. Heterogeneous nucleation is, however, the dominating step controlling the crystallization rate of the PP/CaCO3 nanocomposite blends (Lin et al., 2011).

Besides mechanical properties, thermal contraction and uniformity of molded samples may also be influenced by different concentrations of the nano-CaCO3 in PP matrix. The objective of this research is to determine mechanical performances, shrinkage behavior and at the mean time the injection molding consistency of PP filled with nano-CaCO3 particles. In this research, PP/nano-CaCO3 polymer nanocomposites of different compositions were prepared by using a twin-screw extruder. PP-g-MAH compatibiliser was applied to improve the interfacial interaction between nano-CaCO3 and PP, and to extend the dispersion of nanoparticles in polymer matrix. An injection molding machine was employed to produce the standard specimens. The melt pressure inside the mold cavity was measured during the injection molding process of PP-CaCO3 nanocomposites to assess the production consistency. Dimensional conditions of different samples were characterized in order to determine the effect of nano-CaCO3 inclusion on the shrinkage rates. Morphology was observed and tensile, flexural and impact properties were examined to ascertain the influence of nano-CaCO3 on the mechanical performances.

2. Experimental procedures

2.1 Material used, compounding and sample preparation

Polypropylene (PP500P, SABIC) has melt flow rate of 3.1 (2.16 kg at 230 °C) and density of 905 kg/m^3 was used as matrix resin. Nano-sized synthetic ultrafine surface treated precipitated calcium carbonate (Socal 312, Solvay, France) with mean particle diameter of 70 nm used as filler phase. PP-g-MAH compatibiliser (Priex 20097, Solvay, France) with a maleic anhydride content of 0.05 wt % and MFI of 15 (2.16 kg at 230 °C) was employed to promote the interfacial interaction between nano-CaCO3 and PP, and to extend the dispersion of nanoparticles in polymer matrix. Compounds used as processing materials are listed in the table 1.

The Effects of Adding Nano-Calcium Carbonate Particles on the Mechanical and Shrinkage Characteristics and
Molding Process Consistency of PP/nano-CaCO3 Nanocomposites

131

Material	PP (wt%)	nano-CaCO3 (wt%)	PP-g-MAH (wt%)
PP	100	0	0
P5	94	5	1
P10	88	10	2
P15	82	15	3

Table 1. List of various compounds used in the experimentations

The neat PP and PP-g-MAH were dried in a vacuum oven at 80°C for 6 hours. Melt extrusion technique was applied to produce different compositions of PP/nano-CaCO3 by a ZSK-25 (Coperion Werner Pfliederer-Germany) co-rotating twin-screw extruder (D= 25 mm, L/D = 40) with a barrel temperature profile ranging from 160°C near the hopper to 200°C at the die and a screw speed of 400 rpm.

Molded samples utilized throughout the experimentations comprised standard tensile (ASTM D-638), flexural (ASTM D-790) and impact (ASTM D-256) specimens. An advanced microprocessor control injection molding machine (Poolad-110/380) with clamping force capacity of 110 tones and shot size capacity of 268 grams was employed to produce corresponding samples. The reciprocating screw diameter was equal to 45 millimeters and the ratio of screw length to diameter was 20. The values of molding parameters settings for standard tensile, flexural and impact specimens are given in table 2. The values of different settings were obtained on the basis of dimensional conditions of molds cavities and feed channels and in a manner to produce samples free of defects.

Molding Parameter	Unit	Value for tensile specimens	Value for Flexural and impact specimens
Nozzle Temperature	°C	180	180
Mold Temperature	°C	40	40
Injection Pressure	MPa	57	89
Injection Speed	cm³/s	68	68
Holding Time	s	8	15
Holding Pressure	MPa	32	63
Screw feeding speed	rpm	108	108

Table 2. The values of molding parameters settings for standard tensile, flexural and impact specimens

2.2 Characterization

2.2.1 Molding process consistency

The uniformity of injection molding process of different compounds was characterized by melt pressure measurement inside the mold cavity. A piezo-electric sensor (9221 AA0.6, Kistler, Switzerland) incorporated inside the tensile mold cavity and located behind and in contact to the ejector pin to measure the force exerted to the pin by melt pressure during the cyclic molding process against the time. The influence of nano-CaCO3 inclusion in the PP matrix resin on the sameness of pressure-time profiles and hence the variation of molding process was investigated. Two variables encompassing the maximum force and the force-time integral were chosen as the control parameters to assess the consistency of injection molding process during the molding cycles. For each compound, 10 samples were examined and the mean values of control parameters were considered.

2.2.2 Dimensional conditions

Standard tensile specimens (ASTM D-638) of PP with different contents of nano-CaCO3 were molded and employed to analyze the dimensional conditions and shrinkage rates. Figure 1, indicates the reference dimensions and their locations for shrinkage estimation of corresponding molded parts. The reference dimensions of matching cavity including cavity length, width and depth were equal to the 165.83, 19.10 and 3.20 millimeters respectively. The molded parts reference dimensions including parts length, width and thickness were measured with 0.01 mm accuracy after passing two weeks of molding. The differences of cavity dimensions and molded parts dimensions were considered as shrinkage values. For each compound, 5 samples were tested and the mean values were taken into account.

Fig. 1. Reference dimensions and their locations for shrinkage assessment of part length (L), part width (W) and part thickness (H)

2.2.3 Mechanical properties

Tensile, flexural and impact tests were performed based on the ASTM D-638, ASTM D-790 and ASTM D-256 standards respectively in order to analyze the effect of adding nano-CaCO3 particles on the mechanical performances of PP/CaCO3 nanocomposites. For each compound, 5 samples were tested and the mean values were considered. The specifications of testing equipment and conditions are presented in table 3.

Tensile test	
Manufacturer: Zwick/Roll	Model: TIFR010THA50
Load range: 0.5 g to 3000 kg	Adjusted load sensitivity: 0.5 g
Speed range: 1 to 200 mm/min	Adjusted crosshead speed: 50 mm/min
Impact resistance test	
Manufacturer: Gotech	Model: GT-7045-I
Capacity: 100 kg-cm	Test type: Izod-type
Flexural strength test	
Manufacturer: Gotech	Model: GT-7010A2
Load range: 3000 kg	Speed range: 10 to 200 mm/min
Extension range: 10.00 mm	Span: 30.00 mm
End point: 7.00 mm	Adjusted speed: 10 mm/min

Table 3. The specifications of mechanical properties testing equipment and conditions

2.2.4 SEM

A scanning electron microscope (Tescan VegaII) was employed to observe the dispersion of
nano-CaCO3 particles in PP matrix in the fracture sections of impact specimens. Prior to
SEM observations, the samples were made conductive by gold sputtering.

3. Results and discussions

3.1 Melt flow rate

Figure 2, indicates the melt flow rates of different compounds of PP/nano-CaCO3 with
respect to the nano-CaCO3 contents.

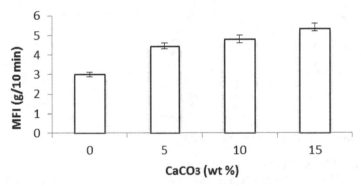

Fig. 2. Melt flow rate versus nano-CaCO3 contents

Inclusion of nano-CaCO3 considerably raised the melt flow rate. The presence of nano-
CaCO3 in the molten PP has a rolling effect which facilitates sliding of melt on the cylinder
wall of MFI tester (Jiang & Huang, 2008). According to Xie et al., spherical nanoparticles
serve as ball bearings, reducing the interlayer interaction of melts (2004). The increment of
MFI rate can facilitate the injection molding of thin-walled parts and can lead to the
reduction of energy consumption of molding process.

3.2 Injection molding uniformity

The results of melt pressure measurement during the molding cycles for different
compounds are indicated in table 4 and figure 3.

Material	Integral (N-s)		Max. Force (N)	
	Ave. Value	Std Dev.	Ave. Value	Std Dev.
PP	4953.29	133.30	843.28	8.75
P5	4923.87	138.26	820.49	11.20
P10	4894.25	50.45	826.49	7.49
P15	4468.54	52.53	800.05	4.46

Table 4. The magnitudes of control parameters and their deviations during the molding
process

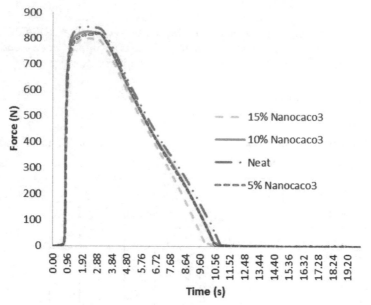

Fig. 3. Force-time traces of different compounds

As the nano-CaCO3 loading was increased, the maximum force and the force-time integral were decreased. This may be attributed to the influence of nano-CaCO3 on melt fluidity and crystallization behavior of PP matrix. By incorporating higher loading of nano-CaCO3 (10 and 15 wt%), the deviation of molding process was significantly declined. By adding 15 wt% of nano-CaCO3, the maximum force standard deviation dropped from 8.75 to 4.46 (N) and the force-time integral standard deviation declined from 133.30 to 52.53 (N-s). According to the figure 3, by increasing the nano-CaCO3 content in the PP matrix, a change in the slope on the right side of force-time profile is observed. This may possibly be related to the increase of PP crystallinity and, in consequence, higher thermal contraction of melt during the molding process, as a result of nano-CaCO3 presence.

3.3 Shrinkage behavior

Table 5, indicates the shrinkage rates along the flow, across the flow, and along the thickness of molded parts with respect to the amount of nano-CaCO3 content.

nano-CaCO3 (wt%)	Ave. Shrinkage along the flow (%)	Ave. Shrinkage across the flow (%)	Ave. Shrinkage along the flow thickness (%)
0	1.64	1.74	5.00
5	1.42	1.50	4.93
10	1.51	1.65	4.87
15	1.53	1.64	4.62

Table 5. Effect of nano-CaCO3 concentration on the shrinkage of PP/nano-CaCO3

The neat PP possessed the highest shrinkage rate. The inclusion of nano-CaCO3 reduced the thermal contraction because of its filling effect and its lower thermal contraction. The molded samples with concentration of 5 wt% nano-CaCO3 possessed the lowest shrinkage along the flow and across the flow directions. Addition of higher values of nano-CaCO3 (10, 15 wt%) to the PP, elevated the shrinkage rates along and across the flow directions. Nano-CaCO3 has nucleating effect and can lead to the increment of crystallization rate in the PP matrix (Chan et al., 2002; K. Yang & Q. Yang, 2006). Addition of higher values of nano-CaCO3 (10, 15 wt%) to the PP, slightly increased the tendency of non-isotropic thermal contraction along and across the flow directions. This can possibly be related to increment of PP crystallinity as a result of higher loadings of nano-CaCO3, and non-isotropic crystal growth of PP Matrix.

3.4 Morphology

Figure 4, indicates the presence and distribution of nano-CaCO3 (5 and 15 wt%) in the PP matrix in the cross sections of impact test samples. Higher concentration of nano-CaCO3 particles is observed in the PP matrix with higher content, i.e. 15wt%, of nano-CaCO3. Due to the surface modification of nano-CaCO3 and application of PP-g-MAH compatibiliser, relatively good dispersions of nanofillers were achieved. According to figure 4-b, more tendency to agglomerate is observed at 15 wt% loading of nano-CaCO3.

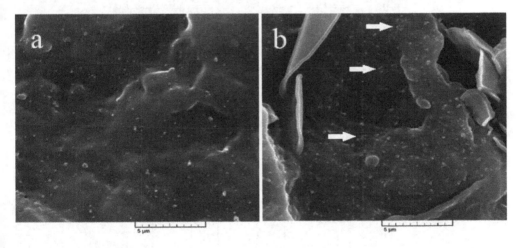

Fig. 4. Morphologies of PP/nano-CaCO3 at two loadings of (a) 5 and (b) 15 wt% of nano-CaCO3

3.5 Mechanical properties

The tensile tests results of various compounds of PP/nano-CaCO3 are indicated in Table 6 and Figures 5 and 6.

Material	Ave. Elastic modulus (Mpa)	Ave. Tensile Strength (Mpa)	Ave. Strength at break (Mpa)	Ave. Elongation at break (%)
PP	960.17	36.04	19.81	75.03
P5	971.65	35.67	11.62	122.88
P10	983.94	34.64	6.81	161.29
P15	990.41	33.80	7.58	118.87

Table 6. The results of tensile tests

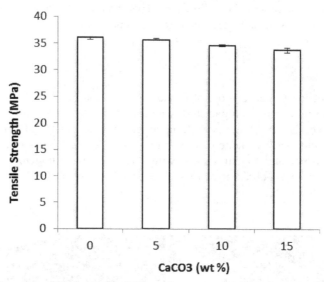

Fig. 5. Tensile strength of PP/nano-CaCO3 against nano-CaCO3 content

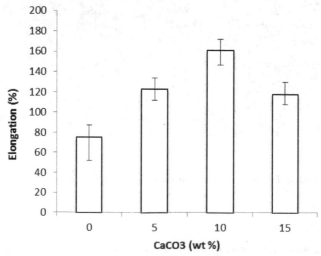

Fig. 6. Elongation of PP/nano-CaCO3 against nano-CaCO3 content

Figure 7, Compares the stress-strain characteristics of median specimens of different compounds.

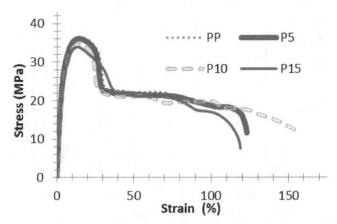

Fig. 7. The influence of adding nano-CaCO3 on the stress-strain behavior of PP/nano-CaCO3

The results of tensile tests revealed that the inclusion of nano-CaCO3 slightly increased modulus and decreased tensile strength and significantly increased the elongation at break. Addition of rigid particles to a polymer matrix can easily improve the modulus since the rigidity of inorganic fillers is generally much higher than that of organic polymers (Fu et al., 2008). The reduction of tensile strength may be attributed to the weakly bonded nanoparticles which promote matrix yielding (Kemal et al., 2009). According to Zhang et al., when CaCO3 particles were introduced, small and imperfect spherulites formed. The reduction of spherulite size and the disappearance of sharp interfaces among spherulites favored the increase of elongation at break for the PP/CaCO3 composites (2004). Xie et al. reported that the increases in elongation at break can be attributed to ellipsoidal voids formation in the matrix surrounding the particles, allowing ductile pull out (2004). At high fraction of nano-CaCO3 (i.e.15 wt%), the elongation at break was declined. This result is consistent with the morphology of corresponding compound which shows agglomeration sites of nano particles in the PP matrix as presented in figure 4-b.

Figure 8 depicts the results of flexural tests.

The incorporation of nano-CaCO3 led to the elevation of flexural strength in PP/nano-CaCO3 compounds. This may also be related to the nucleating effect of nano-CaCO3 in the PP matrix. Figure 9 shows the impact tests results.

Addition of nano-CaCO3 elevated the impact strength significantly. This may be explained by the fact that the presence of nano-CaCO3 in the PP matrix lead to a more uniform distribution of impact energy. Additionally, the presence of nano-CaCO3 may influence the crystallization behavior by reducing the spherulites size and subsequently alter the impact strength. Kemal et al. reported that the raise of toughness and impact energy may be attributed to enhanced micro-void formations initiated by nanoparticles, which locally deform the matrix surrounding the particles and initiate mass plastic deformation (2009).

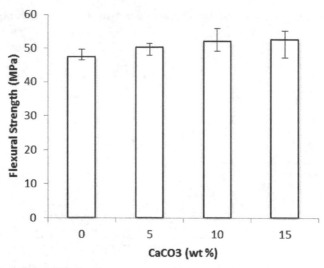

Fig. 8. Flexural strength of PP/nano-CaCO3 versus nano-CaCO3 content

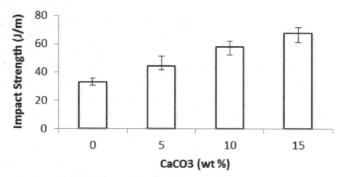

Fig. 9. Impact strength of PP/nano-CaCO3 versus nano-CaCO3 content

4. Conclusion

In this study the influences of nano-CaCO3 on the production consistency, shrinkage and melt flow rates, as well as mechanical properties of PP/nano-CaCO3 nanocomposites were experimentally investigated. PP-g-MAH compatibiliser with a maleic anhydride content of 0.05 wt% was employed to improve the interfacial adhesion between nano-CaCO3 and PP and to extend dispersion of nanoparticles in polymer matrix. Inclusion of nano-CaCO3 raised the melt flow rate as high as 77%. By incorporating higher loading of nano-CaCO3 (10 and 15 wt%), the deviation of molding process was significantly declined. The molded samples with concentration of 5 wt% nano-CaCO3 possessed the lowest shrinkage along the flow and across the flow. The results of tensile tests revealed that the inclusion of nano-CaCO3 slightly increased modulus and decreased tensile strength and significantly increased the elongation at break. At high fraction of nano-CaCO3 (i.e.15 wt%), the elongation at break was declined. Addition of nano-CaCO3 elevated the impact strength as high as 107%. The nano-sized feature, shape and dispersion conditions of nano-CaCO3, played important roles in determining the performances of PP/nano-CaCO3.

5. References

Avella, M.; Cosco, S., Di Lorenzo, M.L., Di Pace, E. & Errico, M.E. Gentile G. (2006). Nucleation activity of nanosized CaCO3 on crystallization of isotactic polypropylene, in dependence on crystal modification, particle shape, and coating. *European Polymer Journal*, Vol. 42, pp. (1548–1557)

Chan, C.M.; Wu, J.S., Lee, J. X. & Chung, Y. K. (2002). Polypropylen/calcium carbonate nanocomposites. *Polymer*, Vol. 43, pp. (2981-2992)

DeBoest, J. F. (1988). Reinforced polypropylenes, In: *Engineering Plastics*, pp. (192-193)

Fu, S.Y.; Feng, X.Q., Lauke, B. & Mai, Y. W. (2008). Effects of particle size, particle/matrix interface adhesion and particle loading on mechanical properties of particulate-polymer composites. *Composites*: Part B, Vol. 39, pp. (933–961)

Fuad, M.Y.A.; Hanim, H., Zarina, R., Mohd Ishak, Z.A. & Hassan, A. (2010). Polypropylene/calcium carbonate nanocomposites – effects of processing techniques and maleated polypropylene compatibiliser. *eXPRESS Polymer Letters*, Vol. 4, pp. (611–620)

Jiang, G. & Huang, H.X. (2008). Online shear viscosity and microstructure of PP/nano-CaCO3 composites produced by different mixing types. *Journal of Materials Science*, Vol. 43, No. 15, pp. (5305–5312)

Kemal, I.; Whittle, A., Burford, R., Vodenitcharova, T. & Hoffman, M. (2009). Toughening of Unmodified Polyvinylchloride through the Addition of Nanoparticulate Calcium Carbonate. *Polymer*, Vol. 50, pp. (4066-4079)

Lam, T. D.; Hoang, T. V., Quang, D. T. & Kim, J. S. (2009). Effect of nanosized and surface-modified precipitated calcium carbonate on properties of CaCO3/polypropylene nanocomposites. *Materials Science and Engineering*: Part A, Vol. 501, pp. (87–93)

Lin, Y.; Chen, H., Chan, C.M. & Wu, J. (2011). Nucleating effect of calcium stearate coated CaCO3 nanoparticles on polypropylene. *Journal of Colloid and Interface Science*, Vol. 354, pp. (570–576)

Mohd Ishak, Z.A.; Kusmono, Chow, W.S., Takeichi, T. & Rochmadi (2008). Effect of Organoclay Modification on the Mechanical, Morphology, and Thermal Propertiese of Injection Molded Polyamide6/Polypropylene/ Montmorillonite Nanocomposites. *Proceedings of the Polymer Processing Society 24th Annual Meeting*, Salerno (Italy), June 2008

Roberts, D. & Constable, R.C. (2003). Chemical coupling agents for filled and grafted polypropylene composites, In: *Handbook of Polypropylene and Polypropylene Composites*, Karian, H. G., pp. (28–68), Marcel Decker, New York

Xie, X.L.; Liu, Q.X., Li, R.K.Y., Zhou, X.P., Zhang, Q.X., Yu, Z.Z. & Mai, Y.W. (2004). Rheological and Mechanical Properties of PVC/CaCO3 Nanocomposites Prepared by In Situ Polymerization. *Polymer*, Vol. 45, pp. (6665-6673)

Yang, K. & Yang Q. (2006). Morphology and mechanical properties of polypropylene/calcium carbonate nanocomposites. *Materials Letters*, Vol. 60, pp. (805-809)

Zhang, Q.X.; Yu, Z.Z., Xie, X.L. & Mai, Y.W. (2004). Crystallization and Impact Energy of Polypropylene/CaCO3 Nanocomposites with Nonionic Modifier. *Polymer*, Vol. 45, pp. (5985-5994)

7

The Influence of Filler Component on Mechanical Properties and Thermal Analysis of PP-LDPE and PP-LDPE/DAP Ternary Composites

Kamil Şirin[1], Mehmet Balcan[2] and Fatih Doğan[3]
*[1]Celal Bayar University, Faculty of Science and Arts,
Department of Chemistry, Manisa,
[2]Ege University, Faculty of Science, Department of Chemistry, İzmir,
[3]Çanakkale Onsekiz Mart University, Faculty of Education,
Secondary Science and Mathematics Education, Çanakkale,
Turkey*

1. Introduction

Composite material is a material system consisting of a mixture or combination of two or more micro-constituents mutually insoluble and differing in form and/or material composition. Particulate-filled thermoplastic composites have proved to be of significant commercial importance in recent years, as industrialists and technologists have sought to find new and cost-effective materials for specific applications (Shonaike & Advani, 2003; Ma et al., 2007). With addition of inorganic filler, various changes occur in the molecular and supermolecular structure of a thermoplastic resin. Composite properties depend on a variety of material-process variables (e.g., polymer matrix structure, filler content, chemical composition, surface activity, particle size and shape, compounding extruder design, mold design, and extruder-molding process conditions). Some changes may be latent or delayed (i.e., occurring later in the service life of the plastic part as a result of surrounding conditions). Reduction of molecular weight, crystal and spherulite size, and molecular mobility are among the most profound effects that solid filler has on the polymer matrix structure (Ayae & Takashi, 2004). The microstructure of the polymer–filler interphase is mirrored by the mechanical integrity of the molded part and long-term durability to extremes of surrounding temperatures and applied stresses. Calcium carbonate is very commonly used filler in the plastics industry. The incorporation of fillers such as calcium carbonate into thermoplastics is a common practice in the plastics industry, being used to reduce the production costs of molded products. Fillers are also used to modify the properties of plastics, such as the modulus and strength. High filler loadings, however, may adversely affect the processability, ductility, and strength of composites (Rai & Singh, 2003). Some polyolefins are prone to chain-scission reactions in the presence of free radicals. PP is degraded due to chain scission in β position to the macroradicals site, while PE is crosslinked, due to macroradical recombination (Braun et al., 1998). The use of organic peroxides for controlled degradation of PP is the most important commercial application of the chain-scission or visbreaking of polyolefin chains and results in the so-called controlled rheology PP grades with enhanced melt flow behavior (Zweifel, 2001). Polypropylene (PP) filled with calcium carbonate is among the more recent development on the polyolefin market and in the last decade have shown impressive growth rates. In polyethylene sector, calcium carbonate fillers now play a role preferably in films and sheets. Low density polyethylene's (LDPE and LLDPE) are usually filled with very pure calcium carbonate grades.

Several studies dealing with the melt rheology, mechanical, deformation, impact behavior and thermal properties of various blend or composites were published during the last decade (Chen et al., 2004; El-Sabbagh et al., 2009; Kolarik & Jancar, 1992; Mishra et al., 1997; Sirin & Balcan, 2010, Zhang et al., 2002). Wang WY (Wang, 2007, 2008) was studied the preparation and characterization of calcium carbonate/Low-Density-Polyethylene and $CaCO_3$/acrylonitrile-butadiene-styrene composites. Tang *et al.* studied rheological properties of nano-$CaCO_3$/ABS composites such as shear viscosity, extension viscosity, and entry pressure dropped by capillary extrusion (Tang & Liang, 2003). Liang investigated the tensile, flow, and thermal properties of $CaCO_3$-filled LDPE/LLDPE composites (Liang, 2007). Effects of coupling agents on mechanical and morphological behavior of the PP/HDPE blend with two different $CaCO_3$ were studied by Gonzalez et. al (Gonzalez et al., 2002) . Also, several researches have reported different properties of ternary composites with calcium carbonate (Jancar & Dibenedetto, 1995; Kim et al, 1993; Premphet, 2000).

This study has focused on the investigation of the changes on thermal, mechanical and morphological properties of PP-LDPE/DAP (90/10 /0.06 wt. %) blend and PP-LDPE/DAP (90/10 /0.0 wt. %) blend when different ratios of 5-10-20 wt. % $CaCO_3$ are added. The blend (PP-LDPE/DAP (90/10/0.06 wt.%) used in this study was prepared in terms of heat sealing strength properties by the results based on our previous work (Sirin & Balcan, 2010) and is optimum as well (Şirin, 2008).

2. Experimental

2.1 Materials

Isotactic polypropylene (PP-MH418) and Low-density polyethylene (LDPE-I 22-19 T) were supplied as pellets by Petkim Petrochemical Company (Aliaga, Izmir, Turkey). The number-average molecular weight (Mn), weight-average molecular weight (Mw) and polydispersity index (PDI) values of PP and LDPE homopolymer were 20300, 213600 g.mol^{-1} and 10.5, and 29600, 157000 g.mol^{-1} and 5.3, respectively. The specific gravity of the PP-MH418 is 0.905 g.cm^{-3} and that of the LDPE-I 22-19 T is 0.919-0.923 g.cm^{-3}, with melt flow index of 4-6 g.10 min^{-1} (2.16 kg, 230 ± 0.5 °C) and 21– 25 g.10 min^{-1} (2.16 kg, 190 ± 0.5 °C), respectively. (2, 5-dimethyl-2, 5-di (tert-butyl peroxy)-hexane, (Sinochem, Tinajin/Chine) was used as dialkly peroxide (DAP). Calcium carbonate filler (AS 0884 PEW) was provided by Tosaf Company (Israel).

2.2 Preparation of blends

In the preparation of the composites, two different procedures were used. In the first procedure, PP-LDPE/$CaCO_3$ composites were prepared without addition of DAP. These composites are called as PC0, PC1, PC2 and PC3. In the second procedure, PP-LDPE/$CaCO_3$

The Influence of Filler Component on Mechanical Properties and Thermal Analysis of PP-LDPE and PP-LDPE/DAP
Ternary Composites

143

composites were prepared with (0.06 % wt.) addition of DAP. These composites are called as PC4, PC5, PC6 and PC7.All compounds were prepared by using single screw extruder (Collin E 30P). The blends were prepared by melting the mixed components in extruder which was set at the extruder diameter: 30 mm, length to diameter ratio: 20, pressure: 9-10 bar, temperature scale composites from filing part to head were 190-250 °C and screw operation speed: 30 rev.min^{-1}. The composites were produced as 70 µm thick and 10 cm wide films. These ratios and their codes are given in Table 1. All of these composites were prepared as samples weighing 1000 grams, while keeping the PP-LDPE (90/10) ratio constant.

	Composition, wt %			
Sample Code	PP	LDPE	CaCO$_3$	DAP
PC0	90	10	-	-
PC1	90	10	5	-
PC2	90	10	10	-
PC3	90	10	20	-
PC4	90	10	-	0.06
PC5	90	10	5	0.06
PC6	90	10	10	0.06
PC7	90	10	20	0.06

Table 1. Nomenclature, components and composition of composites

2.3 Melt flow index (MFI) measurements

Melt flow index measurements of the composites were carried out on a MFI (MP-E) Microprocessor apparatus at 230 °C and under a 2.16 kg weight. The capillary die was 2.095 mm in diameter and 8 mm in length. About 5 g of composite was put into barrel and heated for 5 min to reach the predetermined temperature on the plunger to extrude the melt through the capillary die. After a steady flow state was reached, five samples were cut sequentially and their average weight value was obtained. Experiments were done according to ASTM D-1238.

2.4 Mechanical testing measurements

The tensile properties were determined using a Instron tensile tester (model 4411) following the ASTM D-638 procedure and using type 1 test specimen dimensions. The crosshead speed was set at 50 mm.min^{-1} and 5 samples were tested for each composition. Tensile stress at yield, tensile strength at break and elongation at break were determined from the recorded force versus elongation curve.

2.5 Hardness test measurements

Shore D scale was used to determine the hardness values of all samples. The tests were carried out Zwick/Roell apparatus of out at the room temperature and 76 cm Hg pressure hardness test (Shore D) was performed according to ASTM D 2240. Hardness test measurements were carried out the dimensions of 60x60x4 mm at 50 N, at the room temperature.

2.6 Heat seal tester

Heat sealing testers of samples were carried out with a TP701 (trade name, Tester Sangyo K.K.) apparatus at 1 sec timer and 2 kg/cm² pressure.

2.7 Thermal measurements

Differential scanning calorimetric (DSC) analyses were performed in a Shimadzu DSC-50 thermal analyzer in nitrogen atmosphere. The samples were heated from 25 to 200 °C at 10 °C min⁻¹, cooled to 25 °C at the same rate, and re-heated and cooled under the same conditions. Melting (T_m) and crystallization (T_c) temperatures and enthalpies were determined from the second scan. T_m was considered to be the maximum of the endothermic melting peak from the heating scans and T_c that of the exothermic peak of the crystallization from the cooling scans. The heat of fusion (ΔH_f) and crystallization enthalpy (ΔH_c) were determined from the areas of melting peaks and crystallization peaks.

The crystallinty of composites were calculated with the total enthalpy method [see eq. (1)]; in all calculations, the heats of fusion at equilibrium melting temperature were 209 and 293 Jg⁻¹, for PP and LDPE crystals, respectively (Brandrup & Imergut, 2003)

$$(X_c) = \frac{\Delta H_f}{\Delta H_{crys}} x100 \tag{1}$$

ΔH_f = Heat of fusion (Jg⁻¹)
ΔH_{crys} = 100% crystal polymer crystallization energy (Jg⁻¹)
(X_c) = Crystallinty (%)

The various melting and crystallization parameters which were determined by means of heating and cooling scans for composites are given in Table 3. Thermogravimetric (TG-DTG-DTA) curves were performed on a Seteram Labsys TG-16 thermobalance, operating in dynamic mode, with the following conditions; sample weight ~5 mg, heating rate = 10 °C.min⁻¹, atmosphere of nitrogen (10 cm³.min⁻¹), sealed platinum pan.

2.8 Scanning electron microscopy (SEM)

A Philips XL-305 FEG e SEM model scanning electron microscopy (SEM) was used to examine the morphologies of the composites

3. Conclusion

3.1 Mechanical analysis

CaCO₃ has a high chemical purity, which eliminates a negative catalytic effect on the aging of polymers. In addition, it has high whiteness and low refractive index that can help to reduce consumption of expensive abrasive pigments such as titanium dioxide. On the other

The Influence of Filler Component on Mechanical Properties and Thermal Analysis of PP-LDPE and PP-LDPE/DAP
Ternary Composites

145

hand $CaCO_3$ is very well suited for the manufacture of colorful products. Low abrasiveness, which contributes to low wear of machine parts such as extruder screws and cylinders, is another advantage. These properties and its low cost make $CaCO_3$ a very strong alternative to be considered as filler.

Melt flow index (MFI) analysis of composites are shown in Table 2. MFI values of composites without peroxide showed small differences by increasing amount of $CaCO_3$. Contrariwise, MFI values of composites with peroxide are proportional to the increasing amount of $CaCO_3$. MFI values of composites without DAP are between 9 and 10 g/10 min, however addition of 0.06 %wt. DAP to the composites resulted MFI values to vary between 22 and 26 g/10 min. This increase in MFI values is a result of degradation of PP by the DAP.

Sample Code	MFI / g.10 min^{-1} ± 0.1	Tensile strength at break / kg.cm^{-2} ±10	Tensile strength at yield/ kg.cm^{-2}±10	Elongation at break/ (%)±5	Hardness (ShoreD) ±1	Heat sealing strength/ kg.cm^{-2} ± 0.1 at 145 °C	Heat sealing strength/ kg.cm^{-2} ±0.1 at 150 °C
PC0	9.20	270	322	240	62.10	2.15	3.27
PC1	9.40	171	263	190	62.60	1.52	3.10
PC2	9.60	125	240	160	63.10	1.33	2.50
PC3	9.70	98	210	141	63.20	1.11	2.04
PC4	22.00	290	360	310	63.00	2.53	9.80
PC5	23.80	180	298	287	63.50	1.62	4.74
PC6	24.50	230	336	302	63.90	1.96	5.22
PC7	25.20	146	218	161	63.40	1.44	4.22

Table 2. MFI, heat sealing strength and mechanical analysis values of composites

Mechanical analysis of the composites such as tensile strength at break, tensile strength at yield and elongation at break are shown in Table 2. As shown in Table 2, samples with/without peroxides showed increase and decrease in their mechanical properties of composites in terms of increasing amount of $CaCO_3$. The highest values in mechanical properties were observed in PC6 . Even though PP, LDPE and $CaCO_3$ amounts were same in PC6 and PC2, there was only change in peroxide amounts. In other words, when PC6 and PC2 were compared it was observed that tensile strength at break, at yield, elongation at break values and heat sealing strength values of the composite PC6 showed high peaks. Table 2 summarizes tensile strength at break values for composites which do not contain peroxide displayed decreasing. However, by adding peroxide and increasing the amount of $CaCO_3$, these values showed increasing due to crosslinking of LDPE with the effect of peroxide. Furthermore, the tensile strength of the composites decreased with increasing amount of $CaCO_3$ due to the weak interfacial adhesion and dispersion of the $CaCO_3$ filler to PP-LDPE surface. The decrease of yield stress is likely due to the depending between inorganic fillers and the PP matrix at large deformations. In particular, a higher drop in tensile strength at yield is observed for PP-LDPE/$CaCO_3$, possibly due to the splitting of aggregated particles as well as depending between $CaCO_3$ particles and the PP matrix. In Table 2, with/without peroxide it is clearly shown that elongation at break values of the composites decreases with increasing ratio of $CaCO_3$. On the other hand, while comparing samples with $CaCO_3$ to each other the highest mechanical properties were observed in PC6. The reason for this is that adding DAP effect LDPE by crosslinking and PP by degradation. Shore D values of the composites with/without DAP increase in some degree with increasing amount of $CaCO_3$. As mentioned above, the reasons for increase in shore d values of the composites are addition of DAP effect to LDPE by crosslinking, PP by degradation, and dispersion of $CaCO_3$ in PP and LDPE. Table 2 shows that, increasing amounts of $CaCO_3$ resulted a decrease on heat sealing strength values at 145 °C and 150 °C independent of DAP. Therefore, the decrease of heat sealing strength is owing to the depending between in organic filler and the blends matrix of deformations. PP-LDPE/$CaCO_3$ composites still gave better heat sealing strength results than homopolymer PP and LDPE. In previous studies, we have examined heat sealing strength of homopolymer PP and LDPE (Sirin & Balcan, 2010; Şirin, 2008) Same as in mechanical properties values, heat sealing strength values showed similar trend in composite PC6 for best results.

3.2 Morphology observation

Figure 1 (a-g) shows the scanning electron microscopy (SEM) micrographs of composites reinforced with different amounts of $CaCO_3$ (0, 5, 10, and 20 %wt.) and DAP (0, 0.06 %wt.). From these micrographs, it is clear that $CaCO_3$ fillers were dispersed well in PC4, PC6 and PC7 matrices with DAP of composites. In these matrices, a homogeneous dispersion of reinforcing particles can also be observed. Moreover, the fillers remained intact within the matrix. This indicates that good bonding existed between the $CaCO_3$ particles and matrix. In contrast, the agglomeration of fillers can be observed for the composites containing $CaCO_3$ particles and dialkylperoxide (DAP). In homogeneous dispersion of fillers can cause a loss in the mechanical strength of the composites considerably. At higher loading levels however, $CaCO_3$ will agglomerate and remain confined in the polymer matrix.

3.3 Thermal analysis of composites

The results of the thermal analysis that was carried out by means of differential scanning calorimetry (DSC) are presented in Table 3. Overall, there were increased effect of both the content of the $CaCO_3$ filler and its healing on the melting temperature (T_m) and the crystallization temperature (T_c). However, a decrease in the energy required for the fusion of the crystalline parts was noted when the content of the filler in the composite increased. The melting enthalpy and crystallinity % (X_c) of the $CaCO_3$ containing composites with DAP (0.06 %wt.) were a little higher than the composites without DAP. Yet, the values of heat of fusion remain lower than that of homopolymer PP or LDPE. It can therefore be noted that the filler alters the crystalline phase of the polymer.

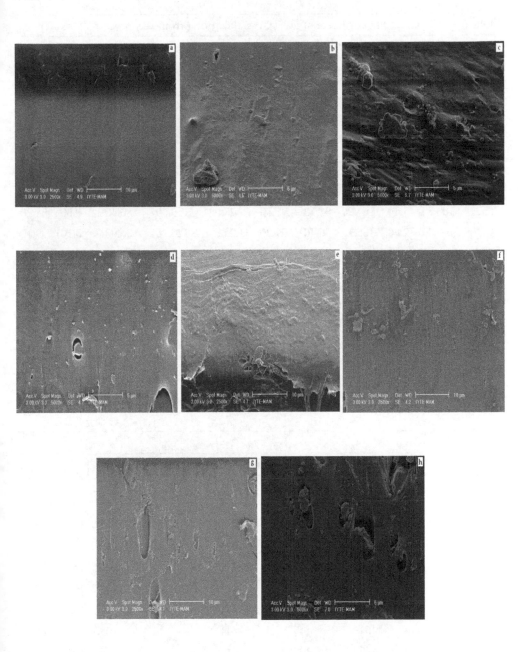

Fig. 1. SEM micrograph of blend and composites (a) PC0 (b) PC1 (c) PC2 (d) PC3 (e) PC4
(f) PC5 (g) PC6 (h) PC7

	Melting (from second heating scans)				Crystallization (from second cooling scans)					
	LDPE		PP		LDPE		PP		LDPE	PP
Sample Code	$T_m/$ °C	$\Delta H_f/$ J.g^{-1}	$T_m/$ °C	$\Delta H_f/$ J.g^{-1}	$T_c/$ °C	$\Delta H_c/$ J.g^{-1}	$T_c/$ °C	$\Delta H_c/$ J.g^{-1}	$X_c/$ %	$X_c/$ %
PC0	102.00	2.90	162.10	91.00	95.00	3.00	118.40	88.90	1.00	43.50
PC1	102.60	2.90	162.90	83.10	95.50	1.40	118.90	90.40	1.00	39.80
PC2	102.80	2.30	162.30	73.20	96.80	1.20	119.30	76.00	0.80	35.00
PC3	103.40	2.20	162.60	70.30	99.40	1.10	120.10	66.30	0.70	33.60
PC4	102.50	3.10	162.30	85.50	94.20	2.90	118.50	87.00	1.10	40.90
PC5	103.20	2.40	164.90	81.70	95.30	2.00	119.00	80.40	0.80	34.20
PC6	106.30	2.80	165.10	86.40	99.20	2.80	120.70	86.90	0.70	39.10
PC7	104.70	1.10	164.60	62.00	97.40	1.30	120.40	76.90	0.40	29.70

Table 3. Thermal properties of PP-LDPE/CaCO$_3$ composites

Thermal behavior of the composites was studied with a thermogravimetric analyzer under a protective nitrogen atmosphere. The temperature was scanned from 30 °C to 700 °C at a heating rate of 10 °C.min^{-1}. Figure 2 and 3 shows the typical thermogravimetric curves for composites with/without DAP and different filler contents are presented. 5 % and 20 % weight loss temperatures ($T_{-5\%}$ and $T_{-20\%}$) and maximum weight loss temperature ([b]$W_{max.}T$), derived from the derivative weight loss and differential thermal analysis curves are tabulated in Table 4.

			TG			DTA
Compounds	[a]T_{on}	[b]$W_{max.}T$	weight loss/ (5 %)	weight loss/ (20 %)	residual	Endo
PC0	403	454	406	438	0.5	105.2, 162.2, 457
PC1	402	453	406	438	3	105.8, 162.6, 458
PC2	405	454	407	440	9	106.6, 163.2, 457
PC3	405	455	407	440	18	106.5, 163.5, 459
PC4	407	456	409	443	6	107.2, 164.4, 459
PC5	409	457	413	444	11	107.9, 165.7, 459
PC6	414	461	415	444	11	108.1, 167.3, 462
PC7	414	459	409	444	26	108.7, 164.5, 460

Table 4. Thermal decomposition values of all the compounds ([a]The onset temperature, [b]Maximum Weight Temperature)

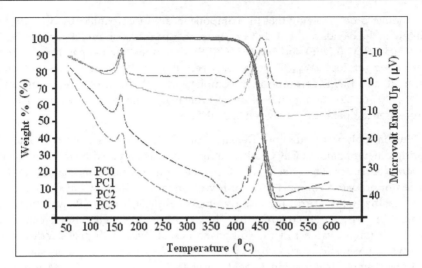

Fig. 2. TG and DTA curves of composites containing the different amounts of CaCO$_3$ (PC0,
PC1, PC2, and PC3)

TG curves for all the compounds exhibits one stage decomposition and a similar
characterizations. The thermal decomposition of composites occurs between 403 and 500 °C.
The thermal stabilities of composites increased usually with increasing CaCO$_3$ content. PC6
has the highest thermal stability among the polymer blends. It is also shown that the rate
curve related to compound shifts to a higher temperature. Also, three endothermic thermal
effects at different temperature in DTA profiles correspond to the melting and the
decomposition of composites. In DTA curves, the first two peaks are two melting peaks.

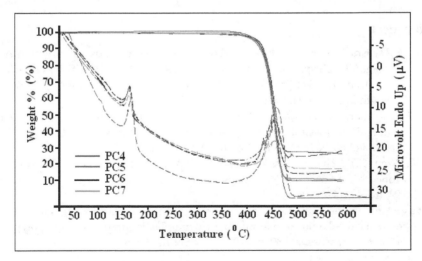

Fig. 3. TG and DTA curves of composites containing the different amounts of CaCO$_3$ (PC4,
PC5, PC6 and PC7)

This is another proof showing that the components are incompatible. But CaCO₃ dispersed completely in the blends and it didn't result another peak. As observed, melting peaks were between 105-108 °C (LDPE) and 162-167 °C (PP), decomposition temperatures related to the maximum weight loss (W_{max}.T) were between 454-459 °C. In the case of composites, although PP and LDPE decomposition completely at 500 °C, according to the amount of CaCO₃ 5, 10, or 20 % amount of mass, remained due to the reason that CaCO₃'s decomposition temperature is between 850-900°C.

In conclusion, polypropylene-Low Density Polyethylene blends with/without DAP and containing different amount CaCO₃ filler component was prepared by melting-blend with a single-screw extruder. The effects of CaCO₃ filler component on mechanical and thermal properties of prepared composites were investigated. Addition of CaCO₃ particles to the polymer matrix with DAP significantly increased MFI values. It was observed that mechanical properties (tensile strength at break, at yield, elongation at break values and heat sealing strength values) of the composite PC6 showed high peaks. With addition of CaCO₃, while mechanical properties of the composites were decreasing shore D values showed increasing. Heat sealing strength at 150 °C increased by increasing amount of CaCO₃ particles in the polymer matrix with DAP. SEM images showed that CaCO₃ particles were well-dispersed in the polymer matrix with DAP. The observation of TG/DTG/DTA curves revealed that the thermal stabilities of composites increased usually by increasing amounts of CaCO₃ and the blends are incompatible.

4. Acknowledgements

This study was carried out in Petkim Petrochemical Holding A.Ş., Turkey. In addition, it was supported by the research funds of Celal Bayar University (Project No: FEF-2006/085).

5. References

Ayae, S. & Takashi K. (2004). Calcium carbonate/polymer composites: polymorph control for aragonite, *Composite Interfaces*, vol.11, No.4, pp.287-295(9), ISSN: 1568-5543

Brandrup, J. & Imergut, E.H.(2003). Polymer Handbook Second Edition, John Wiley and Sons, 4th Edition V-(13–27). ISBN: 978-0-471-47936-9

Braun, D.; Richter, S.; Hellmann, G.P. & Ratzsch, M. (1998). Peroxy-initiated chain degradation, crosslinking, and grafting in PP-PE blends, *Journal of Applied Polymer Science*, vol.68, No.12, pp.2019-2028. ISSN:1097-4628

Chen, N.; Wan, C.Y.; Zhang, Y. & Zhang, Y.X. (2004). Effect of nano-CaCO₃ on mechanical properties of PVC and PVC/Blendex blend. *Polymer Testing*, vol.23, No.2, pp.169-194, ISSN: 0142-9418

El-Sabbagh, A.; Steuernagel, L. & Ziegmann, G.(2009). Processing and modeling of the mechanical behavior of natural fiber thermoplastic composite: Flax/polypropylene. *Polymer Composites*, vol.30, pp.510-519, ISSN:1548-0569

Gonzalez, J.; Albano, C.; Ichazo, M. & Diaz, B.(2002), Effects of coupling agents on mechanical and morphological behavior of the PP/HDPE blend with two different CaCO₃. *European Polymer Journal*, 38(12), 2465-2475. ISSN: 0014-3057

Jancar, J. & Dibenedetto, A.T.(1995). Failure mechanics in ternary composites of polypropylene with inorganic fillers and elastomer inclusions: Part II Fracture toughness. *Journal of Materials Science*, 30(9), 2438-2445, ISSN:1573-4803.

Kim, B.K.; Kim, M.S. & Kim, K.J.(1993). Viscosity effect in polyolefin ternary blends and composites *Journal of Applied Polymer Science*, 48(7), 1271-1278

Kolarik, J. & Jancar, J. (1992). Ternary composites of polypropylene/elastomer/calcium carbonate: effect of functionalized components on phase structure and mechanical properties. *Polymer*, vol.33, pp.4961-4967. ISSN: 0032-3861

Liang, J.Z.(2007), Flow and mechanical properties of polypropylene-low density polyethylene blends. *Journal of Material Processing Technology*, 66, 158-164, ISSN:0924- 0136

Ma,C.G.; Mai, Y.L.; Rong, M.Z.; Ruan, W.H. & Zhang M.Q. (2007). Phase structure and mechanical properties of ternary polypropylene/elastomer/nano-CaCO₃ composites, *Composites Science and Technology*, vol.67, No.14, pp.2997-3005, ISSN: 0266-3538

Mishra, S.; Perumal, G.B. & Naik, J.B. (1997). Studies on Mechanical Properties of Polyvinyl Chloride Composites. *Polymer-Plastics Technology and Engineering*, vol.36, No.4, pp.489-500, ISSN: 1525-6111.

Premphet, K. & Horanont, P.(2000). Polymer, Phase structure of ternary polypropylene /elastomer /filler composites: effect of elastomer polarity, *Polymer*, 41, 9283–9290, ISSN: 0032-3861

Rai, U.S. & Singh, R.K. (2003). Synthesis and mechanical characterization of polymer-matrix composites containing calcium carbonate/white cement filler. *Materials Letters*, vol.58, pp.235– 240, ISSN:0167-577X

Shonaike, G.O. & Advani, S.G.(2003). Advanced Polymeric Materials Structure Property Relationships, CRC Pres, pp. 463-478, ISBN 1-58716-047-1, New York, USA

Sirin, K. & Balcan, M.(2010). Mechanical properties and thermal analysis of low-density polyethylene + polypropylene blends with dialkyl peroxide. *Polymer   Advanced Technology*, vol.21, pp.250–255, ISSN: 1099-1581.

Şirin, K. (2008). Preparation of polymer blends and their composites, and determination of their properties, PhD.Thesis, Ege University, Izmir, Turkey.

Tang, C.Y. & Liang, J.Z.(2003), A study of the melt flow behaviour of ABS/CaCO₃ composites, *Journal of Materials Processing Technology*, vol.138, pp.408–410. ISSN: 0924-0136

Wang, W.Y.; Wang, G.Q.; Zeng, X.F.; Shao, L. & Chen, J.F. (2008). Preparation and properties of nano-CaCO₃/acrylonitrile-butadiene-styrene composites. *Journal of Applied Polymer Science*, vol.107, No.6, pp.3609-3614, ISSN:1097-4628

Wang, W.Y.; Zeng, X.F.; Wang, G.Q. & Chen J.F.(2007). Preparation and Characterization of Calcium Carbonate/ Low-Density-Polyethylene Nanocomposites. *Journal of Applied Polymer Science*, vol.106. No.3, pp.1932-1938. ISSN:1097-4628

Zhang, L.; Wang, Z.H.; Huang, R.; Li, L.B. & Zhang, XY. (2002). PP/elastomer/calcium
        carbonate composites: effect of elastomer and calcium carbonate contents on the
        deformation and impact behavior. *Journal of Materials Science,* vol.37, pp.2615-2621,
        ISSN:1573-4803
Zweifel, H. (2001).Plastic additives Handbook, Ch14, Crosslinking and Controlled
        Degradation of Polyolefin, ISBN: 3- pp.723, 446-19579-3, Switzerland

8

Preparation of Polypropylene Nanocomposites Using Supercritical Technology

Jia Ma and Ton Peijs
Queen Mary University of London
UK

1. Introduction

In recent years polypropylene (PP) nanocomposites have attracted great interest in academia. The attractiveness of this new class of material lies in the large improvements in both mechanical and thermal properties, as well as in gas barrier and fire resistance (Grunes et al., 2003). Nanofillers, being additives of nanometre scale, are dispersed in PP matrix, offering multifunctional and high-performance polymer characteristics beyond those possessed by traditional filled materials.

Many methods have been employed for preparing PP nanocomposites (Wu & Lerner, 1993; Andrews et al., 2002; Zhao et al., 2003; Garcia-Leiner and Lesser, 2004). Among them, supercritical CO_2 technology has been intensively studied in recent years. The main advantages of this technology are that CO_2 is a 'green' non-combustible and non-toxic solvent and the addition of $scCO_2$ into the polymer can cause plasticization, which brings the opportunity to lower polymer viscosity, leading to improved dispersion, and to process the polymer at lower temperatures (Behles & Desimone, 2001; Cansell et al., 2003). Furthermore, CO_2 can act as a transport medium, which facilitates the diffusion of monomers, initiators and molecules into a polymer matrix (Kikic & Vecchione, 2003). The aim of this chapter is to provide a review of the use of supercritical CO_2 technology in the preparation of PP nanocomposites. In this chapter, the different nanofillers used, the processing methods based on using supercritical CO_2 technology and the resulting properties of the nanocomposites will be discussed in detail. The progress and challenges on various aspects of PP nanocomposites will also be discussed.

2. Nanofillers

The term 'nanofillers' is vague and no precise definition exists. Nanofillers are understood, in essence, to be additives in solid form, which differ from the polymer matrix in terms of their composition and structure. Nanofillers are of the order of 100 nm or less in at least one dimension.

Nanofillers are often added to enhance one or more of the properties of polymers. Inactive fillers or extenders raise the quantity and lower the cost price, while active fillers bring about targeted improvements in certain mechanical or physical properties. Common nanofillers include calcium carbonate, ceramic nanofillers, carbon black, carbon nanotubes (CNTs), carbon nanofibres, cellulose nanowhiskers, nanoclays, gold particles, kaolin, mica, silica, silver nanoparticles, titanium dioxide, etc. Because of their impressive intrinsic mechanical properties, nanoscale dimensions and high aspect-ratio, nanofillers such as CNTs or nanoclays are among the most promising due to the fact that small amounts (less than 5 %) of them can provide the resulting nanocomposite material with significant property improvements.

2.1 Carbon Nanotubes (CNTs)

CNTs are considered to be ideal candidates for a wide range of applications in materials science because of their exceptional mechanical, thermal, and electronic properties (Baughman & Zakhidov, 2002). Carbon nanotubes exist as two types of structures: singlewall carbon nanotubes (SWNTs) and multiwall carbon nanotubes (MWNTs). Fig. 1 shows the schematic pictures of different types of carbon nanotubes. SWNT can be considered as graphene sheet rolled cylinders of covalently bonded carbon atoms with very high aspect ratios of 1000 or more. MWNTs consist of a number of graphene cylinders concentrically nested like rings in a tree trunk with an interlayer distance of ~0.34 nm.

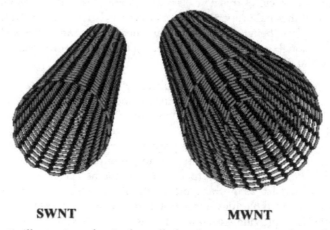

SWNT   **MWNT**

Fig. 1. A schematic illustration of a single-walled carbon nanotube and a multi-walled carbon nanotube (Dresselhaus et al., 2003).

The carbon-carbon covalent bonds in graphite and carbon nanotubes are considered to be one of the strongest atomic bonds in nature. The mechanical properties of CNTs have been extensively studied, both experimentally and theoretically (Yakobson et al., 1996; Lu, 1997). CNTs possess tensile moduli and strengths as high as 1 TPa and 150 GPa respectively. Especially in terms of strength CNTs are exceptional as this value is more than an order of magnitude higher than for high strength carbon fibres. Their density can be as low as 1.3 g.cm^{-3}, which is lower than commercial carbon fibres at 1.8-1.9 g.cm^{-3}. CNTs are thermally stable at up to 2800 ºC in vacuum. Their thermal conductivity is about twice as high as that of diamond, while their electric-current-carrying capacity is 1000 times higher than that of copper wires (Thostenson et al., 2001).

Since the first CNTs and polymer composites were made in 1994 (Ajayan et al., 1994), large amounts of work have been done on polymer/CNT composites. Incorporating nanotubes into plastics can potentially provide structural materials with a dramatic increase in both stiffness and strength. Extensive studies have been carried out producing strong polymer/CNT composites (Coleman et al., 2006; Moniruzzaman & Winey, 2006; Ahir, 2005, Wang et al., 2007). The effective use of CNTs as reinforcements still presents some major difficulties (Wang et al., 2007). The key challenge still remains in breaking down bundles of aggregated CNTs and reaching a fine dispersion in the selected polymer matrix. Much work needs to be done to optimise the conditions required for the potential dispersion of nanotubes, especially at higher CNT loadings, as well as a good interfacial interaction (Moniruzzaman & Winey, 2006).

CNTs have extremely low electrical resistance and their electric-current-carrying capacity is 1000 times higher than copper wires (Thostenson et al., 2001). Devices have been developed using CNTs such as field-emission displays (Fan et al., 1999). CNTs have also been widely used to produce conductive polymer composites (CPCs). It has been reported that the percolation threshold in CPC can be as low as 0.0025 wt% in the case of low viscosity thermosetting resins (Bryning et al., 2005). In thermoplastics the percolation threshold is typically around 1 wt%, but can be significantly lower using latex technology (Lu et al., 2008) and is highly dependent on processing history (Zhang et al., 2009; Deng et al., 2009a and 2009b).

2.2 Nanoclays

Clay is a natural, earthy, fine-grained material and is the main constituent of the sedimentary rocks in marine sediments and in soils. Clay minerals belong to the family of phyllosilicates (or layered silicate) and have particles less than 2 μm in size as stated in ISO 14688 (Moore & Reynolds, 1997). There are three main groups of clays: kaolinite, montmorillonite-smectite and halloysite, and most clay are a mixture of these different types. The shape of clay minerals is of a distinctive character. Montmorillonite, which has irregular flakes, is the most commonly used nanoclay in polymer/clay nanocomposites (Fig. 2a). Recently, sepiolite, a fibrous shaped mineral, has gained increasing attention as nanofiller reinforcement (Y.P. Zheng &Y. Zheng, 2006). Sepiolite is a needle-like shaped nanoclay composed with elemental particles of lengths of 0.2-4 μm, widths of 10-30 nm and thicknesses of 5-10 nm (Fig. 2b). Sepiolite can have a surface area as high as 200-300 m^2g^{-1} and normally it is found stuck together as bundles of fibres which can form micro-agglomerates. Sepiolite is a hydrous magnesium silicate with $[Si_{12}O_{30}Mg_8(OH)_4(H_2O)_4 \cdot 8H_2O]$ as the unit cell formula.

Compared with layered montmorillonite, the morphology of fibre-like sepiolite provides a lower specific surface area and smaller contact surface between the nanoclay particles. Therefore, polymer chains have a better chance not only of interacting with the external surface of the sepiolite, but also of penetrating into the structure, which facilitates a more even dispersion of the clays in the polymer matrix. Sepiolite provides a pseudoplastic and thixotropic behaviour which make it a valuable material in multiple applications to improve the processability, application or handling of the final product. Sepiolite has been successfully used to reinforce different polymers, such as poly (hydroxyethyl acrylate), epoxy, PP, etc (Bokobza et al., 2004; Y.P. Zheng &Y. Zheng, 2006; Bilotti et al., 2008).

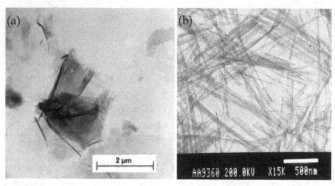

Fig. 2. The structure of (a) montmorillonite (Beermann & Brockamp, 2005) and (b) sepiolite (Bilotti et al., 2008)

3. Polypropylene nanocomposites

For decades we have been dealing with polymer microcomposites, where the length scale of the fillers is in micrometres. In the case of polymer nanocomposites, nanofillers, being additives of nanometre scale, are dispersed in a polymer matrix, offering multifunctional, high-performance polymer characteristics beyond those possessed by traditional filled polymeric materials. Improvements in physical and mechanical properties have been well documented in the literature (Dubois & Alexandre, 2006; Thostenson et al., 2001; Koo, 2006). Apart from mechanical enhancements, the value of polymer nanocomposites comes from providing value-added properties not present in the polymer matrix, without sacrificing the inherent processability and mechanical properties of the matrix. The multifunctional features consist of improved thermal, fire, and moisture resistance, decreased permeability, charge dissipation, and chemical resistance (Moniruzzaman & Winey, 2006; Bokobza et al., 2004; Koo, 2006). Because of PP's good balance between properties and cost and its wide usage in industry, PP nanocomposites have been extensively studied in recent years (Deng et al., 2009, 2010; Bilotti et al., 2008; Andrews et al., 2002; Liu & Wu, 2001).

The properties of nanocomposites can be greatly affected by the dispersion of the nanofillers in the polymer matrix (Ray & Okamoto, 2003). Generally, the better the dispersion, the better is the properties of the final nanocomposite. However, nanofillers are, in essence, agglomerates due to their high surface energy, and it is very difficult to disperse them in most polymers. Due to the lack of polar groups in PP, many efforts have been attempted to improve the dispersion of inorganic fillers such as clay and CNTs into a PP matrix for the preparation of effective PP nanocomposites and enhanced mechanical properties. The achievement of well-dispersed nanofillers is the most investigated research topic worldwide (Coleman et al., 2006; Koo, 2006; Ray & Okamoto, 2000; Lu et al., 2008).

4. Preparation methods

4.1 Traditional processing methods

Polymer nanocomposites are generally prepared by three methods: solution intercalation, in-situ polymerisation or melt compounding.

Solution intercalation has been known for over a century and has proved to be one of the most successful methods for incorporating nanofillers into polymers. Nanocomposites with water-soluble polymers such as poly (ethylene oxide) and poly (vinyl alcohol) and organic solvent-soluble polymers such as HDPE, have been successfully prepared via this method (Wu & Lerner, 1993; Ogata, et al., 1997; Joen et al., 1998; Wang et al., TP, 2007). In terms of PP nanocomposites, the poor solubility of PP in most organic solvents has severely limited the use of this method. Although PP nanocomposites have been reported being produced using xylene, tetrhydronaphthalene and decalin as solvents (Chang et al., 2005; Grady et al., 2002), elevated temperatures are necessary for the evaporation of these high-boiling solvents. Also, their application on an industrial scale is still hindered by the involvement of large quantities of organic solvent.

In-situ polymerisation has been intensively investigated in recent years. The advantage of this process is that the polymer chain can be grafted onto the nanofillers on a molecular scale. This gives excellent dispersion and the potential for good interfacial strength between the nanofillers and the polymer matrix. A relatively good dispersion can be maintained even with high nanotube loading in the matrix. Successful investigations have been reported in the literature from different groups on PP nanocomposites (Ma et al., 2001; Zhao et al., 2003; Koval'chuk et.al, 2008; Funk & Kaminsky, 2007). PP/ MMT nanocomposites were prepared by in-situ polymerization using a Ziegler-Natta catalyst (Zhao et al., 2003). PP/ MWNT nanocomposites have been fabricated by Funck and Kaminsky (2007) using a metallocene/ methylaluminoxane (MAO) catalyst. However, the molecular weight of the polymer is often significantly lower with a wide distribution by comparison with other methods.

Melt compounding is the most common method used to create thermoplastic polymer nanocomposites because it is a cost-effective technology for polyolefin-based systems and is compatible with current industrial practices, such as extrusion, injection moulding, etc (Andrews et al., 2002). However, melt compounding, especially in the case of PP matrices, is generally less effective at dispersing nanofillers such as CNTs and clays, and is limited to low nanofiller loadings due to the high viscosity of the composite systems caused by the addition of nanofillers (Andrews et al., 2002). Moreover, the high shear rates and high temperatures utilised can also cause thermal instability of the molten polymers (Potschke et al., 2003). One approach is using the masterbatch process, where a pre-mixed highly loaded nanofiller composite is diluted with a fresh polymer melt (Prashantha et al., 2009).

Much effort has been expended to facilitate the achievement of good dispersion of nanofillers and efficient stress transfer. The de-agglomeration of CNTs is necessary for dispersing individual nanotubes before mixing them with the polymer matrix. The ultrasonication process is the most common method, but severe sonication may make the tubes shorter (Saito et al., 2002, Inam et al., TP, J. Comp Materials, 2011). Milling and grinding is considered to be a cheap and fast method for industrial processes although it is the most destructive method for CNTs (Pierard et al., 2001). Non-covalent functionalizations, such as surfactants, are often utilised to overcome carbon nanotube entanglements resulting from Van der Waals forces (Bonduel et al., 2005). Covalent functionalization of CNTs helps to break up the CNT bundles and improves the polymer / CNT interfacial adhesion (Liu et al., 2005). Polymer wrapping has been proposed as an alternative method to achieve a good dispersion of the CNTs without destroying their electrical properties (Star et al., 2001). Dissociated CNTs were produced by grafting polymer chains directly onto the CNTs to achieve a homogeneous polymer coating on the CNT surface (Peeterbroeck et al., 2007; Dubois & Alexandre, 2006). High-density polyethylene has been used to coat MWNTs when producing PP nanocomposites and composites in other polymer matrices such as ethylene-vinyl acetate, polycarbonate and polyamide (Deng et al., 2010; Star et al., 2001; Peeterbroeck et al., 2007; Dubois & Alexandre, 2006; Pötschke et al., 2008). Compared with the direct incorporation of CNTs in polymer melts, enhanced dispersion and improved properties have been reported.

There are two ways to modify the surface of hydrophilic clays in order to improve their dispersion in the polymer matrix. The first one is to modify the surface with cationic surfactants to make the silicate surface organophilic via ion exchange reactions (Lan & Pinnavaia, 1994; Shi et al., 1996). The second method is based on grafting polymeric molecules through covalent bonding to the hydroxyl groups existing on the particles. However, for polyolefin polymers such as PP and PE, they are non-polar and incompatible with silicate surfaces even after modifying them with non-polar long alkyl groups. Therefore, a compatibilizer is often needed to facilitate the interaction between the polymer and the clays (Hasegawa et al., 2000; Garcia-Lopez et al., 2003). Compatibilizers are usually polar functional oligomers providing both a hydrophobic part (which can be easily mixed with a polymer) and a hydrophilic part (which is compatible with clay). Maleic anhydride grafted polypropylene (PP-g-MA) is a commonly used compatibilizer to aid the dispersion of clay or CNTs in PP matrix (Bilotti et al, 2008 and 2010).

4.2 The supercritical fluid technique

In recent years, supercritical technology, especially supercritical carbon dioxide ($scCO_2$), has been widely applied in the processing of polymer nanocomposites. A supercritical fluid is defined as "any substance, the temperature and pressure of which are higher than their critical values, and which has a density close to, or higher than, its critical density" (Darr & Poliakoff, 1999). Fig. 3 shows a schematic representation of the density and organization of molecules of a pure fluid in solid state, gas state, liquid state and the supercritical domain. No phase separation occurs for any substance at pressures or temperatures above its critical values. In other words, the critical point represents the highest temperature and pressure at which gas and liquid can coexist in equilibrium.

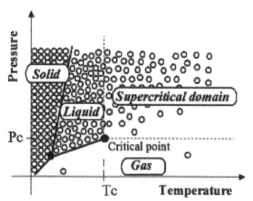

Fig. 3. A schematic representation of the microscopic behaviour of a pure fluid in the P-T plane phase diagram (Cansell et al., 2003)

Supercritical fluids are unique solvents with a wide range of interesting properties. Supercritical fluids have high diffusivities, similar to gas which allows them to effuse through solids, while also having liquid-like densities that allow them to act as effective solvents for many compounds. In addition, small changes in pressure and temperature greatly affect the density of a supercritical fluid and therefore many properties of SCF can be 'fine-tuned'. The unique properties of supercritical fluids allow them to be widely exploited in materials processing. The most promising developments are the processing of fine powders, core-shell particles, the processing and / or impregnation of aerogels, foams, surface modifications and the processing of polymers. (Cansell et al., 2003)

Among all the supercritical fluids, the use of $scCO_2$ is the most desirable for polymer processing because of its environmental compatibility as well as the following properties:

- CO_2 has relatively easily accessible critical points of 31.06 °C and 1070 psi (7.38 MPa) (Hyatt, 1984)
- The density of $scCO_2$ is easily tunable (Fig. 4). Therefore, the solvent strength and processes can be easily controlled
- CO_2 is non-combustible and non-toxic in contrast to most of the organic solvents suitable for supercritical applications
- It is also easily available because it occurs naturally as well as being the by-product of many industrial processes; therefore CO_2 is relatively inexpensive
- As CO_2 is a gas at ambient temperatures and pressures it can be easily removed, leaving no solvent residues in the processed material (Behles & Desimone, 2001)

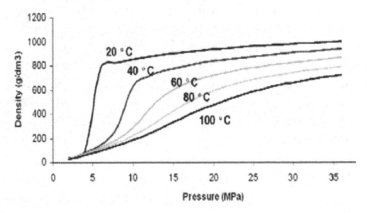

Fig. 4. The density of CO_2 as a function of pressure for a range of temperatures

$ScCO_2$ has found widespread application in industrial processes, including the extraction of metals or organic material, the decaffeination of green coffee beans, dry cleaning and degreasing, nano- and micro-particle formation, impregnation and dyeing, the processing and synthesis of polymers and composites, tissue engineering scaffolds, drug delivery, etc (Quirk et al., 2004; Hyde et al., 2001). The important fields most relevant to this Chapter are in the $scCO_2$ processing of polymers and composites which will be discussed in greater detail below.

ScCO$_2$ processing, as one of the new and cleaner processing methods for polymer nanocomposites, has recently received increasing attention. The solubility of polymers in scCO$_2$ is poor for many high molecular weight polymers. Table 1 gives a representative sample of the available literature data for the solubility of CO$_2$ in a variety of polymers. The low solubility is a result of the lower density of scCO$_2$ and the weak interaction between the CO$_2$ molecules and the non-polar groups of many polymers (Cansell et al., 2003). However, even for polymers which are not soluble in scCO$_2$, the CO$_2$ is still able to permeate resulting in substantial and sometimes dramatic changes in the properties of these polymers. The permeation of scCO$_2$ into a polymer causes it to swell. Aided by their zero surface tension, the addition of scCO$_2$ into the polymer phase gives the chains a greater mobility. The CO$_2$ molecules act as lubricants, which reduces the chain-chain interactions as it increases the inter-chain distance and free volume of the polymer. This is called plasticization. The physical properties of the polymer are changed dramatically, including the depression of the glass transition temperature (T_g), the lowering of interfacial tension and a reduction of the viscosity of the polymer melt. ScCO$_2$ may increase the crystallinity of the polymers because the polymer chains are given more freedom to align themselves into a more favourable order.

Polymer	Method[a]	Pressure [atm]	Temp. [°C]	Solubility
poly(methyl methacrylate)	GM-M	50	65	46 SCC/cm^3
	GM-D	204	70	10.5 wt.%
polystyrene	GM-M	13.2	25	14.5 SCC/cm^3
high-impact polystyrene	GM-D	204	70	0.5 wt.%
polycarbonate	GM-M	13.2	25	24 SCC/cm^3
	GM-D	68	25	13 g/100g
poly(ethylene terephthalate)	GM-D	136	40	1.5 wt.%
poly(vinyl chloride)	GM-D	68	25	8 g/100g
	GM-D	136	40	0.1 wt.%
poly(vinyl acetate)	GM-D	54.4	25	29 g/100g
low density polyethylene	GM-D	68	40	0.2 wt.%
high density polyethylene	GM-D	68	40	0.1 wt.%
polypropylene	GM-D	68	40	0.1 wt.%
	GM-D	204	25	0.1 wt.%
	BM	73	160	1.59 g/100g
	BM	61.2	200	1.09 g/100g
Nylon 66	GM-D	68	40	1.8 wt.%
polyurethane	BM	136	40	2.2 wt.%
Teflon	GM-D	68	40	0.0 wt.%

[a]Method: GM-M, gravimetric method (microbalance); GM-D, gravimetric method (desorption); BM, barometric method. Units: SCC/cm^3, cm^3(STP)/cm^3 of polymer; g/100g, g of CO$_2$/100g of polymer.

Table 1. The solubility of CO$_2$ in polymers (Tomasko et al., 2003)

Garcia-Leiner and Lesser (2004) have studied the use of $scCO_2$ in the processing of a variety of polymers (s-PS, FEP, and PTFE) using a modified single-screw extruder with a CO_2 injection (Fig. 5). The present of CO_2 significantly enhanced the processability of the polymer–CO_2 system by plasticization effect and the hydrostatic contribution. This presented an effective alternative to process intractable or high melt viscosity polymers.

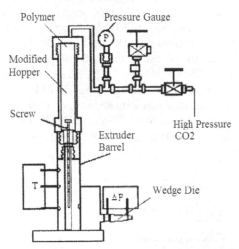

Fig. 5. Modified extrusion system for CO2-assisted polymer processing (Garcia-Leiner & Lesser, 2004)

A wide range of opportunities have opened up for $scCO_2$ that have an impact on the plastics industry (Tomasko et al., 2003). These include usages for extraction, foaming and impregnation. Extraction occurs merely by removing the soluble extractant material, such as any unreacted monomer, while leaving the insoluble substrate. Foaming occurs when rapid decompression forms gaseous CO_2 inside the polymer, leaving the polymer in the form of a porous (micro) material. This is very important in supporting the growth of blood vessels and collagen fibres in the matrix of biodegradable polymers or when the final product is intended to be used as a catalyst. As for impregnation, CO_2 can act as a transport medium facilitating the diffusion of monomers, initiators and molecules to impregnate a polymer, while the CO_2 can be cleanly removed afterwards. A better dispersion of the molecules has been provided within the polymer matrix. Substances impregnated into polymers have included dyes, fragrances, drugs for controlled release, anti-microbial and anti-fungal agents, and nanoparticles (Kikic & Vecchione, 2003).

The effective dispersion of the fillers in the polymer matrix and the improvement of polymer-filler interactions are two key challenges in the field of polymer nanocomposites. The development of polymer processing technologies in $scCO_2$ has enabled the synthesis of very complex polymer nanocomposites. Zerda et al. (2003) developed poly (methyl methacrylate) / montmorillonite composites via in-situ polymerization. $ScCO_2$ was used as the reaction medium to distribute homogeneously the monomer and the initiator and allowing polymerization under lower viscosity conditions. The modulus of the nanocomposites increased 50% with 40 wt % of clay content. A significant improvement in physical properties has also been reported by Green et al. (2000) for $scCO_2$ processed poly (methyl methacrylate) / silicate nanocomposites, in which scCO$_2$ acted as both a plasticizer for the polymer matrix and a carrier for the monomer. Polystyrene / clay has also been produced via in-situ polymerization by Li et al. (2006), where styrene monomer and initiator were directly intercalated into organomontmorillonite (OMMT) with the aid of sCCO$_2$.

For PP nanocomposites, PP will become plasticized when treated with CO$_2$. Varma-Nair et al. (2003) have suggested that 1% CO$_2$ could be dissolved in PP at 50 °C at 180 psi. Both decreases in T_m and T_c of PP in presence of CO$_2$ have been reported in the literature (Varma-Nair et al., 2003; Garcia-Leiner & Lesser, 2003). Studies in PP have indicated improvements in the dispersion of nanofillers in the matrix via melt compounding in scCO$_2$ (Garcia-Leiner & Lesser, 2003, Ma et al, 2007, Zhao & Huang, 2008). Zhao & Huang (2008) improved a twin-screw extruder with the aid of scCO$_2$ for the preparation of PP / clay nanocomposites (see Fig. 6). A CO$_2$ injection system was connected to the extruder and several kneading and reverse conveying elements were added to prevent CO$_2$ from leaking. This continuous extrusion process of PP nanocomposites using scCO$_2$ has set good examples for scCO$_2$ usage in an industrial environment.

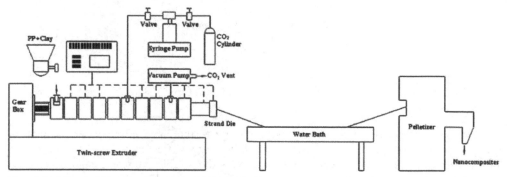

Fig. 6. ScCO$_2$ used in the extrusion process for the preparation of PP/ clay composites (Zhao & Huang, 2008)

5. Morphology and properties of PP nanocomposites

The preparation methods have crucial impact on the dispersion of the nanofillers and the final properties of the nanocomposites. In this section, the morphology of the PP nanocomposites and various nanocomposite properties will be discussed.

5.1 Morphology

Polymer-grafted CNTs were used in the study of Yang et al. (2008). They reported that PP-g-MWNTs were dispersed individually in the PP matrix at 1.5 wt%, but difficulty was found at higher CNT contents (Yang et al., 2008). Deng et al. (2010) reported that large bundles of MWNTs were present in their PP/MWNT samples using traditional melt compounding methods and better dispersions of MWNTs were observed for the HDPE coated MWNTs samples. The scanning electron microscopy (SEM) images of fracture surfaces of the nanocomposites are presented in Fig. 7. The HDPE coats the MWNTs and reduces the tendency of the nanotubes to aggregate and is partly miscible with the PP matrix. Ma et al.

(2010) studied the use of scCO$_2$ to assist melt compounding of PP nanocomposites. The nanocomposites are prepared in an autoclave. The autoclave was filled with liquid CO$_2$ and held at 2175 psi and 200 °C under stirring for 30 min. In their work, pristine MWNTs without HDPE coating were observed to exhibit good dispersion in the PP matrix under scCO$_2$ assisted mixing (Fig. 7c). This indicates the great efficiency of scCO$_2$ being a processing lubricant between the polymer chains enhancing the polymer diffusion.

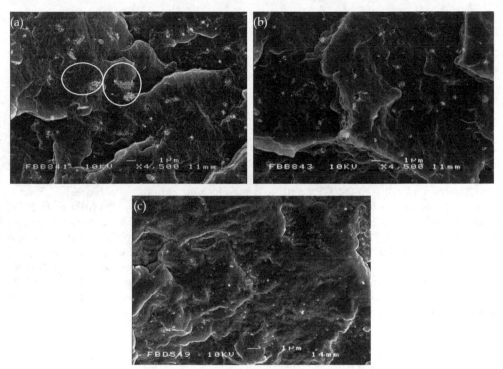

Fig. 7. SEM images of freeze-fractured samples of PP nanocomposites with 0.5 wt.% MWNT loading (a) melt-compounded MWNT (b) melt-compounded PP/coated MWNT (Deng et al., 2010) (c) scCO$_2$ PP/MWNT (Ma et al., 2010)

In Zhao & Huang' (2008) work, a better dispersion of clays in PP matrix was also confirmed with scCO$_2$ assisted mixing. Fig. 8 shows that the sample prepared with 2.5% CO$_2$ has more uniform distributed clay and aggregations of clay are much thinner.

Ma et al. (2007) also reported a good dispersion of sepiolite clays in PP matrix with scCO$_2$ assisted mixing. Without the use of any compatibilizer, the dispersion of sepiolite in the PP matrix improved significantly in the scCO$_2$ assisted mixing compared to melt mixing. This indicates that compatibilizers such as PP-g-MA are not needed to achieve good dispersions in the case of scCO$_2$ assisted processing of polyolefins. This is of particular relevance because the role of compatibilizers, such as PP-g-MA, is generally regarded as essential for the creation of well dispersed nanoclays composites based on polyolefins using traditional melt compounding methods.

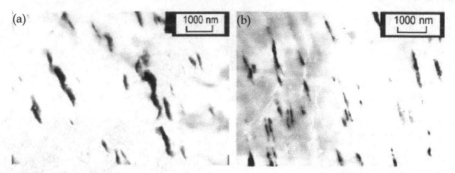

Fig. 8. TEM images of PP/clay nanocomposites prepared (a) without and (b) with 2.5% CO_2 (Zhao & Huang, 2008)

5.2 Mechanical properties

The mechanical properties including the tensile strength and Young's modulus of the PP nanocomposites can be significantly enhanced by the incorporation of nanofillers. Andrews et al. (2002) fabricated PP / MWNT composites by a shear mixer and found a modulus increase from 1.0 GPa to 2.4 GPa with a relatively high nanotube content of 12.5 wt%. However, this was at the expense of a reduction in yield stress from 30 MPa to 18 MPa. Similarly, Wang and Sheng (2005) found that the modulus of the PP nanocomposites increased from 788 MPa to 908 MPa by the addition of 7 wt% Attapulgite clay, but at the expense of a reduction in strength from 32 MPa to 26 MPa. Deng et al. (2010) investigated the effects of a HDPE coating onto MWNTs on the mechanical properties of PP / MWNT composites produced by melt compounding. A property increase from 1.42 GPa to 1.79 GPa for Young's modulus and from 34.5 MPa to 37.9 MPa for tensile strength, was achieved at relatively low loadings (0.5 wt%) of coated MWNTs.

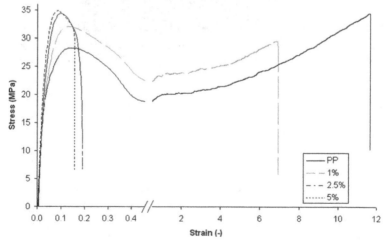

Fig. 9. Stress-strain curves of $scCO_2$ PP / sepiolite nanocomposites with different clay contents (Ma et al., 2007)

The elongation at break of the nanocomposites is often decreased. The increase of nanofiller content compromises the ductile nature of the PP matrix. Yielding followed by drawing still occurred in the 1 wt% sepiolite samples; while 2.5 wt% and 5 wt% sepiolite samples show brittle fractures at low strains (Fig. 9).

The effect on the modulus and yield stress of the scCO$_2$ processed PP nanocomposites was studied for the addition of MWNTs and sepiolite clays (Ma et al., 2007 and 2010). The results were compared with traditional melt compounded nanocomposites.

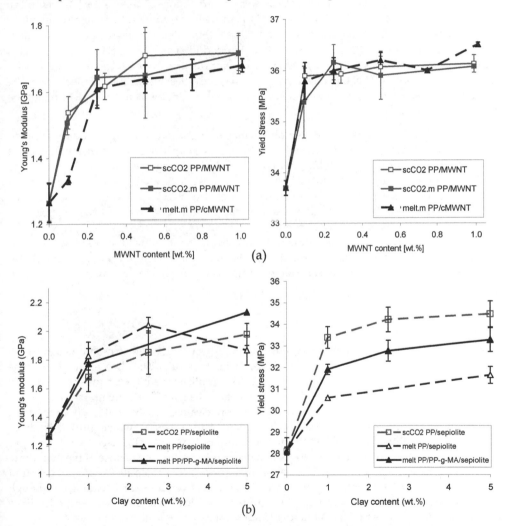

Fig. 10. The Young's modulus and yield stress of scCO$_2$ processed and melt-compounded PP nanocomposites (a) PP / MWNT nanocomposites (b) PP / sepiolite nanocomposites. The label, melt.mPP/cMWNT, in the figure represents melt-compounded and masterbatch based HDPE-coated MWNT PP nanocomposites (Ma et al., 2007; Ma et al., 2010)

Increases of both properties were observed in all systems with the nanofiller content as expected. The melt-compounded and masterbatch based PP/coated MWNT composites show a continuous modulus increase up to 1.7 GPa by the addition of 1.0 wt% CNT loading (Fig. 10a). This is as a result of the de-aggregation of the MWNT bundles by the HDPE coating (Ma et al., 2010). ScCO$_2$ processed nanocomposites with pristine MWNTs also showed a competitive increase in Young's modulus. Furthermore, the use of masterbatch in the scCO$_2$ assisted mixing method did not show much improvement in terms of Young's modulus. Similarly, increases in yield stress were observed for scCO$_2$ PP / MWNT samples, with a significant 6 % increase at 0.1 wt% CNT loading followed by roughly constant values across the rest of the CNT loading range up to 1 wt% (Fig. 10a). The degree of reinforcement of pristine MWNTs is as good as for melt-compounded PP/coated MWNT composites. This indicates that scCO$_2$ assisted mixing achieves a better dispersion of CNTs without the need for a HDPE coating. The enhanced mechanical properties are also benefited from the relatively low shear forces involved in scCO$_2$ assisted mixing, which causes less nanotube breakage and damage.

In Ma et al.' (2007) work about PP/sepiolite nanocomposites, the Young's modulus of melt compounded PP nanocomposites (no PP-g-MA used as compatibilizer) shows a reduction at 5 wt% sepiolite loading (Fig. 10b). This corresponds to the aggregation of clay at higher loadings with the absence of PP-g-MA. This is in contrast to the analogous material with PP-g-MA, which shows a 68 % increase in modulus from 1 to 5 wt% clay content. The modulus for scCO$_2$ processed PP nanocomposites without PP-g-MA shows a steady increase in modulus (by up to 56 % for 5 wt% sepiolite). In terms of yield stress, scCO$_2$ processed PP nanocomposite without PP-g-MA shown the highest increase in yield stress of up to 23 % for 5 wt% sepiolite, which was benefited from the good dispersion of the sepiolite and also the benefit of preserving the clay fibre length when using scCO$_2$ assisted mixing.

5.3 Characterization of PP

5.3.1 Thermal behaviour

Nanofillers increase the crystallization temperature (T_c) of PP. It has been observed for both CNT and clay nanocomposites (Ma et al., 2007; Bilotti et al., 2008; Lee et al., 2008). The nanofillers act as nucleating agents enabling PP to crystallize at a higher temperature during the cooling process. This increase in T_c is dependent on the nanofiller content, which suggests that higher loadings of nanofiller can provide higher nucleating efficiency. It has been reported that SWNTs are more effective nucleating agents than MWNTs, which provided larger increase in T_c (Miltner et al., 2008). Furthermore, pristine MWNTs also give a stronger nucleation effect than HDPE coated MWNTs, which suggests that the surface of the nucleating agents plays an important role in the crystallization process (Ma et al., 2010). It was also found that a more pronounced nucleating effect was observed for very low CNT loadings. Although nucleating agents significantly increase the number of nucleation sites, above certain loadings the introduction of more CNTs may hinder chain mobility and retard crystal growth.

Apart from the T_c, the crystallinity of PP has also been increased by the addition of nanofillers (Bilotti et al., 2008). The Young's modulus of the composites has been reported to increase with polymer matrix crystallinity in the work of Coleman et al. (Coleman et al., 2004). Similar results were also shown in other studies that changes in the crystallinity and the crystalline morphology of the polymer matrix can have pronounced effects on the mechanical properties of nanocomposites (Bhattacharyya et al., 2003; Wang et al., 2007). Hence, the mechanical reinforcing effects observed are a combination of the modification of the PP matrix through increased crystallinity as well as true reinforcing effects from CNT fillers.

The effect of nanofillers on the melting temperature (T_m) of PP is not clear. Yang et al. (2008) and Bikiaris et al. (2008) have observed an increase in T_m. Lee et al. (2008) has reported a decrease in T_m of PP. Other authors reported that T_m remains unchanged (Seo et al., 2005; Jin et al., 2009).

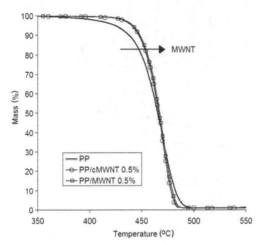

Fig. 11. The TGA of PP / MWNT nanocomposites processed using scCO$_2$ assisted mixing, showing a slightly retarded thermal degradation of nanocomposites by the presence of MWNTs (Ma et al., 2010)

The thermal stability of PP is also slightly improved by the addition of the nanofillers. TGA shows that the thermal degradation of PP is retarded by the presence of CNTs (Fig. 11). PP / MWNT nanocomposites showed a higher onset decomposition temperature (T_{onset}). The Reason is that the CNTs hinder the flux of degradation products and improves the heat dissipation within the composites (Huxtable et al., 2003). Seo & Park (2004) have observed a T_{onset} increase from 278 °C to 352 °C at the addition of 5 wt% of MWNTs. The effect of the clay surface modification on the thermal stability of PP has been studied by Tartaglione et al. on the melt-compounded PP / sepiolite composites (Tartaglione et al., 2008).

5.3.2 Microstructure of PP

PP can crystallize in three crystalline modifications: monoclinic (α), hexagonal (β), and orthorhombic (γ) (Bruckner et al., 1991). These phases can be examined by XRD. The XRD patterns for all the PP nanocomposites shows peaks corresponding to PP α-phase at $2\theta = 14$, 17, 18.5, 21.5, 21.9 and 25.4 ° for six major reflections: (110), (040), (130), (111), (041) and (060) plane, respectively. Most studies showed that only α phase of PP was present in the PP nanocomposites and no other forms of PP crystallites were detected. This indicates that the addition of sepiolite does not affect the crystal modification of the final PP molecules

However, Wang & Sheng (2005) did report that the relative intensities of 040 peaks and 110 peaks can change (Fig. 12) and an increase in the I_{040} / I_{110} ratio was found. This preferential orientation of PP crystal growth has also been found with the addition of MWNTs and sepiolite clays (Ma et al., 2007 and 2010). Data of relative intensities are listed in Table 2. This increase indicates that the addition of sepiolite promotes a preferential orientation of PP crystals growing with (040) planes parallel to the sepiolite surface and the b-axis perpendicular to it (Ferrage et al., 2002; Wang & Sheng, 2005).

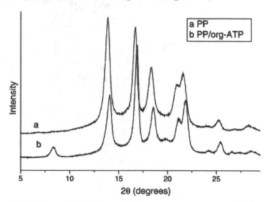

Fig. 12. XRD patterns of (a) PP and PP/org-ATP nanocomposites (Wang & Sheng, 2005)

Sepiolite [wt.%]	0	1	2.5	5
PP/sepiolite	0.80	2.41	2.40	2.24
PP/PP-g-MA/sepiolite	0.80	1.40	3.00	2.71

Table 2. Ratio of intensity between diffraction peaks of (040)a and (110)a in XRD patterns (Ma et al., 2007)

6. Conclusion

Researches on nanofillers such as CNTs and nanoclays reinforced PP nanocomposites have attracted wide attention. The properties of the nanocomposites can be significantly improved by the addition of nanofillers. The challenge in producing high performance PP nanocomposites is to achieve homogeneous dispersion of nanofillers thus obtaining efficient stress transfer. Among the nanocomposite preparation methods, scCO$_2$ assisted mixing is an interesting alternative to produce PP nanocomposites as it overcomes some of the issues in the traditional melt compounding process, such as high viscosities, high temperatures and high shear stresses involved, which can all lead to polymer degradation. The plasticization of scCO$_2$ could reduce the PP viscosity, decrease the T_m of the PP, and hence increase the processablity of nanocomposites and improve the dispersion of nanofillers. ScCO$_2$ assisted mixing also benefits from a better preservation of nanotube or nanofibre lengths. Good nanofiller dispersion and mechanical properties were achieved without using the aid of compatibilizers such as PP-g-MA or polymer coated nanofillers, which are commonly used aiding the dispersion of nanofillers in the melt compounding process. This is all very encouraging from an economical point of view.

7. References

Ahir, S. (2005) Polymer Containing Carbon Nanotubes: Active Composite Materials. In: *Polymeric Nanostructures and Their Applications*. Nalwa, S., American Scientific Publishers, ISBN: 1-58883-068-3 California

Ajayan, P.; Stephan, O.; Colliex, C. & Trauth, D. (1994) Aligned Carbon Nanotube Arrays Formed by Cutting a Polymer Resin-Nanotube Composite. *Science.* Vol.265, No.5176, pp.1212-1214, ISSN 0036-8075

Andrews, R.; Jacques, D.; Minot, M. & Rantell, T. (2002) Fabrication of Carbon Multiwall Nanotube/Polymer Composites by Shear Mixing. *Mater. Eng.* Vol.287, No.6, pp.395-403, ISSN 1438-7492

Baughman, A. & Zakhidov, W. (2002) Carbon Nanotubes - The Route Toward Applications. *Science.* Vol.297, No.5582, pp.787-792, ISSN 0036-8075

Beermann, T. & Brockamp, O. (2005) Structure analysis of montmorillonite crystallites by convergent-beam electron diffraction. *Clay Miner.* Vol.40, No.1, pp.1-13, ISSN 0009-8558

Behles, J. & DeSimone, J. (2001) Developments in CO_2 Research. *Pure Appl. Chem.* Vol.73, No.8, pp.1281-1285

Bhattacharyya, A.; Sreekumar, T.; Liu, T.; Kumar, S.; Ericson, L.; Hauge, H. & Smalley, R. (2003) Crystallization and Orientation Studies in Polypropylene/Single Wall Carbon Nanotube Composite. *Polym.* Vol.44, No.8, pp.2373-2377, ISSN 0032-3861

Bilotti, E.; Fischer, H. & Peijs, T. (2008) Polymer nanocomposites based on needle-like sepiolite clays: Rffect of functionalized polymers on the dispersion of nanofiller, crystallinity, and mechanical properties. *J. Appl. Polym. Sci.* Vol.107, No.2 pp.1116-1123, ISSN 0021-8995

Bilotti, E.; Ma, J. & Peijs, T. (2010) Preparation and Properties of Polyolefin/Needle-Like Clay Nanocomposites. In: *Advances in Polyolefin Nanocomposites*. Mittal V., Taylor & Francis (CRC Press), USA.

Bokobza, L.; Burr, A.; Garnaud, G.; Perrin, M. & Pagnotta, S. (2004) Fibre Reinforcement of Elastomers: Nanocomposites Based on Sepiolite and Poly(hydroxyethyl acrylate). *Polym. Int.* Vol.53, No.8, pp.1060-1065, ISSN 0959-810

Bonduel, D.; Mainil, M.; Alexandre, M.; Monteverde, F. & Dubois, P. (2005) Supported Coordination Polymerisation: A Unique Way to Potent Polyolefin Carbon Nanotube Nanocomposites. *Chem. Commun.* Vol.14, No.6, pp.781-783

Bruckner, S.; Meille, S.; Petraccone, V. & Pirozzi, B. (1991) Polymorphism in Isotactic Polypropylene. *Prog. Polym. Sci.* 16, No.2-3, pp.361-404

Bryning, M.; Islam, M.; Kikkawa, J. & Yodh, A. (2005) Very Low Conductivity Threshold in Bulk Isotropic Single-Walled Carbon Nanotube–Epoxy Composites. *Adv. Mater.* Vol.17, No.9, pp.1186-1191

Cansell, F.; Aymonier, C. & Loppinet-Serani, A. (2003) Review on Materials Science and Supercritical Fluids. *Curr. Opin. Solid State Mater. Sci.* Vol.7, No.4-5, pp.331-340

Chang, t.; Jensen, L.; Kisliuk, A.; Pipes, R.; Pyrz, R. & Sokolov, A. (2005) Microscopic mechanism of reinforcement in single-wall carbon nanotube / polypropylene nanocomposites. *Polym.* Vol.46, No.2,pp.439-444

Coleman, J.; Cadek, M.; Blake, R.; Nicolosi, V.; Ryan, K.; Belton, C.; Fonseca, A.; Nagy, J.; Gun'ko, Y. & Blau, W. (2004) High Performance Nanotube-Reinforced Plastics:

Understanding the Mechanism of Strength Increase. *Adv. Funct. Mater.* Vol.14, No.8, pp.791-798

Coleman, J.; Khan, U. & Gun'ko, Y. (2006) Mechanical Reinforcement of Polymers Using Carbon Nanotubes. *Adv. Mater.* Vol.18, No.6, pp.689-706

Darr, J. & Poliakoff, M. (1999) New Directions in Inorganic and Metal-Organic Coordination Chemistry in Supercritical Fluids. *Chem. Rev.* Vol.99, No.2, pp.495-542

Deng, H.; Skipa, T.; Zhang, R.; Lillinger, D.; Bilotti, E.; Alig, I. & Peijs T. (2009a) Effect of melting and crystallization on the conductive network in conductive polymer composites. *Polym.* Vol.50, No.15, pp.3747-3754

Deng, H.; Zhang, R.; Reynolds, C.; Bilotti, E. & Peijs, T. (2009b) A Novel Concept for Highly Oriented Carbon Nanotube Composite Tapes or Fibres with High Strength and Electrical Conductivity. *Macromol. Mater. Eng.* Vol.294, No.11,pp.749-755

Deng, H.; Bilotti, E.; Zhang, R. & Peijs, T. (2010) Effective Reinforcement of Carbon Nanotubes in Polypropylene Matrices. *J. Appl. Poly. Sci.* Vol.118, No.1, pp.30-41

Dresselhaus, M.; Lin, Y.; Rabin, O.; Jorio, A.; Souza Filho, A.; Pimenta, M.; Saito, R.; Samsonidze Ge, G. & Dresselhaus, G. (2003) Nanowires and nanotubes. *Mater. Sci. Eng.* Vol.23, No.1, pp.129-140

Dubois, P. & Alexandre, M. (2006) Performant Clay/Carbon Nanotube Polymer Nanocomposites. *Adv. Eng. Mater.* Vol.8, No.3, pp.147-154

Fan, S.; Chapline, M.; Franklin, N.; Tombler, T.; Cassell, A. & Dai, H. (1999) Self-oriented regular arrays of carbon nanotubes and their field emission properties. *Science.* Vol.283, No.5401, pp.512-514

Ferrage, E.; Martin, F.; Boudet, A.; Petit, S.; Fourty, G.; Jouffret, F.; Micoud, P.; De Parseval, P.; Salvi, S.; Bourgerette, C.; Ferret, J.; Saint-Gerard, Y.; Buratto, S. & Fortune, J. (2002) Talc as Nucleating Agent of Polypropylene: Morphology Induced by Lamellar Particles Addition and Interface Mineral-matrix Modelization. *J. Mater. Sci.* Vol.37, No.8, pp.1561-1573

Fleming, G. & Koros, W. (1986) Dilation of Polymers by Sorption of Carbon Dioxide at Elevated Pressures. 1. Silicone Rubber and Unconditioned Polycarbonate. *Macromol.* Vol.19, No.8, pp.2285-2291

Funck, A. & Kaminsky, W. (2007) Polypropylene carbon nanotube composites by in situ polymerization. *Compos. Sci. Technol.* Vol.67, No.5, pp.906-915

Garcia-Leiner, M. & Lesser, A. (2003) Melt Intercalation in Polymer-Clay Nanocomposites Promoted by Supercritical Carbon Dioxide. *Polym. Mater. Sci. Eng.* Vol.89, pp.649-650

Garcia-Leiner, M. & Lesser, A. (2004) CO_2-assisted polymer processing: A new alternative for intractable polymers. *J. Appl. Polym. Sci.* Vol.93, No.4, pp.1501-1511

Garcia-Lopez, D.; Picazo, O.; Merino, J. & Pastor, J. (2003) Polypropylene Clay Nanocomposites: Effect of Compatibilizing Agents on Clay Dispersion. *Eur. Polym. J.* Vol.39, No.5, pp.945-950

Grady, B.; Pompeo, F.; Shambaugh, R. & Resasco, D. (2002) Nucleation of polypropylene crystallization by single-walled carbon nanotubes. *J. Phys. Chem. B.* Vol.106, No.23, pp.5852-5858

Green, J.; Rubal, M.; Osman, B.; Welsch, R.; Cassidy, P.; Fitch, J. & Blanda, M. (2000) Silicon-Organic Hybrid Polymers and Composites Prepared in Supercritical Carbon Dioxide. *Polym. Adv. Technol.* Vol.11, No.8-11, pp.820-825

Grunes, J.; Zhu, J. & Somorjai, G. (2003). Catalysis and Nanoscience. *Chem. Commun.* Vol.7, No.18, pp.2257-2260

Hasegawa, N.; Okamoto, H.; Kato, M. & Usuki, A. (2000) Preparation and Mechanical Properties of Polypropylene-Clay Hybrids based on Modified Polypropylene and Organophilic Clay. *J. Appl. Polym. Sci.* Vol.78, No.11, pp.1918-1922

Huxtable, S.; Cahill, D.; Shenogin, S.; Xue, L.; Ozisik, R.; Barone, P.; Usrey, M.; Strano, M.; Siddons, G.; Shim, M. & Keblinski, P. (2003) Interfacial Heat Flow in Carbon Nanotube Suspensions. *Nat. Mater.* Vol.2, No.11, pp.731-734

Hyatt, J. (1984) Liquid and Supercritical Carbon Dioxide as Organic Solvents. *J. Org. Chem.* Vol.49, No.26, pp.5097-5101

Hyde, J.; Licence, P.; Carter, D. & Poliakoff, M. (2001) Continuous Catalytic Reactions in Supercritical Fluids. *Appl. Catal., A.* Vol.222, No.1-2, pp.119-131

Jin, S.; Kang, C.; Yoon, K.; Bang, D. & Park, Y. (2009) Efect of compatibilizer on morphology, thermal, and rheological properties of polypropylene/functionalized multi-walled carbon nanotubes composite. *J. Appl. Polym. Sci.* Vol.111, No.2, pp.1028-1033

Joen, H.; Jung, H.; Lee, S. & Hudson, S. (1998) Morphology of polymer/Silicate Nanocomposites High Density Polyethylene and a Nitrile Copolymer. *Polym. Bull.* Vol.41, No.1, pp.107-111

Kikic, I. & Vecchione, F. (2003) Supercritical Impregnation of Polymers. *Curr. Opin. Solid State Mater. Sci.* Vol.7, No.4-5, pp.399-405

Koo, J. (2006) *Polymer Nanocomposites: Processing, Characterization, and Applications.* McGraw-Hill, ISBN 0071458212, 9780071458214, New York

Koval'chuk, A.; Shevchenko, V.; Shchegolikhin, A.; Nedorezoca, P.; Klyamkina, A. & Aladyshev, A. (2008) Effect of carbon nanotube functionalization on the structural and mechanical properties of polypropylene/MWCNT composites. *Macromol.* Vol.41, No.20, pp.7536-7542

Lan, T. & Pinnavaia, T. (1994) Clay-Reinforced Epoxy nanocomposites. *Chem. Mater.* Vol.6, No.12, pp.2216-2219

Lee, G.; Janannathan, S,; Chea, H.; Minus, M. & Kumar, S. (2008) Carbon nanotube dispersion and exfoliation in polyperopylene and structure and properties of the resulting composites. *Polym.* Vol.49, No.7, pp.1831-1840

Li, J.; Xu, Q.; Peng, Q.; Pang, M.; He, S. & Zhu, C. (2006) Supercritical CO_2-Assisted Synthesis of Polystyrene/clay Nanocomposites via in situ Intercalative Polymerisation. *J. App. Polym. Sci.* Vol.100, No.1, pp.671-676

Liu, L.; Barber, A.; Nuriel, S. & Wagner, H. (2005) Mechanical Properties of Functionalized Single-Walled Carbon-Nanotube/Poly(vinyl alcohol) Nanocomposites. *Adv. Funct. Mater.* Vol.15, No.6, pp.975-980

Liu, X. & Wu, Q. (2001) PP/Clay Nanocomposites Prepared by Grafting-Melt Intercalation. *Polym.* Vol.42, No.25, pp.10013-10019

Lu, J. (1997) Elastic Properties of Carbon Nanotubes and Nanoropes. *Phys. Rev. Lett.* Vol.79, No.7, pp.1297-1300

Lu, K; Grossiord, N; Koning, C.; Miltner, H.; Mele, B. & Loos, J. (2008) Carbon Nanotube/Isotactic Polypropylene Composites Prepared by Latex Technology: Morphology Analysis of CNT-Induced Nucleation. Macromol. 41, pp. 8081

Ma, J.; Bilotti, E.; Peijs, T. & Darr, J. (2007) Preparation of Polypropylene/Sepiolite Nanocomposites Using Supercritical CO2 Assisted Mixing. Eur. Polym. J. Vol.43, No.12, pp.4931-4939

Ma, J.; Deng, H. & Peijs, T. (2010) Processing of Polypropylene/Carbon Nanotube Composites Using ScCO2-Assisted Mixing. Macromol. Mater. Eng. Vol.295, No.6, pp.566-574

Ma, J.; Qi, Z. & Hu, Y. (2001) Synthesis and characterization of polypropylene/clay nanocomposites. J. Appl. Polym. Sci. Vol.82, No.14, pp.3611-3617

Moniruzzaman, M. & Winey, K. (2006) Polymer Nanocomposites Containing Carbon Nanotubes. Macromol. Vol.39, No.16, pp. 5194-5205

Moore, D. & Reynolds, R. (1997) X-Ray Diffraction and the Identification and Analysis of Clay Minerals (2nd edition). University Press, ISBN 0195087135, New York

Ogata, N.; Kaawakage, S. & Ogihara, T. (1997) Poly(vinyl alcohol)-Clay and Poly(ethylene oxide)-Clay Blends Prepared Using Water as Solvent. J. Appl. Polym. Sci. Vol.66, No.3, pp.573-581

Peeterbroeck, S.; Laoutid, F.; Taulemesse, J.; Monteverde, F.; Lopez-Cuesta, J.; Nagy, J.; Alexandre, M. & Dubois, P. (2007) Mechanical Properties and Flame-Retardant Behavior of Ethylene Vinyl Acetate/High-Density Polyethylene Coated Carbon Nanotube Nanocomposites. Adv. Funct. Mater. Vol.17, No.15, pp.2787-2791

Pierard, N.; Fonseca, A.; Konya, Z.; Willems, I.; Van-Tendeloo, G. & Nagy, J. (2001) Production of short carbon nanotubes with open tips by ball milling. Chem. Phys. Lett. Vol.335, No.1-2, pp.1-8

Potschke, P.; Bhattacharyya, A.; Janke, A. & Goering, H. (2003) Melt Mixing of Polycarbonate/Multi-wall Carbon Nanotube Composites. Compos. Interfaces. Vol.10, No.4-5, pp.389-404

Pötschke, P.; Pegel, S.; Claes, M & Bonduel, D. (2008) A Novel Strategy to Incorporate Carbon Nanotubes into Thermoplastic Matrices. Macromol. Rapid Commun. Vol.29, No.3, pp.244-251

Prashantha, K.; Soulestin, J.; Lacrampe, M.; Krawczak, P.; Dupin, G. & Claes, M. (2009) Masterbatch-Based Multi-Walled Carbon Nanotube Filled Polypropylene Nanocomposites: Assessment of Rheological and Mechanical Properties. Compos. Sci. Technol. Vol.69, No.11-12, pp.1756-1763

Quirk, R.; France, R.; Shakesheff, K. & Howdle, S. (2004) Supercritical Fluid Technologies and Tissue Engineering Scaffolds. Curr. Opin. Solid State Mater. Sci. Vol.8, No.3-4, pp.313-321

Ray, S. & Okamoto, M. (2003) Polymer/Layered Silicate Nanocomposites: A review from Preparation to Processing. Prog. Polym. Sci. Vol.28, No.11, pp.1539-1641

Saito, T.; Matsushige, K. & Tanaka, K. (2002) Chemical Treatment and Modification of Multi-Walled Carbon Nanotubes. Physica B. Vol.323, No.1-4, pp.280-283

Seo, M. & Park, S. (2004) A kinetic study on the termal degradation of multi-walled carbon nanotubes-reinforced poly(propylene) composites. Macromol. Mater. Eng. Vol.289, No.4, pp.368-374

Seo, M.; Lee, J. & Park, S.(2005) Crystallizaiton kinetics and interfacial behaviors of polypropylene composites reinforced with multi-walled carbon nanotubes. *Mater. Sci. Eng.* Vol.404, No.1-2, pp.79-84

Shi, H.; Lan, T. & Pinnavaia, T. (1996) Interfacial Effects on the Reinforcement Properties of Polymer-Organoclay Nanocomposites. *Chem. Mater.* Vol.8, No.8, pp.1584-1587

Star, A.; Stoddart, J.; Steuerman, D.; Diehl, M.; Boukai, A.; Wong, E.; Yang, X.; Chung, S.; Choi, H. & Heath, J. (2001) Preparation and Properties of Polymer-Wrapped Single-Walled Carbon Nanotubes. *Angew. Chem. Int. Ed.* Vol.40, No.9, pp.1721-1725

Tartaglione, G.; Tabuani, D,; Camino, G. & Moisio, M. (2008) PP and PBT composites filled with sepiolite: Morphology and thermal behavior. *Compos. Sci. Technol.* Vol.68, No.2, pp.451-460

Thostenson, E.; Ren, Z. & Chou, T. (2001) Advances in the science and technology of carbon nanotubes and their composites: a review. *Compos. Sci. Technol.* Vol.61, No.13,pp.1899-1912

Tomasko, D.; Li, H.; Liu, D; Han, X.; Wingert, M.; Lee, L. & Koelling, K. (2003) A Review of CO_2 Applications in the Processing of Polymers. *Ind. Eng. Chem. Res.* Vol.42, No.25, pp.6431-6456

Varma-Nair, M.; Handa, P.; Mehta, A. & Agarwal, P. (2003) Effect of Compressed CO_2 on Crystallization and Melting Behavior of Isotactic Polypropylene. *Thermochim. Acta.* Vol.396, No.1-2, pp.57-65

Wang, L. & Sheng, J. (2005) Preparation and properties of polypropylene / org-attapulgite nanocomposites. *Polm.* Vol.46, No.16, pp.6243-6249

Wang, Z.; Ciselli, P. & Peijs, T. (2007) The Extraordinary Reinforcing Efficiency of Single-Walled Carbon Nanotubes in Oriented Poly(vinyl alcohol) Tapes. *Nanotechnol.* Vol.18, No.45, 455709

Wu, J. & Lerner, M. (1993) Structural, Thermal, and Electrical Characterization of Layered Nanocomposites Derived from Sodium-Montmorillonite and Polyethers. *Chem. Mater.* Vol.5, No.6, pp.835-838

Yakobson, B.; Brabec, C. & Bernholc, J. (1996) Nanomechanics of Carbon Tubes: Instabilities beyond Linear Response. *Phys. Rev. Lett.* No.76, No.14, pp.2511-2514

Yang, B.; Shi, J.; Pramoda, K. & Goh, S. (2008) Enhancement of the mechanical properties of polypropylene using polypropylene-grafted multiwalled carbon nanotubes. *Compos. Sci. Technol.* Vol.68, No.12, pp.2490-2497

Zerda, A.; Caskey, T. & Lesser, A. (2003) Highly Concentrated, Intercalated Silicate Nanocomposites: Synthesis and Characterisation. *Macromol.* Vol.36, No.5, pp.1603-1608

Zhao, H.; Zhang, X.; Yang, F.; Chen, B.; Jin, Y.; Li, G.; Feng, Z. & Huang, B. (2003) Synthesis and characteriztion of polypropylene/montmorillonite nanocomposites via an in-situ polymerization approach. *Chin. J. Polym. Sci.* Vol.21, No.4, pp.413-418

Zhao, Y. & Huang, H. (2008) Dynamic rheology and microstructure of polypropylene/clay nanocomposites prepared under Sc-CO_2 by melt compounding. *Polym. Test.* Vol.27, No.1, pp.129-134

Zheng, Y. & Zheng, Y. (2006) Study on Sepiolite-Reinforced Polymeric Nanocomposites. *J. App Polym Sci.* Vol.99, No.5, pp.2163-2166

Zhang, R.; Dowden, A.; Deng, H.; Baxendale M. & Peijs T. (2009) Conductive network formation in the melt of carbon nanotube/thermoplastic polyurethane composite. *Compos. Sci. Technol.* Vol.69, No.10, pp.1499-1504

9

Morphology and Thermo Mechanical Properties of Wood/Polypropylene Composites

Diene Ndiaye[1,*], Bouya Diop[1], Coumba Thiandoume[2],
Papa Alioune Fall[1], Abdou Karim Farota[1] and Adams Tidjani[2]
[1]Universite Gaston Berger de Saint-Louis,
[2]Universite Cheikh Anta Diop de Dakar,
Senegal

1. Introduction

Because of the future scarcity of fossil raw material and taking into account current environmental concerns, the development of eco-materials occupies a large number of research centers. Wood polymer composites (WPC) made from wood flour and polymer matrices, are part of this logic. In today's world, the growing needs of the population and the growing technological innovation are pushing industrial and researchers to move towards so-called new generation products such as wood polymer composites whose production increases considerably from year to year. In recent years, wood–fibers have gained significant interest as reinforcing material for commercial thermoplastics. They are now fast evolving as a potential alternative to inorganic fillers for various applications. These composites made from blends of thermoplastics and natural fibers have gained popularity in a variety of applications because they combine the desirable durability of plastics with the cost effectiveness of natural fibers as fillers or reinforcing agents and several advantages like low density, high specific properties, non-abrasive to processing equipment, low cost and most importantly biodegradability (Timmons et al., 1971). In tropical countries, fibrous plants are available in abundance; these fibers with high specific strength improve the mechanical properties of the polymer matrix. Wood is renewable, recyclable and biodegradable, characteristics well appreciated by environmentalists. For these reasons, combining the plastic timber produces a more accepted material. However, consumers want more and more natural materials. But in practice there are a lot of waste that can be exploited in combination with polymers to form composites that are resistant thermoplastic timber, recyclable and can be burned for energy recovery. The addition of wood wood flour, into the polymer matrix leads to an improvement in the stiffness of the composite and decreases in the abrasiveness on processing equipment and density of the product compared to mineral fillers. Because of these attributes, Wood/polypropylene composites (WPCs) are used in a variety of innovative applications. A composite material is a blend of at least two different elements. The new material thus formed, has properties that the elements alone do not possess. Wood polymer composite consists of a wooden frame

---

* Corresponding Author

called reinforcement (load) that provides the mechanical strength and protection called matrix that is the plastic (thermosetting or thermoplastic resin), which ensures the cohesion of the structure and transmission efforts towards the reinforcement while ensuring the cohesion of the material, gives it its final form and provides the interface with the mechanical environment and additives (accounting, anti-UV, antioxidants, fire retardants) which give the composite properties particular requirements for durability and performance of these materials for outdoor use. However the primary drawback of using wood–fibers for reinforcement is the poor interfacial adhesion between polar-hydrophilic wood–fibers and non polar-hydrophobic plastics (Diène et al., 2008).   The WPC are used in four areas: Building materials account for 75% of production. The products concerned are mainly patios, fences, doors, windows and moldings decorative. The consumer and industrial products account for 10% of this market. In this domain, WPC are used in making furniture, cabinets, floor, pallet handling, brackets, boxes and containers. Motor vehicles occupy 8% of this sector. WPC are like interior components of vehicles such as door panels, components of trunk, empty-pockets, cargo cover and so on. Other applications include mainly municipal infrastructure, marine applications, etc... They account for 7% of the production. Among the products manufactured facilities parks, picnic tables, modules, games, etc... Interfacial interactions are very weak in wood/polymer composites , because the surface free energy of both the filler and the polymer is very small (Maldas & Kokta, 1993). As a consequence adhesion must be improved practically always to achieve acceptable properties. Various techniques are used or at least tried for the improvement of interfacial adhesion including the treatment of the wood with sodium hydroxide (Ichazo et al. 2001; Cantero et al. 2003), coupling with functional silanes (Ichazo et al. 2001), or the coating of wood flour with stearic acid (Stark N.M., 1999; Raj R.G. & Kokta B.V., 1991). However, polymers functionalized with maleic anhydride (Kazayawoko et al., 1999; Bledzki A. K. & Faruk O., 2002). The functional groups of these polymers were shown to interact strongly or even react chemically with the surface of wood (Lu et al., 2005; Kazayawoko et al. 1997), while the long alkyl chains diffuse into the matrix making stress transfer possible. The aim of this work is to develop composite materials from polyolefin (polypropylene) and wood flour with coupling agent. One of most popular methods to improve the durability of wood is chemical modification by some small chemical reagents. Among these reagents, acid anhydrides, inorganic acid esters, acid chlorides, aldehydes, lactones, reactive vinyl compounds, epoxides and isocyanates are most useful compounds. In this research, maleic anhydride was selected for its active ring-anhydride group, which is capable of easily reacting with hydroxyl groups on wood without reversed effect on environment and resultantly reducing amounts of hydroxyl groups. Consequently, it's a promising way to improve the wood durability. A first series of specimen was obtained by blending wood flour and polymeric material. In a second series, polypropylene grafted- maleic anhydride was added to the previous ingredients. The first key point for the production of acceptable WPC is the compatibility between wood and polymer host matrix. Wood is hydrophilic in nature (high surface tension), which lowers the compatibility with hydrophobic polymeric material (low surface tension) during composite preparation; this leads to WPCs with poor dispersions of wood fibers [Kazayawoko et al., 1999, 1997; Woodhams et al., 1984; Li Q. & Matuana LM., 2003). Scanning the literature, one can find different surface treatments that have been experienced to improve wood/polymer adhesion in composites. Remind that the level of adhesion and/or the dispersion state of wood are the key points for the improvement of mechanical properties of the composites. Indeed, the wood particles which have high strength and modulus – with good adhesion and uniform dispersion – can impart better mechanical properties to the host polymer in order to obtain a composite with better properties than those of the unfilled polymer. Substantial research has been carried out on the surface modification of wood fibers with coupling agents to improve the strength properties of WPCs (Woodhams et al., 1984; Li Q. & Matuana L. M., 2003), among these, the addition of maleated polypropylene (MAPP) in polypropylene (PP)-based WPCs has been shown to appreciably improve the dispersion of fibers in the matrix and the mechanical properties of WPCs because of the formation of linkages between the OH groups of wood and maleic anhydride. Many authors (Kazayawoko et al., 1999; Woodhams et al., 1984) in-depth studies have elucidated the mechanisms of adhesion between MAPP treated wood fibers and the PP matrix that cause the improvement. In our study the effects of the incorporation of wood particles with and without a compatibilizing agent on the processing and properties of WPC and the effects of wood flour concentrations on the mechanical properties of the composites were investigated and the results are discussed.

2. Experimental

2.1 Materials

The wood flour particles of 425 microns (40 mesh) in size were kindly donated by American Wood fibers (Schofield, WI) and are constituted predominantly with ponderosa pine, maple, oak, spruce, southern yellow pine, cedar. The wood was oven dried at 100°C for 24 h before processing to remove moisture. The isotactic polypropylene matrix (PP) has a density of 0.9 g/cm^3 and a melt flow index of 2.5 g/10 min, it was provided by Solvay Co. Polypropylene grafted with maleic anhydride (MAPP) with an approximate maleic anhydride (MA) content of 3 wt. % was purchased from Aldrich Chemical Company, Inc. (Milwaukee, WI). All ingredients were used as received.

2.2 Compounding and processing

Before compounding, the wood flour was dried in an oven for at least 48 h at 105°C to a moisture content of less than 1%. The dried wood flours were stored in a sealed plastic container to prevent the absorption of water vapor. The PP matrix, dried wood flour, MAPP, were added to a high-intensity mixer (Papenmeier, TGAHK20, Germany) and dry-blended at room temperature for 10 min. After blending, the compounded materials were stored in a sealed plastic container. Several formulations were produced with various contents of wood flour, PP and MAPP (table 1). For the mechanical property experiments, test specimens were molded in a 33-Cincinnati Milacron reciprocating screw-injection molder (Batavia, OH). The nozzle temperature was set to 204°C. The extrudate, in the form of strands, was cooled in the air and pelletized. The resulting pellets were dried at 105°C for 24 h before they were injection-molded into the ASTM test specimens for flexural, tensile (Type I, ASTM D 638), and Izod impact strength testing. The dimensions of the specimens for the flexural tests were 120x 3x 12 mm^3 (Length x Thickness x Width). The different samples and their code are listed in table 1.

2.3 Electron and optical microscopy

The state of dispersion of the wood inside the polymeric matrix was analyzed using optical microscopy on samples of 100–200 lm thick. Scanning electron microscopy (SEM) was used to obtain microphotographs of the fracture surfaces of the wood composites. These fractures have been performed in liquid nitrogen to avoid any deformation. SEM has been performed using a FEI Quanta 400 microscope working at 30 kV. The polymer surface was examined with LEICA optical microscope working in a transmission mode. Samples were thin enough that no special preparation of the samples was needed for their observations with the optical microscope.

2.4 Differential scanning calorymeter (DSC)

Wood, natural and synthetic polymers are subject to a degradation of the mechanical properties under the influence of increased temperatures (Munker M., 1998). It is very important to have knowledge about the effect of the processing temperatures in relation to the processing duration because there is always thermal stress during the manufacturing of WPC. Important properties concerning the thermal stability of the WPC are obtained from the differential scanning calorimetric (DSC). DSC is widely used to characterize the thermal properties of WPCs. DSC can measure important thermoplastic properties, including the melting temperature (Tm), heat of melting, degree of crystallinity [$x(\%)$] crystallization, and presence of recyclates/ regrinds, nucleating agents, plasticizers, and polymer blends (the presence, composition, and compatibility). Thermal analysis of the WPC samples was carried out on a differential scanning calorimeter (PerkinElmer Instruments, Pyris Diamond DSC, Shelton, Connecticut) with the temperature calibrated with indium. All DSC measurements were performed with powdered samples of about (8–10)±0.5mg under a nitrogen atmosphere with a flow rate of 20 ml/min. Three replicates were run for each specimen. All samples were subjected to the same thermal history with the following thermal protocol, which was slightly modified from the one reported by (Valentini L. et al., 2003):

First, the samples were heated from 40 to 180°C at a heating rate of 20°C/min to erase the thermal history.

Second, the samples were cooled from 180 to 40.00°C at a cooling rate of 10°C/min to detect the crystallization temperature (Tc).

Finally, the samples were heated from 40 to 180°C at a heating rate of 10°C/min to determine Tm. Tm and the heat of fusion (ΔH_m) were calculated from the thermograms obtained during the second heating. The values of (ΔH_m) were used to estimate $x(\%)$, which was adjusted for each sample.

2.5 Mechanical tests

Tensile tests (tensile strength and tensile strain) and three-point flexural tests (flexural strength) were carried out on an Instron 5585H testing machine (Norwood, MA) with crosshead rates of 12.5 and 1.35 mm/min according to the procedures outlined in ASTM standards D 638 and D 790, respectively eight replicates were conducted to obtain an average value for each formulation. Before each test, the films were conditioned in a 50% relative humidity chamber at 23°C for 48 h. The Izod impact strength was measured with an

Instron impact pendulum tester (model PW5) according to ASTM D 256 with acutely notched specimens (notch depth = 2 mm) at room temperature. Each mean value represented an average of eight tests. The impact strength is defined as the ability of a material to resist the fracture under stress applied at a high speed. The impact properties of composite materials are directly related to their overall toughness. In the Izod standard test, the only measured variable is the total energy required to break a notched sample. Specimens for the test had the following dimensions 50 x 12.7 x 3.2 mm^3. Eight replicates for each composition were tested for impact strength.

Sample	PP (%)	Wood (%)
WPPC0	100	0
WPPC1	95	5
WPPC3	75	25
WPPC4	50	50

(a)

Sample	PP (%)	Wood (%)	MAPP (%)
WPPC0*	95	0	5
WPPC1*	90	5	5
WPPC3*	70	25	5
WPPC4*	45	50	5

(b)

Table 1. Composition and code of the wood/polymer composites (percentage is in weight). The star (*) denotes a composite with Wood/PP and 5% wt MAPP.

3. Results and discussions

The results are discussed in several sub-sections in accordance with the goals of the study.

3.1 Structure and morphology (SEM)

Figure 1 and figure 2 show the micrographs of WPCs. It is well known that the properties of wood polymer composites are highly dependent on the wood dispersion and adhesion with the polymer matrix. Figure 1a shows the SEM image of the pure PP fracture surface which was smooth and featureless. As for the WPC, the wood's particles are detected as white dots in figure 1b, 2a and 2b. These figures reveal more separate wood chips and polymer areas for the non compatibilized system. The micrograph in Fig.1c and fig.2c, show that with MAPP there is better dispersion and less voids than without MAPP in fig.1b, fig.2a and fig.2b. On the other hand, between the fig.2a and 2b, it is visible that the dispersion is better in figure 2b; where there's less wood, these two figures show that dispersion decrease with wood loading. When the wood content was enough higher, the particles were uniformly distributed in the PP matrix. They exhibited many single disperse particles and aggregates integrated with particles. The matrix is not enough to encapsulate the solid micro particles of wood. However, large aggregates were found, and the aggregate size increased substantially in these micrographs with higher with higher wood loading wood. The copolymers in the blends can act as a compatibilizer decreasing the interfacial tension between the blend components of the mixture while enhancing the dispersion of dispersed phase in the matrix (Moon H.S., 1994). MAPP improves interfacial adhesion and prevents the debonding of even very large particles. Interfacial interactions and the strength of adhesion determine micromechanical deformation processes and the failure mode of the composites (Renner et al., 2009). The SEM micrographs taken from the surface of broken specimens provide indirect information about the failure mode and interfacial adhesion. Fig.1c and fig.2c present the fracture surface of specimens prepared with MAPP. The coverage of the wood with the polymer and the relatively small number of holes related to debonding or fiber pull out indicate good adhesion. On the other hand, the opposite is observed in composites prepared without MAPP. The number of debonded particles is quite large, the contours of particles remaining on the surface are sharp, and adhesion seems to be poor, at least compared to MAPP modification.

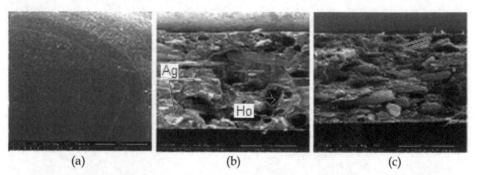

(a)   (b)   (c)

Fig. 1. SEM micrographs of (a) pure PP and composites containing (b) wood/PP (25/75) and (c) wood/PP/MAPP (25/70/5). Ho seems holes and Ag seems aggregate.

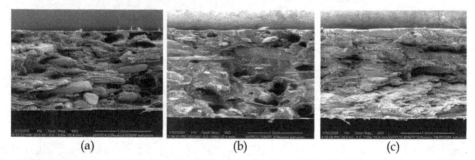

(a)   (b)   (c)

Fig. 2. SEM micrographs (a) Wood/PP (50/50), (b) Wood/PP (25/75) and (c) Wood/PP /MAPP (25/70/5).

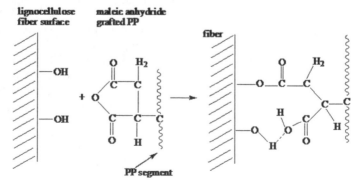

Fig. 3. Schematic description of the grafting of maleic anhydride with wood (Gauthier et al., 1999).

Figure 3 shows the summary of the grafting reaction of MAPP with wood.

3.2 Thermal and crystallization behavior

In this work, the thermal stability, the process of crystallization and melting of PP in its composites with wood (red pine) are studied by SEM analysis and differential scanning calorimetry, respectively, as a function of the wood content and coupling agent.

The physical properties of the WPC could be significantly affected by the crystallization characteristics of PP. Table 2 summarizes the results obtained from this heating run for all of the samples. The measurements were performed immediately after the melt-quenching thermal and crystallization behavior. The physical properties of the WPC could be significantly affected by the crystallization characteristics of PP. The measurements were performed immediately after the melt-quenching scans, so the samples had the same thermal history without an aging cycle. The curves revealed the following thermal events with increasing temperature: the cold crystallization process characterized by T_c and the cold crystallization enthalpy ΔH_c, and the melting process with following characteristics melting temperature (T_m) and melting enthalpy (ΔH_m). Comparing the thermograms and calorimetric parameters collected in table 2, one can see that with a filling of wood, as shown in this table, neat PP represented a tiny broad exothermic peak at 120°C, which indicated a rather low cold crystallization capability. However, in the case of WPC, this peak was sharper and appeared at much higher temperature, and the crystallization enthalpies increased correspondingly.

Sample code	$T_c(°C)$	$-\Delta H_c(J/g)$	$T_m(°C)$	$\Delta H_m(J/g)$	$\chi(\%)$	$\chi_{corr}(\%)$
WPPC0	120.8	87.9	160.1	89.2	37.9	37.9
WPPC1	123.4	94.9	160.8	95.7	34.3	38.2
WPPC3	124.7	92.7	162.5	93.0	30.1	40.1
WPPC4	125.1	89.0	162.7	90.7	20.6	41.2

Table 2. Thermal and Crystalline Properties of the neat PP and WPCs

The double endothermic melting peaks are visible on the thermograms, as shown in fig.4. Due to the reorganization during heating; the composite appears to slide to a more stable phase that melts at a higher temperature. We know that for pure PP one endothermic peak of melting occurs at 160°C corresponding to melting of its crystalline phase. On the thermograms of samples with a rate of wood more than 25%, we notice a slight peak at 110°C after the main crystallization peak around 120°C (cooled from the melt) and another small peak at 145°C (heating) before the main melting peak of the composite which arrived at 160°C. This corroborates the heterogeneity of the material. Scanning the literature, some authors have noticed these very specific peaks. Maillard and co-author (Maillard et al., 2008) observed these peaks at the same temperatures. Some authors (Quilin et al., 1993) believe that this is not a real effect, but simply a result in heat capacities between the samples of pure PP and those impregnated wood, causing an apparent shift in temperature measurement. According to the data listed in table 2, the Tc values of all the WPC were greatly increased compared to that of neat PP.

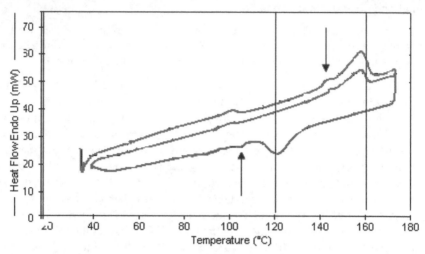

Fig. 4. DSC curve of heating and cooling of WPPC3

In the meantime, ΔH_c were decreased gradually with increasing the rate the rate of wood incorporated. This phenomenon between reinforcing material and polymer matrix has been reported in many other articles (Nam et al., 2001; Sinha et al., 2002). Generally, with the addition of lower content of wood, the polymer in the WPC, formed crystals much more easily because of the nucleation effect of the wood particles. However, when it reached certain content, the appearance of some aggregates restricted the crystallization behavior of PP. These results suggested that the incorporation of wood enhanced the cold crystalline ability of PP. The crystallization arrives sooner in WPC than in pure PP; this phenomenon was ascribed to a nucleating effect of the wood, which accelerated the crystallization speed of PP. The addition of wood flour had the effect of shifting Tm to higher temperatures. This increase was accompanied by an increased of crystallinity $x(\%)$. We corrected the degree of crystallinity of the composites $x_{cor}(\%)$ in the equation (1) by taking into account the wood-flour concentration:

$$x_{cor}(\%) = \frac{\Delta H_m}{(1\text{-MF}) \cdot \Delta H_o} \tag{1}$$

where ΔH_m and ΔH_o are respectively the melting enthalpies of the composite and polymer with 100% crystallinity and MF is the mass fraction of the wood in the composite.

This result can be explained by the agglomeration of the wood's particles. More wood's particles were added, and more aggregates were formed. In general, larger aggregates contributed to the crystallization of PP. This significantly conformed by the gradually decrease of ΔH_c that corresponded to the increase in wood content. With wood loading, the T_m increases for all WPC; this was ascribed to the poor thermal conductivity of wood. In the composite, wood flour acted as an insulating material, hindering the heat conductivity. As a result, the WPC compounds needed more heat to melt. Similar findings were previously reported by Matuana and co-author (Matuana L. M. & Kim J.W., 2007) for PVC based wood–plastic composites. They found that the addition of wood flour to the PVC resin caused significant significant increase in the temperature and energy at which fusion between the particles started. The delayed fusion time observed in rigid PVC/wood flour composites was attributed to the poor thermal conductivity of the wood flour; this decreased the transfer of heat and shear throughout the PVC grains. These phenomena were consistent with the results of this study.

3.3 Mechanical properties

Tensile strength, flexural strength, module of elasticity (MOE) and elongation at break provide an excellent measure of the degree of reinforcement provided by the fiber to the composite (Mueller D.H. & Krobjilowski A., 2003). It can be seen from fig.5a and fig.5b respectively that the tensile strength and the tensile modulus increase with wood content.

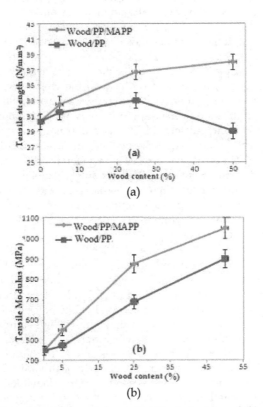

(a)

(b)

Fig. 5. Tensile strength and tensile modulus respectively in (a) and (b) versus wood content.

Tensile strength increases over WPC. Whereas, without compatibilizer, the tensiles strength of the composites are in the range of 33.5-28.5 N/mm^2 at wood loading from 5% to 50%, suggesting that, there is little stress transfer from the matrix to the fibers irrespective of the amount of wood   present. When MAPP was incorporated into the WPC, the wood was relatively well dispersed and the interaction has occurred between the wood and matrix that was corresponding to the improved tensile strength. The use of MAPP improves interaction and adhesion between the fibers and matrix leading to better matrix to stress transfer. Similar observations were reported by Felix and Gatenholm (Felix J. M. & Gatenholm P., 1991) where tensile strength of the composites increased linearly with fiber content when MAPP treated fibers were used instead of untreated fibers. Myers et al. (Myers et al., 1993) reported 21% increase in tensile strength for a 50:50 wood flour polypropylene composites when MAPP was used as a compatibilizer. Stark and Rowlands (Stark N. M. & Rowlands R. E., 2003) also reported a 27% increase in tensile strength of composite prepared with 40% wood-fiber and 3% MAPP. At lower filler content the tensile modulus does not seem to be affected by improved adhesion (Felix J. M. & Gatenholm P., 1991). However at higher filler loading the tensile modulus of the composites with compatibilizer was much superior to that of the composites without compatibilizer. Compatibilizers can change the molecular morphology of the polymer chains near the fiber-polymer interphase. Yin et al. (Yin et al., 1999) reported that the addition of coupling agent (MAPP) even at low levels (1-2%) increases the nucleation capacity of wood- fibers for polypropylene, and dramatically alters the crystal morphology of polypropylene around the fiber. When MAPP is added, surface crystallization dominates over bulk crystallization and a transcrystalline layer can be formed around the wood-fibers. Crystallites have much higher moduli as compared to the amorphous regions and can increase the modulus contribution of the polymer matrix to the composite modulus (Sinha et al., 2002). The elongation at break (Fig.6) decreases steadily with the wood-fiber content. There is no significant difference in elongation at break for composites with and without compatibilizer so we have only represented the stress versus strain for only WPC without coupling agent. The steep decline in elongation immediately on filler addition is obvious, because wood-fibers have low elongation at break and restrict the polymer molecules flowing past one another. This behavior is typical of reinforced thermoplastics in general and has been reported by many researchers (Felix J.M. & Gatenholm P., 1991). Adding a suitable interface modifier will promote the stability of the morphology in wood plastic composite (Snijder M.H.B. & Bos H.L., 2000). The crystallinity results can help to explain the results from the mechanical testing. The strength and modulus are increasing with wood content in WPC at the same time, in the DSC curves (not represented here), the degree of crystallinity of these samples increases. Introducing wood filler modifies the mechanical behavior by making the material stiffer, which is characterized by the significantly decreased failure strain. The addition of wood decreases the compatibility between the hydrophilic wood material and hydrophobic matrices (binders) and entanglement between the PP and its molecules. This results in poor interphase properties and a lower strain to failure of the composite. The percentage elongation of all the composites with wood decreased with wood loading until 50% as compared to neat polymer. From the stress-strain curve of the composites with different wood loading (fig.6a), it is found that the stress-strain curve of pure polypropylene is similar to that of brittle materials. The behavior is perfectly elastic, the stress increases linearly with strain. However, addition of wood makes the matrix more ductile. This is evident from the elongation at break values of the composites. The flexural strength (fig.6b) increase with wood loading and the coupling agent make this phenomenon more pronounced.

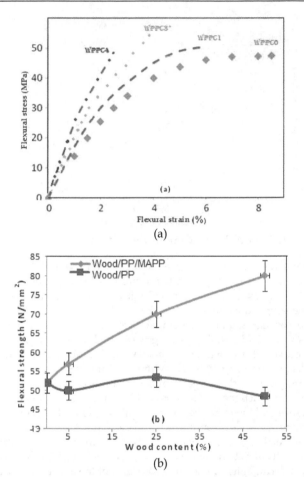

Fig. 6. Curves of: (a) Flexural stress/strain and (b) flexural strength function of wood content.

Figure 7 exhibits the curves of Izod impact strength function of wood incorporated. It is important to note that the impact test machine used in this study did not provide enough energy to break the neat PP because of the high flexibility of the PP matrix. By contrast, all of the composites with and without MAPP broke completely. For the impact property testing, composites with high wood fiber content possess low impact strength and notched impact strength decreases with the increasing of wood flour content. The positive effect of the coupling agent reduces this phenomenon. The main factors influencing impact strength is the size of the disperse phase. Impact strength of WPC blends decreases as a function of dispersed phase (Albano et al., 2002). SEM micrographs (Figures 1b, 1c and 2) show some fracture surfaces of WPC where the fibers are still covered by the polymer matrix; this result indicates that matrix cohesive failure was the dominant failure mode. Wood flour is kind of stiff organic filler, comparing to PP, so adding wood flour could decrease the impact strength of composite. The scanning electron microscope illustrated that the polymer intimately associated in the wood structure altered the mode of fracture. Composite treated with MAPP like interface modifier exhibited better impact strength than the untreated ones

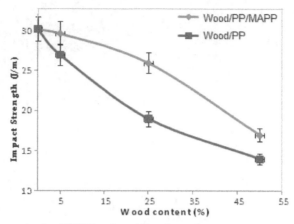

Fig. 7. Izod impact strength of WPC as function of wood content

since the debonding behavior between the interface of wood flour and PP matrix absorb larger impact energy in modified composites than the unmodified ones. Decreased impact strength of wood/ PP/MAPP composites was explained with the stronger interfacial adhesion and rigidity created by the larger functionality of MAPP used (Hristov et al., 2004). It was clear that the addition of wood particles impaired the impact strength of the wood polymer composites. The wood's particles were rigid with high strength and high modulus, which cannot generate deformation when impact is exerted on them. So, they cannot absorb impact energy by terminating the cracks or producing craze. Therefore, the brittleness of the composites increased and the impact strength decreased. A larger number of bonds may form on a unit surface of the wood in the case of larger functionality and this could lead to better stress transfer, but this was contradicted by our results. However, the filler/coupling agent interaction is only one side of the interphase forming in these composites and we must consider also the coupling agent/polymer interface. Large functionality leads to more reactions with the wood and shorter free chains, which cannot entangle with the polymer as efficiently as longer molecules. This leads to smaller deformability of the interphase, and of the entire composite.

4. Conclusion

The effects of the incorporation of wood particles both with and without a compatibilizing agent on the processing and properties of WPC composites were investigated. The morphology (SEM) indicates that the composites treated with MAPP modifiers exhibit much better bonding between flour and matrix. Heat deflection temperature increases with the increasing of wood flour content. When exposed to a source of chemical degradation, the morphology of WPC may be change, and in some cases disruption of the crystalline order occurs as detected by reduction in the fractional crystallinity of PP. Tensile strength, tensile modulus and flexural strength are significantly increased with increasing wood flour content. Mechanical properties, measured in tensile and flexural tests, demonstrated that the wood flour used in this work act as effective reinforcing agents for PP. Addition of wood flour, at all levels, resulted in more rigid and tenacious composite, but had lower impact energy and lower percentage of elongation as compared to the polymer matrix. The increase in mechanical properties demonstrated that MAPP is an effective compatibilizer for wood polypropylene composites. The presence of wood in the composite generated imperfect adhesion between the components of the composite; this increased the concentration of stress and decreased the impact strength. Increasing the wood content in the composites led to an increased stress concentration because of the poor bonding between the wood flour and the polymer. Although crack propagation became difficult in the polymeric matrix reinforced with filler. The results show that the presence of wood flour in the composite was accompanied by an evolution of the crystallization. This study demonstrated that wood flour could be successfully used as a reinforcing material in a polypropylene matrix. MAPP coupling agent improves the compatibility between wood flour and PP resin. Taking these advantages into account, wood flour reinforced composites can be manufactured successfully using injection moulding. They represent a suitable material which is an alternative to glass fibers reinforcements for lots of applications in the range of lower mechanical loads. The manufacture of wood polymer composites allows not only to recover the waste wood from the forest industry, but also to reduce the use of fossil resources.

5. References

Albano C.; Reyes J.; Ichazo M.; Gonzalez J.; Brito M. & Moronta D. Analysis of the mechanical, thermal and morphological behavior of polypropylene compounds with sisal fibre and wood flour, irradiated with gamma rays. *Polymer Degradation and Stability*, vol.76, (2002), pp. 191-203.

Amash A. & Zugenmaier P. Morphology and properties of isotropic and oriented samples of cellulose fibre-polypropylene composites. *Polymer, vol.* 41, No.4, (2000), pp. 1589-1596.

Bledzki A. K.; Faruk O. & Huque M. Physico-mechanical studies of wood fiber reinforced composites. *Plastics Technology and Engineering*, vol.41, No.3, (2002), pp. 435-451.

Cantero G.; Arbelaiz A.; Mugika F.; Valea A. & Mondragon I. Mechanical behavior of wood/polypropylene composites: Effects of fiber treatments and ageing processes. *Journal of Reinforced Plastics and Composites*, vol. 22, (2003), pp. 37-50.

Diène N.; Fanton E.; Morlat-Therias S.; Vidal L.; Tidjani A. & Gardette J.L. Durability of wood polymer composites: Part 1. Influence of wood on the photochemical properties. *Composites Science and Technology*, vol. 68, (2008), pp. 2779-2784.

Felix J. M. & Gatenholm P. The nature of adhesion in composites of modified cellulose fibers and polypropylene. *Journal of Applied Polymer Science*, vol.42, (1991), pp. 609-620.

Gauthier R.; Gauthier H. and Joly C. Compatibilization between lignocellulosic fibers and a polyolefin matrix *Proceedings of the Fifth International Conference on Woodfiber-Plastic Composites, Forest Products Society, Madison, WI, May 1999, p.* 153

Hristov V. N.; Krumova M.; Vasileva St. & Michler G.H. Modified polypropylene wood flour composites. Fracture deformation and mechanical properties. *Journal of Applied Polymer Science,* vol.92, No.2, (2004), pp. 1286-1292.

Ichazo MN, Albano C, Gonzalez J, Perera R, Candal MV. Polypropylene/wood flour composites: treatments and properties. Compos Struct 2001; 54:207-14

Kazayawoko M.; Balatinecz J.J. & Woodhams R.T. Diffuse reflectance Fourier transform infrared spectra of wood fibers treated with maleated polypropylenes. *Journal of Applied Polymer Science*, vol.66, No.6, (1997), pp. 1163-1173.

Kazayawoko M.; Balatinecz J.J. & Matuana L.M. Surface modification and adhesion mechanism in wood fiber-polypropylene composites. *Journal of Mater Science*, vol. 34, No.24, (1999), pp. 6189-992.

Li Q. & Matuana L. M. Surface of Cellulosic Materials Modified with Functionalized Polyethylene Coupling Agents. *Journal of Applied Polymer Science*, vol.88, No.2, (2003), pp. 278-286

Lu J. Z.; Negulescu I. I. & Wu Q. Maleated wood-fiber/high densitypolyethylene composites: coupling mechanisms and interfacial characterization. *Composite Interfaces*, vol. 12, No. 1-2, (2005), pp. 125–140.

Maillard D. & Prud'homme R. E. The crystallization of ultrathin films of polylactides Morphologies and transitions. *Canadian Journal of Chemistry*, vol.86, No.6, (2008), pp. 556-563.

Maldas D. & Kokta B.V. Interfacial adhesion of lignocellulosic materials in polymer composites: an overview. *Composites Interfaces*, vol.1, No.1, (1993), pp. 87–108.

Matuana L.M. & Kim J.W. Fusion Characteristics of Wood-Flour Filled Rigid PVC by Torque Rheometry, *Journal of Vinyl & Additive Technology*, vol;13, No.1, (2007), pp. 7-13.

Meyrs E.G.; Chahyadi I.S.; Gonzalez C. & Coberly C.A. Wood flour and polypropylene or high-density polyethylene composites: influence of maleated polypropylene concentration and extrusion temperature on properties. *In: Walcott MP, editor. Wood fibres/polymer composites: fundamental concepts, processes, and material options. Madison,USA: Forest Product Society*, (1993), pp. 49 –56.

Moon H. -S.; Ryoo B. -K. & Park J. -K., J. *Polymer Science, Part B: Polymer Physic*, vol.32 (1994), pp. 1427-1435.

Mueller D.H. & Krobjilowski A. New Discovery in the Properties of Composites Reinforced with Natural Fibers. *Journal of Industrial Textiles*, vol.33, No.2, (2003), pp. 111–130.

Munker M. Werkstoffe in der Fertigung, 3, (1998), 15.

Nam P. H.; Okamoto M.; Kotaka T.; Hasegawa N. & Usuki, A. A hierarchical structure and properties of intercalated polypropylene/clay nanocomposites, *Polymer*, vol.42, (2001), 9633-9640

Quilin D. T.; Caulfield D. F. & Koutski J. A. Crystallinity in the polypropylene/cellulose system and crystalline morphology, *Journal of Applied Polymer Science*, vol.50. (1993), pp. 1187-1194

Raj R.G. & Kokta B.V. Reinforcing high density polyethylene with cellulosic fibers I: The effect of additives on fiber dispersion and mechanical properties. *Polymer Engineering Science*, vol. 31, No. 18, (1991), pp. 1358–1362.

Renner K.; Móczó J. & Pukánszky B. Deformation and failure of wood flour reinforced composites: effect of inherent strength of wood particles. *Composites Science and Technology*, vol.69, No.10 (2009), pp. 1653–1659.

Sinha R. S.; Maiti P.; Okamoto M.; Yamada K.. & Ueda K. New Polylactide/Layered Silicate Nanocomposites. 1. Preparation, Characterization, and Properties, Macromolecules, vol.35, (2002), pp. 3104-3110.

Snijder M.H.B. & Bos H.L. Reinforcement of polypropylene by annual plant fibres: optimization of the coupling agent efficiency. *Composites Interfaces*, vol.7, No.2, (2000), pp. 69-79.

Stark N. M. Wood fiber derived from scrap pallets used in polypropylene composites. *Forest Products Journal*, vol.49, No 6, (1999), pp. 39 – 46.

Stark N. M., Rowlands RE. Effects of wood fiber characteristics on mechanical properties of wood/polypropylene composites. *Wood Fiber Science*, vol.35, No.2, (2003), pp. 167-174.

Timmons T. K; Meyer J. A. & Cote W. A. Polymer location in the wood polymer composite. *Wood Science*, vol. 41, No. 1, (1971), pp. 13–24.

Woodhams, R. T.; Thomas G. &. Rodgers D. K. Wood fibers as reinforcing fillers for polyolefins. *Polymer Engineering Science*, vol.24, No.15, (1984), pp. 1166-1171.

Yin S.; Rials T.G. & Wolcott M.P. Crystallization behavior of polypropylene and its effect on woodfiber composite properties. *In: Fifth international conference on wood fiber–plastic composites, Madison, WI, Forest Products Society, (1999 May 26–2), pp. 139–146.*

10

Solidification of Polypropylene Under Processing Conditions – Relevance of Cooling Rate, Pressure and Molecular Parameters

Valerio Brucato and Vincenzo La Carrubba
Dipartimento di Ingegneria Chimica, Gestionale, Informatica,
Meccanica Università di Palermo, Palermo
Italy

1. Introduction

Polymer transformation processes are based on a detailed knowledge of material behaviour under extreme conditions that are very far from the usual conditions normally available in the scientific literature. In industrial processing, for instance, materials are subjected to high pressure, high shear (and/or elongational) rates and high thermal gradients. These conditions lead often to non-equilibrium conformational states, which turn out to be very hard to describe using classical approaches. Moreover, it is easy to understand that the analysis of the relationships between the processing conditions and the morphology developed is a crucial point for the characterisation of plastic materials. If the material under investigation is a semicrystalline polymer, the analysis becomes still more complex by crystallisation phenomena, that need to be properly described and quantified. Furthermore, the lack of significant information regarding the influence of processing conditions on crystallization kinetics restricts the possibilities of modelling and simulating the industrial material transformation processes, indicating that the development of a model, capable of describing polymer behaviour under drastic solidification conditions is a very complex task.

However, new innovative approaches can lead to a relevant answer to these scientific and technological tasks, as shown by some recent developments in polymer solidification analysis (Ding & Spruiell, 1996, Eder and Janeschitz-Kriegl, 1997, Brucato et al., 2002) under realistic processing conditions. These approaches are based on model experiments, emulating some processing condition and trying to identify and isolate the state variable(s) governing the process.

So far, due to the experimental difficulties, the study of polymer structure development under processing conditions has been mainly performed using conventional techniques such as dilatometry (Leute et al., 1976, Zoller, 1979, He & Zoller, 1994) and differential scanning calorimetry (Duoillard et al., 1993, Fann et al., 1998, Liangbin et al., 2000). Investigations made using these techniques normally involve experiments under isothermal conditions. However experiments under non isothermal conditions have been limited to cooling rates several orders of magnitude lower than those experienced in industrial processes, which often lead to quite different structures and properties. Finally, in the last years, experiments revealing the crystallinity evolution by measures of crossing light scattering, have been conducted at intermediate cooling rate (Strobl, 1997, Piccarolo, 1992).

For the sake of completeness, it should be conceded that the complexity of the investigation concerning polymer solidification under processing conditions is even greater if the wide latitude of morphologies achievable is considered, especially when dealing with semicrystalline polymers. This would have to take into account also the complexity introduced by the presence of the crystallization process (Eder & Janeschitz-Kriegl, 1997).

Generally speaking, polymer crystallization under processing conditions cannot be considered an "equilibrium" phenomenon, since it is not possible to separate the thermodynamics effects on the processes from the kinetic ones. Furthermore, crystallization of polymeric materials is always limited by molecular mobility, and very often leads to metastable phases, as recently shown by Strobl (Strobl, 1997). Further evidences of the formation of metastable phases under drastic conditions (high cooling rates and/or high deformation rates) have been widely reported for iPP (Piccarolo, 1992, Piccarolo et al., 1992a). Choi and White (Choi & White, 2000) described structure development of melt spun iPP thin filaments, obtaining conditions under which different crystalline forms of iPP were obtained as a function of cooling rate and spinline stresses. On the basis of their experimental results together with many others available in literature, the authors have constructed a diagram, which indicates the crystalline states formed at different cooling rates in isotropic quiescent conditions. Continuous Cooling Transformation curves (CCT) have been reported on that diagram. According to the authors, at low cooling rates and high stresses, the monoclinic a-structure was formed, whereas at high cooling rates and low stresses a large pseudo-hexagonal/smectic ("mesophase") region was evident.

The formation of metastable phases normally takes place in a cooling rate range not achievable using the conventional techniques mentioned above; nevertheless it is worth reminding that the behaviour of a given semi-crystalline polymer is greatly influenced by the relative amount of the constitutive phases. From this general background the lack of literature data in this particular field of investigation should not be surprising, due to the complexity of the subject involved. The major task to tackle is, probably, to identify the rationale behind the multiform behaviour observed in polymer solidification, with the aim of finding the basic functional relationships governing the whole phenomenon. Therefore a possible approach, along this general framework, consists of designing and setting-up model experiments that could help to isolate and study the influence of some experimental variables on the final properties of the polymer, including its morphology. Thus a systematic investigation on polymer solidification under processing conditions should start on the separate study of the influence of flow, pressure and temperature on crystallization.

Due to experimental difficulties, there are only a few reports on the role of pressure in polymer crystallization, especially concerning its influence on the mechanical and physical properties. Moreover, the majority of studies made at high pressure have concentrated only on one polymer, polyethylene, dealing with the formation of extended chain crystals, as shown by Wunderlich and coworkers (Wunderlich & Arakaw, 1964, Geil et al., 1964, Tchizmakov, 1976, Wunderlich, 1973, 1976, 1980, Wunderlich & Davison, 1969, Kovarskii, 1994). The pressure associated with such investigations tends to be extremely high (typically 500 MPa) with respect to the pressures normally used in industrial processes. Furthermore, the experimental conditions normally investigated were quasi-isothermal. This implies that

Solidification of Polypropylene Under Processing Conditions – Relevance of Cooling Rate, Pressure and
Molecular Parameters

191

the obtained results may not be applied to conventional polymer processing, involving very high thermal gradients.

The purpose of this chapter is to provide a general experimental route for studying the crystallization behaviour of isotactic polypropylene under high cooling rates and pressure.

In this respect, two complementary devices were used. The first involves a special equipment that has been developed, and widely tested, to quench polymeric samples at atmospheric pressure in a wide range of cooling rates (from 0.1 up to ca 2000 °C/s) under quiescent conditions and with the use of which it has been possible to collect much information about the influence of cooling rate on the final properties. The second was an innovative equipment specifically designed to evaluate the combined effect of typical injection moulding pressures (up to 40 MPa) and temperature gradients (up to a maximum of ca 100 °C/s), with the aid of a modified injection moulding machine. The results show that the influence of pressure on polymer crystallization is not as obvious as one may expect. An increase of cooling rate generally determines a transition from crystalline to non-crystalline (or pseudo-crystalline) structures. As for the influence of pressure, in iPP an increase in pressure results into a decrease of crystallinity, owing to kinetic factors, such decreased mobility related to the increased Tg.

In the last part of the chapter, a discussion on the influence of molecular parameters on the crystallization kinetics of iPP under processing conditions is presented. As a matter of fact, the crystalline structure of iPP quenched from the melt is affected not only by cooling rate, or generally by processing conditions, but also by molecular parameters like molecular mass (Mw) and molecular mass distribution (Mwd). Different configurations (isotacticity and head-to-tail sequences) or addition of small monomeric units and nucleating agents can also influence the final structure (De Rosa et al., 2005, Foresta et al., 2001, Sakurai et al., 2005, Nagasawa et al., 2005, Raab et al., 2004, Marigo et al., 2004, Elmoumni, 2005, Chen et al., 2005).

Influence of molecular weight on polymer crystallization is controversial. Stem length indeed interferes with entanglement density, thus determining a rate controlled segregation regime of topological constraints in non crystalline regions. Very low molecular weight tails of the distribution are shown to positively affect crystallization kinetics although their thermodynamic action should not favour perfection of crystallites (Strobl, 1997).

It is known from the literature that crystallization kinetics of semicrystalline polymers is influenced by the presence of contaminants. The main effect of the addition of a nucleating agent is an increase of the final crystallinity level together with a higher final density and a finer and homogeneous crystal size distribution. This typical effect of enhancement of the overall crystallization kinetics allows one to infer that crystallization kinetics are nucleation-controlled, being the nucleation step the rate determining one whilst the growth rate remains almost unaffected (Nagasawa, 2005, Raab, 2004).

On the other hand, the incorporation of a small content of ethylene units in the polypropylene chains has an influence on the regularity of the molecular structure. In fact, a change in tacticity induced by the shortening of isotactic sequences was observed (Zimmermann, 1993). Although this has a negative influence on crystallisation kinetics, an opposite effect should come from the enhanced mobility due to the presence of the ethylene sequences. As a result of these counteracting effects, a relatively narrow window of cooling rates exists in which an enhancement of crystallization kinetics sets in (Foresta et al., 2001).

A better understanding of the relation between processing and properties can be achieved if the absolute crystallinity during transformation can be predicted as a function of processing conditions. This prediction has to be supported by a crystallization kinetics model; here a modified two-phase non-isothermal form of the Kolmogoroff-Avrami-Evans model was used to describe the crystallization kinetics (Avrami, 1939, 1940, 1941, Evans, 1945, La Carrubba et al., 2002a, Brucato et al., 1993). The main purpose of this analysis is to underline the relevance of thermal history resulting from various cooling conditions on the crystallization kinetics of different grades of iPP containing various additives such as nucleating agents and small content of ethylene.

More specifically, the discussion attempts to identify relevant material parameters determining quiescent non isothermal crystallization kinetics simulating polymer solidification under processing conditions. One has obviously to cope with commercially relevant grades, which implies constraints in the span they cover. Therefore limitations arise not only due to the intrinsic poor significance of material parameters to crystallization kinetics but also owing to the limitation on the grades one can recover on the market. Finally, one of the main issues of this part of the chapter is the appropriate comparison among the investigated iPP samples in order to outline, when possible, the influence on the crystallization kinetics of average molecular mass, molecular-mass distribution, isotacticity, copolymerization with small amount of ethylene units and the addition of nucleants.

2. Description of the experimental procedure

2.1 Rapid cooling experiment at atmospheric pressure

A schematic drawing of the experimental set-up is shown in **fig. 1 a**. The sample, properly enveloped in a thin aluminium foil, so as to avoid leakage of material while in the molten state, (see **fig. 1 b**), and sandwiched between two identical flat metallic slabs, is heated to a suitable high temperature in a nitrogen fluxed environment.

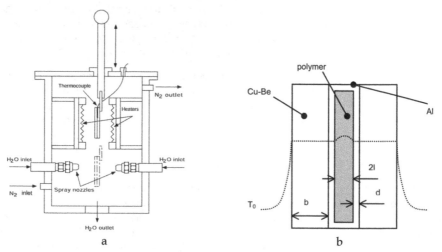

Fig. 1. a. Scheme of the experimental set-up for quench experiments; b. Sample assembly and temperature profiles. b=1-2 mm; l=50-100 μm; d=10 μm

A fast response, 12.5 µm thick (Omega type CO2), thermocouple buried inside one of the slabs allows to record the whole thermal history by a data acquisition system.

A Cu-Be alloy was chosen for the production of the metallic slabs, owing to its high Young's Modulus coupled with a high thermal conductivity (see Goodfellow Catalogue, 1996).

After keeping the sample system at a temperature above the equilibrium melting temperature for a time sufficient to erase memory effects (Alfonso & Ziabicki, 1995, Ziabicki & Alfonso, 1994), the sample assembly was moved to the lower zone of the container where it was quenched by spraying a cooling fluid on both faces through two identical nozzles positioned symmetrically opposite to each face of the sample assembly (**fig. 1a**).

The cooling rate was varied by changing the cooling fluid, its flow rate and temperature, or by changing the thickness of the sample assembly. However, the coolant temperature may not be crucial if it is sufficiently lower than the polymer solidification temperature.

Once the sample reached the final temperature it was immediately removed from the sample assembly and kept at low temperature (-30°C) before further characterization.

Three typical thermal histories (i.e. variation of temperature with time) obtained using this device are shown in **fig. 2**. Results of an extended set of experiments are reported in **fig. 3** as recorded variation of cooling rate with sample temperature. The data in **fig. 3** represent the range of variation of cooling rate covering five orders of magnitude (0.01-1000°C/s). This result is particularly significant when compared to standard DTA or DSC runs which cover only the lowest two decades of this cooling rate range (0.01-1°C/s). It is worth noting that for crystallization kinetics the high cooling rates are very informative, especially for fast crystallizing polymers, such as polyolefins. However, the high cooling rates severely restrict the possibility to detect the structural modifications taking place during solidification. The latter is the main constraint with respect to the real-time information provided by DTA and DSC measurements.

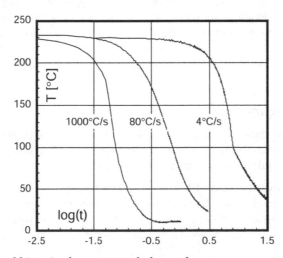

Fig. 2. Typical thermal histories for spray cooled samples

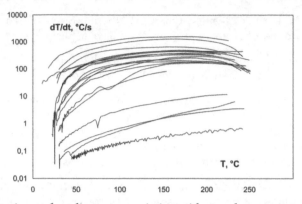

Fig. 3. Typical experimental cooling rates variation with sample temperature

With respect to the thermal histories in **fig. 2**, one will note that there is no temperature plateau associated with crystallization, the process occurring during cooling. This is due to the fact that temperature was measured on the metal slabs and not in the bulk of the polymer sample, albeit the latter has a negligible mass and volume relative to the size of the metal slabs. Furthermore, the very high heat flux to which the polymer was subjected masks the effect of the latent heat of crystallization. So, only the temperature-time history is recorded and, therefore, at the end of the cooling process one gets a thin polymeric film with a known thermal history. Sample structure depends on its thermal history and this relationship can be experimentally assessed if the "length scale" of structural features developed is small compared to the sample thickness and if the final structural features are uniform throughout the whole sample (Titomanlio et al., 1997, Titomanlio et al., 1988a, Titomanlio et al., 1988b). The sample homogeneity is thus crucial to the method envisaged since the recorded thermal history is the only available information for the determination of the final structure of the sample. The proposed model experiment is addressed to design a method for the characterization of the non-isothermal solidification behaviour encompassing typical cooling conditions of polymer processing. Only temperature history determines the structure formed as the melt solidification takes place in quiescent conditions. A discussion on the temperature distribution in a mono-dimensional heat exchange regime and the evaluation of structure distribution obtained along the thickness follows.

2.1.1 Cooling mechanism

We will consider now the effect of the applied heat flux on the temperature distribution of the metal in the sample assembly. Later in the next section the temperature distribution across the sample in contact with the metal will be examined.

The shape of the temperature profile in a flat slab having the following characteristics, thickness $2b$ and thermal conductivity k, and conditions, initial temperature T_i, suddenly exposed to a cooling medium at temperature T_0 and draining heat from the slab with a heat exchange coefficient h it is determined by the dimentionless Biot number:

$$Biot = h \times b / k \qquad (1a)$$

For our experimental conditions the highest value of $Biot$ number is estimated to be 0.3. Although this value does not fulfil the classical requirements for a flat temperature profile distribution within the slab (which requires $Biot < 0.1$), the slab "cooling time" is practically unaffected by slab conductivity, therefore the so-called "regular regime" conditions still apply (Isachenko et al., 1987). In other words, the maximum Biot number for achieving a flat temperature profile is:

$$Biot_{max}=0.1 \tag{1b}$$

On the other hand, an estimate of the response time of the slabs assembly can be easily taken as the time needed for the mid plane to undergo 99% of a sudden drop of the wall temperature. The solution of such transient heat conduction problem gives the characteristic time τ_R as (Carslaw et al., 1986, Bird et al., 1960):

$$\tau_R = 2b^2 / \alpha \tag{2}$$

Where α and b are thermal diffusivity and half thickness of the slab respectively. Using the values of $\alpha = 2.6 \ 10^{-5} \ m^2/s$ (copper-beryllium 2% alloy – Goodfellow Catalogue, 1996) and b=0.001m in **equation (2)** gives $\tau_R = 0.07$ s. Note that the fastest cooling rate in our experiments has a characteristic time $\tau_A = 0.33$ s, which is about five times τ_R. Furthermore, since the real wall boundary thermal condition on the slab is not as sharp as the assumed stepwise drop of the wall temperature, the heat conduction inside the assembly does not affect the cooling history to any appreciable extent. Applying a more realistic boundary condition, i.e. a wall temperature depending on the heat flux, does not lead to a sudden wall temperature drop, and the ratio τ_A / τ_R becomes larger.

In the experiments water sprays were used to drain heat from the slab, therefore the associated heat transfer coefficient depends very much on the flow rate of the cooling medium, as shown in **figs. 4 a-b**. Here the heat flux was evaluated according to the lumped temperature energy balance on a slab of volume $V=Sx2b$, having a heat capacity c_p and density ρ.

$$\rho c_p V \ dT / dt = h \ 2S \ (T_0 - T) = -h2S(T - T_0)$$

$$dT / dt = (T_0 - T) / \tau_l = -(T - T_0)/ \tau_l \qquad \tau_l = \rho c_p b / h \tag{3}$$

Where S is the slab surface, h heat transfer coefficient, T_0 the coolant temperature and T the lumped sample temperature. By assuming that the heat exchange coefficient h is constant, then slope of the cooling rate versus temperature curve is also constant, while the slab temperature decays exponentially with time.

Fig. 4a shows that below the maximum and using smaller nozzles, giving lower mass flow rates, there are two heat transfer regimes separated by the Leidenfrost temperature, i.e. by the onset of temperature for the production of a boiling layer nucleated by the surface of the slab. In **fig. 4b** the increase of coolant mass flow rate results in the disappearance of the Leidenfrost temperature and brings about an extension of the linear dependence of heat flux to a higher temperature range up to the maximum (Ciofalo et al., 1998).

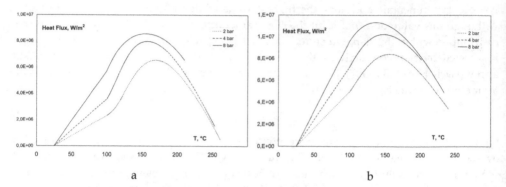

a   b

Fig. 4. Heat Flux variation with sample assembly temperature for two different (a, small nozzles and b, large nozzles) spray nozzles

As long as the heat flux depends on temperature linearly, a constant heat transfer coefficient can be successfully used. This condition is well identified in the low driving force (low temperature difference) region. This result can be understood considering that the heat transfer of convection induced by the liquid drops impacting onto the solid surface is similar to that of nucleated boiling, since it promotes the renewal of the liquid layer close to the solid surface. Indeed the two mechanisms take place in parallel and the spray cooling effectiveness can be varied by changing the mass flow rate of the coolant and, at high values of the mass flow rate, the same value of the heat exchange coefficient is attained in a temperature range spanning from ambient temperature to about 150°C. This last point is particularly relevant for fast crystallizing polymers since high heat transfer coefficients are required at low temperatures to quench them effectively, as in the case of iPP.

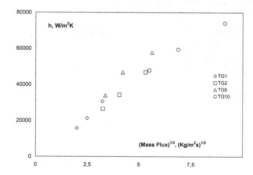

Fig. 5. Heat exchange coefficient vs. coolant mass flux for four different spray nozzles

The relationship between the liquid convection heat transfer coefficient, h, and the mass flow rate is summarized in **fig. 5** for all the nozzles used in this work. Within an error of ±10% there is a square root dependence of h on mass flow rate (Ciofalo et al., 1998).

The time constant, τ_l, obtained from **equation (3),** attains a minimum value of about 0.05 s. A comparison of the values of $5 \times \tau_l$ and τ_R (98.5% of the overall temperature drop) shows

Solidification of Polypropylene Under Processing Conditions – Relevance of Cooling Rate, Pressure and
Molecular Parameters

197

that the driving force (i.e. the temperature drop) is larger in the fluid than in the Cu-Be slab, i.e. the heat transfer is mainly controlled by the fluid heat transfer. At the same time and the definition of τ_l suggests that another way to change linearly the slope of the cooling curves of **figs. 4 a-b** is by modifying the slab thickness. Moreover **equations (2)** and **(3)** show that the ratio τ_l / τ_R is proportional to the inverse of the thickness, suggesting that one should use the thinnest possible slab to achieve a more uniform temperature distribution through the thickness.

In principle the time constant, τ_l, drawn from **figs. 4 a-b** could be used as a parameter to rigorously identify the overall cooling process (Ding & Spruiell, 1996). When the solidification temperature of the polymer falls in a range in which there is a change of the heat transfer regime, the heat transfer coefficient will also change with temperature while the use of τ_l becomes meaningless, as it is no longer constant. On the other hand, the value of τ_l changes slightly when the temperature range where solidification takes place is quite narrow (of the order of 10°C). Although an average value of τ_l could be used, it is preferred to use an equivalent parameter to identify the cooling process, which is the average cooling rate in the range of temperatures within which the polymer solidifies (Brucato et al., 2002, Piccarolo, 1992, Piccarolo et al., 1992a, Piccarolo et al., 1992b, Brucato et al., 1991a, Brucato et al., 1991b, Piccarolo et al. 1992, Piccarolo et al., 1996, Brucato et al., 2009). This parameter, indeed, imposes not only the experimental time to be constant, but also the characteristic range of temperatures in which a given polymer solidifies. For iPP, the average cooling rates at around 70°C (Piccarolo et al., 1992a, Brucato et al., 2002, Brucato et al., 2009) has been chosen, as the parameters characterizing the cooling effectiveness for that polymer. Although this is a semi-quantitative measure of cooling effectiveness, the whole thermal history is available to compare experimental results with predictions from non isothermal kinetic models (Piccarolo et al., 1992a, Brucato et al., 1991a). Furthermore, if the kinetic constant vs. temperature relationship is mapped to the temperature vs. time profile, it is clear that an underestimate of the effective cooling rate is obtained only at low cooling rates. With an exponential temperature decay most of the solidification takes place around the maximum of the kinetic constant, i.e. in the chosen temperature interval.

2.1.2 Temperature distribution in the polymer sample

The solution of **equation (3)**, introducing the dimensionless temperature of the Cu-Be slab Θ_{Cu} with boundary conditions $T=T_i$ for $t=0$ and constant heat exchange coefficient h, is:

$$\Theta_{Cu} = \exp(-t / \tau_l) \qquad (4)$$

where T_i and T_0 are the initial and final temperatures respectively.

If sample thickness is very small compared to that of the slab, **equation (4)**, representing the time dependence of the slab temperature (i.e. the temperature at the sample surface), becomes an exponential decay equation with a time constant defined by **equation (3)**. Furthermore, in the case of very high cooling rates, this dependence of temperature on time extends to high temperatures. The smallest characteristic times are then obtained in the largest temperature range.

An estimate of the temperature profile in the polymer sample under these cooling conditions is, therefore, conservative and may well provide a case for achieving the maximum cooling rates with this technique, aiming to achieve a homogeneous thermal history throughout the entire sample thickness. As it has been previously pointed out, this condition must be satisfied in order to devise a direct relationship between structure obtained and the associated thermal history.

The temperature distribution in the solid polymer sample can well be approximated by the Fourier equation for transient heat conduction within a medium of constant thermal diffusivity, i.e.

$$dT_{pol} / dt = \alpha \times \partial^2 T_{pol} / \partial x^2 \tag{5a}$$

Or, in dimensionless form, i.e.

$$d\Theta_{pol} / dFo = \partial^2 \Theta_{pol} / \partial \xi^2 \tag{5b}$$

Where: $\xi = x / l$, dimensionless half depth; l =slab half depth; $Fo = \alpha t / l^2$ =Fuorier number;

With boundary conditions:

1.   When $Fo = 0$ then $\Theta_{pol} = 0 \ \forall \xi$ (flat temperature profile before cooling);

2.   For $\xi = 0$, $\partial \Theta_{pol} / \partial \xi = 0 \ \forall Fo \geq 0$ (symmetry.)

The cooled wall boundary condition is an exponential decay of temperature according to experimental observation:

3.   For $\xi = 1$, $\Theta_{pol} = \phi(Fo) = \exp(-t / \tau) \ \forall Fo \geq 0$ (τ = exponential time constant, s)   (6)

However, **Equation (5.b)** neglects the heat generated by the latent heat of crystallization. An analytic solution of **equation (5.b)** with the boundary conditions given by **equations (6)** is provided in some texts (Luikov, 1980), i.e.

$$\Theta_{pol}(\xi, Fo) = \frac{\cos(\xi \cdot \sqrt{Pd})}{\cos \sqrt{Pd}} \cdot \exp(-Pd \cdot Fo) + \sum_{n=0}^{\infty} \frac{2\cos[(2n+1)\pi/2 \cdot \xi]}{(\pi/2 + n\pi)\sin[(2n+1)\pi/2]} \exp(-\mu_n^2 \cdot Fo) \tag{7}$$

With $Pd = l^2 / \alpha \cdot \tau$, *Predvotitelev* number (dimensionless time constant) and $\mu_n = \pi/2 + n\pi$.

A similar analytic solution is also provided in other texts (Carslaw & Jaeger, 1986), but (probably due to a misprint) any attempt to use the reported solution has failed.

Prediction of temperature profiles for an iPP slab (Brangrup & Immergut, 1989, van Krevelen, 1972) cooled with an exponential decay from T_i=230°C to T_0=5°C, are summarized for the case of two sample thicknesses (0.2 and 0.1 mm) in **figs. 6** and **7** respectively. The smallest time constant, $\tau_l = 0.05s$, corresponding to the fastest experiment performed, is considered. While diagrams **a)** of **figs. 6** and **7** show the calculated temperature distribution across the thickness (only half sample is considered), diagrams **b)** shows the calculated dependence of cooling rate on temperature at different sample depth along the thickness

Solidification of Polypropylene Under Processing Conditions – Relevance of Cooling Rate, Pressure and Molecular Parameters

199

direction. One can observe that for a sample thickness of 0.1 mm (**fig. 7**) the temperature distribution is almost flat across thickness. Significant deviations on the cooling rate versus temperature dependence are observed only at temperatures significantly higher than the range of solidification for most polymers.

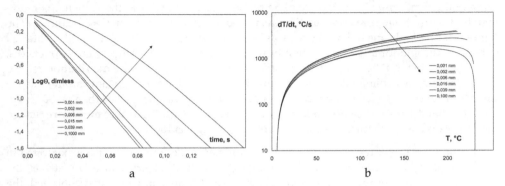

a

b

Fig. 6. iPP film (half thickness=100µm) cooled from 230 to 5°C with an exponential decay with time for characteristic time η=0.05s. a) calculated temperature distribution across the thickness; b) calculated cooling rate vs. temperature at different sample depth

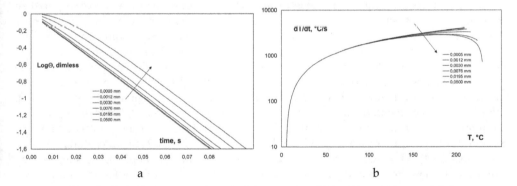

a

b

Fig. 7. iPP film (half thickness=50µm) cooled from 230 to 5°C with an exponential decay with time for characteristic time η=0.05s. a) calculated temperature distribution across the thickness; b) calculated cooling rate vs. temperature at different sample depth

Although this result may appear to be in contradiction with the constraint expressed by **equation (1b),** the analysis of the regimes involved in transient heat conduction, reported in advanced textbooks (Isachenko et al., 1987), provides a consistent explanation. When a solid is suddenly exposed to a coolant kept at a constant temperature T_0, the temperature profile could experience two regimes: an initial one corresponding to dissimilar temperature profiles, and a second one, called "regular", whereby the temperature profiles are almost parallel to each other and self similar at different times. Depending on the *Biot* number the second regime may also not take place and the condition *Biot*<10 determines the onset of the second regime controlling the transient heat conduction for most of the cooling time. The condition expressed by **equation (1b)** may thus be seen to be more restrictive than it is necessary, determining that the flat temperature is the controlling factor for most of the time during cooling of the solid from T_i to T_0.

In the regular regime of transient heat conduction, the onset of almost parallel temperature profiles determines a condition by which at different times the slope of the profile is the same in different sample positions, leading to the same cooling rate at the same temperature and to a correct interpretation of the calculated results reported in **figs. 6** and **7**. The temperature profile may thus be seen as a perturbation propagating from the external surface to the interior as the calculation of **figs. 6** and **7** shows. For larger sample thickness, as in **fig. 6**, although the temperature profile is not flat, the temperature distribution regime is regular and the cooling rate at lower temperatures is still almost constant throughout the sample. This is not so for cooling rates evaluated at temperatures higher than ca 80°C. On the other hand, for small sample thickness, as is shown in **figs. 7**, the heat transfer regime in the sample is always regular even at temperatures as high as 180°C.

As for the influence of the latent heat on the temperature distribution, which is neglected in **equation (5.b)**, one can observe that, although the heat of crystallization affects the temperature profile of the sample, and/or the thermal history to which it is subjected, the overall effect is only moderate. Indeed an estimate of the increase in sample temperature due to latent heat of crystallization, with the assumption that heat release takes place adiabatically, only produces maximum values of about 40°C, if the polymer sample crystallizes to the maximum allowable extent. Although this value may seem large when compared to the effect of temperature on the crystallization kinetics, it must be remembered that at low cooling rates, crystallization takes place over longer time intervals, and consequently does not affect appreciably the temperature of the sample since the heat is being released slowly. With respect to adiabatic conditions, a smaller temperature increase will, indeed, take place during cooling. Furthermore, the "heat sink" effect played by the metal slabs on the polymer film makes the temperature increase negligible. At high cooling rates, on the other hand, the heat is released in a shorter time interval, however in this case the temperature of the sample is controlled by very high heat fluxes and, consequently, the temperature is not affected very much either (Brucato et al., 2002, Brucato et al., 2000, La Carrubba, 2001). Moreover, if very drastic cooling conditions are applied, the sample only experiences a low degree of crystallization and, therefore, releases smaller amount of heat, which affects the temperature even less.

2.2 Rapid cooling under pressure

In order to evaluate the combined effect of typical injection molding pressures and temperatures, a new equipment was designed as a natural extension of the previously described apparatus. A standard Negri Bossi NB25 injection moulding machine was used as a source of molten polymer supplied at a pre determinable and maintainable pressure at which the polymer can be injected into a preheated mould cavity.

A special injection mould has been designed such that samples could be cooled at a known cooling rate and under a known pressure (Brucato et al., 2002, Brucato et al., 2000, La Carrubba, 2001). This heated mould consists of a conical cavity (the sprue), which is located in the fixed platen of the injection molding machine, coupled to a "diaphragm". The front of the cavity is sealed with a high tensile, high thermal conductivity copper-beryllium

"diaphragm", which is spray cooled on the opposite side when the quench starts. A
schematic representation of the apparatus is shown in **fig. 8**. The cavity is located within a
brass block where eight cartridge heaters with a total power of 2 kW are inserted. The
diaphragm is located in the moving platen of the machine. The whole apparatus (cavity and
Cu-Be diaphragm) has been designed in such a way that it can be placed in and removed
from an injection moulding machine as a normal injection mould tool. A cooling channel,
which allows the diaphragm to be spray cooled by pressurized water (at 8 bars) on one side,
is also located in the mobile part. A thermocouple (type E, diam.=0.05 mm) inserted inside
the diaphragm close to the wall facing the polymer sample records the thermal history
during cooling, whilst a pressure sensor (type Dynisco PT46) mounted in the cavity allows
measurement of pressure during the experiment. The pressure sensor and the thermocouple
are connected to a data acquisition system, constituted by a National Instrument card LAB-
LC coupled with an Apple–Macintosh LC computer.

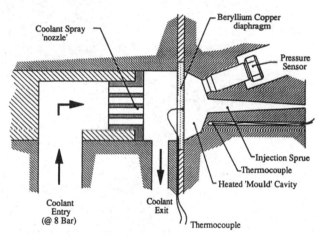

Fig. 8. Apparatus for solidification under pressure fitted to the modified injection molding
machine

The experimental methodology of recording the thermal history experienced by the surface
of rapidly cooled samples and then analyzing the resulting sample morphology has been
adopted. Using the above described configuration, a thin layer in contact with the
diaphragm solidifies under a known recorded thermal history and under a constant
recorded pressure history. Internal layers of the polymer are cooled with different cooling
rates, which can be calculated by solving the transient heat transfer equation (7). In order to
relate thermal history to the structure formed, the relationship between cooling rate
evaluated at 70°C (characteristic temperatures of iPP) and depth in the sample can be
calculated based on the conduction heat transfer problem (**eq. 7**), as shown in **fig. 9**. This is a
sort of "transformation function" or "mapping function", which converts the depth in the
sample in an equivalent value of cooling rate, thus enabling the physical data to be mapped
as a function of cooling rate rather than of the sample depth. This transformation functions
allow the effect of pressure superimposed to that of cooling rate to be properly identified
and quantified. Thin slices (50 to mm) microtomed across a direction parallel to the cooled

surface are then used for post-solidification characterization methods (Brucato et al., 2000, La Carrubba, 2001), being each slice characterized by a well defined cooling rate (averaged across slice thickness).

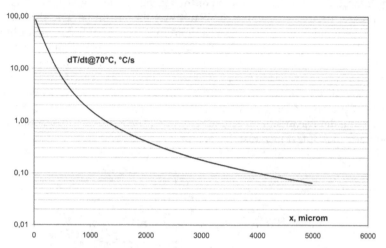

Fig. 9. Depth-Cooling Rate "mapping function"

3. Materials and characterization

Several iPP grades were analysed, with the aim of encompassing a wide latitude of crystallization behaviour and to highlight the influence of molecular parameters on iPP crystallization. The main features of the different grades of iPP tested are listed in **Table 1**.

Material name	Mw	Mwd	Xs[(*)]	notes
HPB	430000	6.6	2.9	
M2	208000	3.5	4.5	
M6	391000	5.6	4.6	
M7N	379000	5.3	3.4	+ Talc 1000 ppm
M9	380000	3.8	5.0	Copolymer 0.5% ethylene
M12	252000	5.4	13.9	
M14	293000	7.3	5.2	Copolymer 3.1% ethylene+DBS
M16	293000	7.3	5.2	Copolymer 3.1% ethylene
iPP1	476000	6		
iPP2	405000	26		bimodal MWD
iPP3	489000	9.7		
iPP4	481000	6.4		

Table 1. Main Characteristics of the iPPs examined. [(*)]Xylene soluble weight percentage

Since cooling rate in the present devices is too fast for recording any macroscopic change
during the solidification process, only the final structure of the solidified sample was
evaluated. The final features of the samples, analyzed by suitable macroscopic probes, such
as powder Wide Angle X-Ray Diffraction (WAXD) patterns and density measurements were
related to thermal history.The X-ray diffraction measurements have been performed by a
Philips vertical diffractometer equipped with a Philips PW1150 generator. The Cu-Ka Nickel
filtered radiation was detected in the interval 6-45°, applying steps of 0.05° in the interval
10-35° and of smaller steps of 0.2° elsewhere with a counting time of 60 s per step
throughout. The gradient column technique was used for density measurements.

4. Results

4.1 Crystallization of iPP at atmospheric pressure

The results of the correlation for iPP3 between the structural features of quenched samples,
assessed through the macroscopic probes cited above (WAXD and density), and thermal
history, identified by the relevant cooling rate in the range of temperatures where the polymer
solidifies (70°C for iPP), are shown in **figs. 10a** and **b** for density and WAXD patterns
dependence on cooling rate, respectively. Such results point out the features of the proposed
method of characterization already reported by the authors (Piccarolo, 1992, Brucato et al.,
1993), with respect to change of structure and density with cooling rate. A broad range of
density were identified as well as extreme structural features in the WAXD patterns. The
WAXD patterns reported in **fig. 10b** show that at low cooling rates only the stable a
monoclinic phase is observed with small amounts of the β phase, as identified by the reflection
at 2θ=16.1°. The crystalline order, determined by the width of the peaks, continuously
decreases on increasing cooling rate up to a point where only two broad diffraction peaks are
observed, showing the presence of the so called mesomorphic phase of iPP (Corradini et al.,
1986, Guerra et al., 1990). At intermediate cooling rates, the coexistence of the two phases is
revealed, over a narrow range of cooling rates, by the superposition of the two broad peaks of
the mesomorphic phase and the faint residues of the peaks related to the a monoclinic phase.

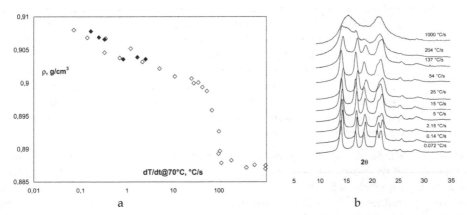

Fig. 10. a. Density dependence on cooling rate (measured at 70°C) of iPP3. Open symbols:
rapid cooling experiments; filled symbols: standard constant cooling rate experiments (DSC)
b. Dependence of WAXD powder patterns of iPP3 on cooling rate (measured at 70 °C)

Furthermore, changes of the WAXD patterns agree with the density measurements, making the two methods consistent and comparable. Although a qualitative cross check can be made for the data in **fig. 10b**, a quantitative comparison can only be obtained by WAXD deconvolution (Martorana et al., 1997). This last has been extensively used to determine the phase content and its dependence on cooling rate. This dependence of phase content on cooling rate, in turn, has been used for the determination of non-isothermal crystallization kinetics. The model adopted was based on the crystallization kinetics of two phases competing for the transformation from melt to solid (Brucato et al., 1998, Piccarolo et al., 1992a, Brucato et al., 1993). The reason for two parallel crystallization mechanisms stems from the WAXD patterns dependence on the cooling rate. In the case of iPP, for example, the patterns show that the stable phase disappears while the mesomorphic phase content increases with increasing cooling rate.

The density versus cooling rate curve of **fig. 10.a** shows three zones characterized by different features related to the WAXD based phase content dependence on cooling rate reported in **fig. 11**:

i.   At low cooling rates where only stable phases are formed, the density decreases to a small extent with the log of the cooling rate. Below ca 5°C/s a slight decrease of density is observed, related to the formation of small amount of β phase formed (Piccarolo, 1991).

ii.   At the highest cooling rates, a low-density plateau is observed related to the mesomorphic phase set-in, since a limiting packing condition has been approached. The nature of the mesomorphic phase is not well known, the most acknowledged hypothesis being a packing very similar to the α-monoclinic phase but with a low range order (Corradini et al., 1986). The most significant feature of this phase, indeed, is that it transforms to the stable α-monoclinic phase upon ageing, which is relevant for the post processing behaviour of iPP. Previous studies point out that the kinetics of this transformation to be measurable only above 80°C (Struik, 1978). More recent annealing experiments, discussed elsewhere (Gerardi et al., 1997) show that such transformation can take place at much lower temperatures and can cause significant density changes.

iii. In an intermediate cooling rate range the material density shows a very high sensitivity to changes in cooling rate. In this zone the stable phase content decreases while that of the mesomorphic phase increases as the cooling rate is increased. This transition is strongly dependent on the material characteristics, e.g. nucleating agents and molecular weight (Sondegaard et al., 1997). Solidification under these intermediate cooling rates shows the effect of the competition between the a monoclinic and the mesomorphic phases in the transformation from melt to solid. The slope of the density curve in this region is a measure of the sensitivity of the crystallization kinetics towards the cooling rate for a given polymer.

To sum up, although the mapping of the structural features provides a general understanding of the relationship between the thermal history and the associated structure formed during a quenching experiment, the density dependence on cooling rate provides an immediate, quantitative information on the non isothermal crystallization behaviour of the polymer. In this respect the identification of the narrow range of cooling rates at which the transition from a monoclinic to mesomorphic phase takes place provides quantitative information on the material non-isothermal crystallization kinetics.

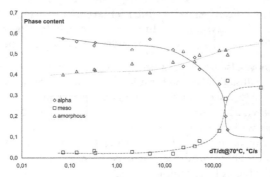

Fig. 11. WAXD deconvolution of iPP3. Phase content vs. cooling rate (measured at 70 °C)

A model-based interpretation of such transitional cooling rates performed on the crystallization kinetics parameters has been published recently (La Carrubba et al., 2002a).

4.2 Crystallization of iPP under pressure

The results of the density measurements made on iPP4 samples solidified under pressure are reported in **figs. 12.a** and **b**. Four different pressure conditions have been explored: 0.1 MPa, 8 MPa, 24 MPa, 40 MPa using two different diaphragm size 3.5 mm and 8 mm thick, (see **fig. 8**). In **fig. 12.a** is shown the density depth profile for the 3.5 mm thick diaphragm, and in **fig. 12.b** the density depth profile for the 8 mm thick diaphragm. Samples obtained with the 3.5 mm thick diaphragm were subjected to an experimental surface cooling rate (measured at 70°C -Brucato et al., 2002) of about 100°C/s. Samples solidified using the 8 mm thick diaphragm experienced a surface cooling rate of about 20°C/s. It is worth noting that the surface cooling rate depends on the coolant heat transfer and on the diaphragm thermal inertia. Changing the diaphragm thickness is, indeed, a simple and reliable way to tune the surface cooling rate (La Carrubba, 2001).

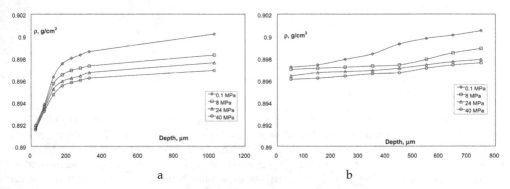

a   b

Fig. 12. iPP4 Density depth profile for different solidification pressures from 0.1 to 40 MPa: a. diaphragm 3.5 mm thick; b. diaphragm 8 mm thick

The curves in **fig. 12a** and **b** show that for both experiments and at all pressures, density increases from the surface to the bulk of the sample. This behavior can be related to the expected increase in crystallinity since the internal layers are cooled at progressively lower rates. In **fig. 12a** and **b** it is shown that the highest level of the density increase takes place in the vicinity of the surface, and that this is independent of the applied pressure.

Both **figs. 12a** and **b** show also somewhat unexpected results, in so far as a density decrease with pressure occurs at all depths in the sample. The reduction of density due to pressure is minimum at the sample surface and grows with depth. Furthermore, the majority of the density change is observed by varying the pressure from 0.1 to 8 MPa, which is quite low especially if compared with the typical pressure values used in injection moulding. This experimental result may be relevant for modeling the shrinkage and the internal stress distribution in injection molded products. Particularly important is the fact that this effect is more pronounced in the bulk of the sample.

Fig. 13 is obtained by plotting the density data in **figs. 12a** and **b** against the cooling rate calculated at 70°C by using a transient heat conduction model (equation (7). The value of the calculated cooling rate was averaged across every slice (50μm thick). The use of the transient model was also validated by overlapping the data referred to a surface cooling rate of 100 and 20 °C/s. In **fig. 13** it is also shown that at constant cooling rate the final density decreases with pressure. The same trend is obtained with respect to cooling rate, indicating that the density drop above 10-20 °C/s is independent of the solidification pressure. Finally, **fig. 13** shows that the decrease of density with pressure vanishes with increasing cooling rate, implying that the influence of pressure is more pronounced in the bulk of the sample. This is a very important information in process simulation.

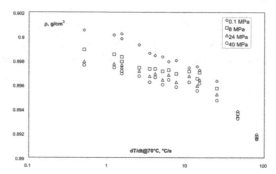

Fig. 13. iPP4 Density versus cooling rate evaluated at 70°C for different solidification pressures from 0.1 to 40 MPa

A similar pressure dependence of the density has also been observed by He and Zoller (He and Zoller, 1994) using a standard dilatometer and measuring the specific volume of this sample during crystallization from the melt. A constant slow cooling rate (2.5 °C/min) under constant pressure was used, bringing the sample back to a fixed pressure at the end of the test. It is worth noting that the majority of experiments have provided information on specific volume under pressure, whereas in our work we have measured the density at ambient pressure after solidification under pressure. He and Zoller observed an increase of specific volume with increasing crystallization pressure in the case of iPP, i.e. the specific volume in the solid phase at the end of the solidification curve is slightly higher than the one measured at the beginning of the melting curve. This behavior demonstrates that during

Solidification of Polypropylene Under Processing Conditions – Relevance of Cooling Rate, Pressure and
Molecular Parameters

207

solidification under pressure some structural transformations take place giving final lower density values. Although He and Zoller have attempted to explain the reduction in density with the formation of the γ phase, which is less dense than the α phase, the samples in our study did not show any evidence of the presence of the γ phase (La Carrubba et al., 2000). The experiments performed by He and Zoller are in agreement with our results, which show that the final density (measured at atmospheric pressure) of samples solidified under pressure is, in fact, lower than that of the samples solidified at atmospheric pressure. We have repeated the PVT measurements on iPP and have published the results in a recent paper (La Carrubba et al., 2002b), where a comparison between specific volumes of samples crystallized at different pressures and/or cooling rates has revealed a decrease in density with increasing cooling rate and pressure. Thereafter, WAXD experiments were performed on slices cut in the transverse direction with respect to the direction of the heat flux. All experiments were performed by the synchrotron radiation source of the DESY center in Hamburg. A very long accumulation time (5 frames of 1 minute) was applied in order to achieve statistically significant results and a good reproducibility.

A qualitative analysis of the diffraction patterns has lead to the conclusion that the alpha phase content decreases on increasing cooling rate, for all the adopted pressures used (Brucato et al., 2000, La Carrubba et al., 2002a, La Carrubba et al., 2000). The data have shown that increase in pressure decreases the alpha phase content. This is better shown by WAXD data after a deconvolution procedure that has already been discussed elsewhere (Martorana et al., 1997). The program employed (implemented on MATLAB) uses a best-fitting procedure to calculate the positions and the intensity and of the alpha phase including mesomorphic phase peaks and that of the amorphous halo.

In **fig. 14** are shown plots of the phase content of the samples against pressure at four different values of cooling rates, ranging from 1.5 to 80 °C/s. A decrease of alpha phase is noticed, which is in agreement with the data from density and micro hardness measurements, showing that the change in the alpha phase content with pressure is highest within the first 10 MPa. By examining the data in **fig. 14** one notes that the decrease of the alpha phase with pressure tends to vanish when the cooling rate increases, particularly for cooling rates above 20°C/s. Additionally, the reduction of the alpha phase is mostly balanced by an increase of the mesomorphic phase content while the amorphous phase seems to be only slightly affected by pressure.

This last point is also very relevant, in so far it shows that as the main effect of pressure is to replace the alpha phase by the mesomorphic phase, leaving almost unaffected the amorphous content. It has been already shown, in fact, that the main effect of increasing the cooling rate at ambient pressure is the substitution of the alpha phase with the mesomorphic one (Piccarolo, 1992, La Carrubba et al., 2002a). In other words, the qualitative effect of pressure (at a constant cooling rate) on the final structure appears to be the same as an increase of cooling rate alone at a constant pressure (La Carrubba et al., 2000). This results is also illustrated in **fig. 15**, reporting the phase fraction, as calculated from the WAXD deconvolution of the samples, crystallized at atmospheric pressure using the rapid quenching apparatus. One can easily see how the decrease of the alpha phase with increasing cooling rate is accompanied by an increase of the mesomorphic phase (**fig. 14**). This observation is a further confirmation of the possibility to adopt a master curve approach to describe the bahaviour of iPP under pressure and high cooling rates.

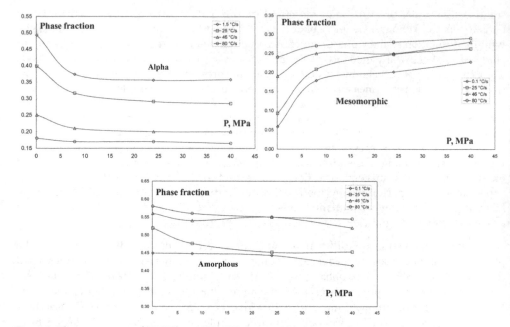

Fig. 14. Phase content of iPP4 from WAXD deconvolution as a function of pressure

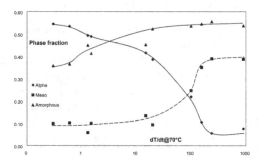

Fig. 15. WAXD deconvolution of iPP4 at 0.1 MPa, showing the phase relative content

4.3 Crystallization kinetics model

When dealing with crystallization of iPP, the numerous crystalline modifications of this material must be accounted for, since α, β or γ crystals may form upon solidification from the melt. The resulting complex frame can be simplified based on some experimental evidences, supported by several references (Foresta et al., 2001, Nagasawa et al., 2005, Raab et al., 2004, Marigo et al., 2004). As for the β phase, it basically shows up only if specific β nucleants are added, therefore for commercial non-b-nucleated iPP's it does not form; traces of γ form crystals are often present, but always in minor amount and in a narrow window of operating conditions (i.e. cooling rates), hence its presence is neglected without affecting the reliability of the results.

Solidification of Polypropylene Under Processing Conditions – Relevance of Cooling Rate, Pressure and Molecular Parameters

209

Under the aforementioned hypotheses, as two different crystalline phases are formed (α and mesomorphic), at least two kinetic processes take place simultaneously. The simplest model is a parallel of two kinetic processes non-interacting and competing for the available molten material. The kinetic equation adopted here for both processes is the non-isothermal formulation by Nakamura et al. (Nakamura et al., 1973, Nakamura et al., 1972) of the Kolmogoroff Avrami and Evans model (Avrami, 1939, 1940, 1941, Evans, 1945).

The model is based on the following equation:

$$X(t) / X_\infty = 1 - \exp\left[-E(t)\right] \tag{8}$$

Where $X(t)$ and X_∞ are the crystallized volume fraction at time t and in equilibrium conditions, respectively. For simplicity and for the sake of generalization X_∞ is here assumed to be a material constant, although it has been reported its dependence upon the crystallization history (crystal size distribution and degree of perfection, Ziabicki, 1976).

$E(t)$ is the expectancy of crystallized volume fraction if no impingement would occur. A different formulation of the model can be easily obtained by differentiation of equation (8):

$$d\xi / dt = (1 - \xi)\dot{E}(t) \tag{9}$$

Where:

$$\xi = X(t) / X_\infty \tag{10}$$

Since in the case of interest two crystalline phases develop, the simplest extension of the present model is to assume that those phases grow independently in parallel, competing each other for the residual fraction of available melt. Under this hypothesis the rate equation, for the general case of m crystalline phases developing simultaneously, becomes:

$$d\xi_i / dt = (1 - \sum_i \xi_i)\dot{E}_i(t) \qquad \text{for i= 1...m} \tag{11}$$

The following function, suggested by several authors (Ziabicki, 1976, La Carrubba et al., 2002a), can be adopted for the expression of the time derivative of the expectancy, leading to a rate equation proportional to the fraction of untransformed material times the current value of the kinetic constant, in which nucleation and growth rates have been lumped together (nucleation and growth are therefore isokinetic):

$$d\xi_i / dt = (1 - \sum_i \xi_i) n_i \ln 2 \left[\int_0^t K_i(T) ds\right]^{n_i - 1} K_i(T) \qquad \text{i= 1...m} \tag{12}$$

The form adopted in equation (12) for the time derivative of the expectancy reduces to the classical *Avrami* form, with a dimensionality index n_i for the ith phase, if an isothermal experiment is considered. As for the dependence of the rate constant K_i on temperature, the simplest expression that one can consider is a Gaussian shaped curve:

$$K_i(T) = K_{0,i} \exp\left[-4\ln 2(T - T_{\max,i})^2 \cdot D_i^{-2}\right] \qquad i = 1 \ldots m \qquad (13)$$

where D_i, $T_{\max,i}$ and $K_{0,i}$ are the half width, the temperature where the maximum of K_i is attained and the maximum value of K_i itself, respectively (Ziabicki, 1976).

The governing equations with reference to two phases (alpha and mesomorphic phase) are:

$$d\xi_\alpha / dt = (1 - \xi_\alpha - \xi_m)n_\alpha \ln 2\left[\int_0^t K_\alpha(T)ds\right]^{n_\alpha - 1} K_\alpha(T) \qquad (14)$$

$$d\xi_m / dt = (1 - \xi_\alpha - \xi_m)n_m \ln 2\left[\int_0^t K_m(T)ds\right]^{n_m - 1} K_m(T) \qquad (15)$$

Where α and m indices stand for the alpha monoclinic and the mesomorphic phases respectively. This system of two coupled ordinary differential equations can be integrated with the appropriate initial conditions ($\xi_\alpha = \xi_m = 0$ for $t=0$). The integration leads to crystallinity development with time under any temperature history.

Fig. 16a shows a typical $K(T)$ curve for the two different phases, and **fig. 16b** outlines the influence of two main parameters, the product of K_0*D (nearly the area under the $K(T)$ curve sometimes called crystallizability, Ziabicki, 1976) and the Avrami index n. This latter is representative of the sensitivity of the crystallization kinetics to the cooling rate, a larger n leads in fact to a faster dependence of final crystallinity on cooling rate, the curves crossover is however always the same, i.e. about one half of the maximum attainable crystallization at an abscissa of K_0*D. The crystallizability is a cooling rate scaling factor of crystallization kinetics; as a matter of fact, a larger value of K_0*D leads to a shift along the abscissa of the curve, i.e. along the cooling rate, such that the larger the crystallizability the more pronounced the material tendency to crystallize.

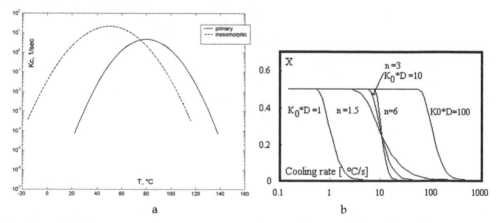

Fig. 16. a. Kinetics constant versus temperature for a and mesomorphic phase; b. Crystallinity volume fraction as a function of cooling rate for various values of n (Avrami index) and K0*D (crystallizability).

Solidification of Polypropylene Under Processing Conditions – Relevance of Cooling Rate, Pressure and
Molecular Parameters

211

Before discussing the results obtained concerning the sensitivity of the cited parameters on polymer composition, it is worth to point out the intrinsic limitations of the approach adopted related to its empirical nature. They depend on the origin of the KAE equation describing the nucleation and growth without diffusivity constraints and without accounting for the possible non isokinetic contribution of each mechanism, with a simple mathematical extension to the non isothermal conditions and finally without accounting for the complexity of crystallization in polymer melts, clearly a multistage process (Strobl, 1997). A slightly different modelling is represented by the so-called "Schneider rate equations" (Schneider et al., 1988); Schneider et al underline that their approach consists in an application of Avrami's (and Tobin's) impingement model leading to a different mathematical and more easy-to-handle formulation, based on a set of differential equations instead of dealing with integral equations. In other words, although their formulation enhances the applicability to process modelling, the physics behind it is completely described by the Avrami model. Therefore the use of Schneider's approach is more advisable when dealing with "non-lumped" problem, to be solved by coupling of transport equations.

All things considered, an analysis of the literature studies on polymer crystallization kinetics shows that the isokinetics approach is the most widely adopted (see the recent review by Pantani et al – Pantani et al., 2005); moreover, the limitations imposed by the isokinetic hypothesis do not weaken the self consistency and the abundance of information here provided. In any case, the limits of the aforementioned approach can be overcome by recalling the original Kolmogoroff's model (which accounts for the number of nuclei per unit volume on spherulitic growth rate) and determining the average radius of spherulites based on geometrical considerations (i.e. counting the number of nuclei), as shown by Zuidema et al. (Zuidema et al., 2001) and Pantani et al. (Pantani et al., 2002). This approach has however some limitations since it can be applied only to conditions where a recognizable spherulitic morphology is formed thus either low cooling rates or conditions where the spherulites are dispersed in a non crystalline matrix as in the case of mesomorphic iPP phase (Piccarolo, 1991). This possible refinement of the analysis is however far beyond the scope of the present work both due to its limitations and to the macroscopic approach adopted aiming to describe crystallization kinetics parameters in the broadest possible range of quiescent solidification conditions, i.e. under conditions emulating, but for the role of orientation and pressure, polymer processing.

4.4 Density data and crystallization kinetics model parameters for various iPPs

Figs. 17.a and b and 18.a and b show a comparison of the density dependence upon cooling rate for the iPP grades studied, whereas **Table 2** reports the crystallization kinetics model parameters calculated by a best fitting procedure not only on the basis of the final monoclinic and mesomorphic content of the quenched samples, taken from the deconvolution of the WAXD patterns, but also accounting for results which provide the time and the temperature at the maxima of the crystallization rate (isothermal tests and DSC measurements) respectively. For this purpose a multiobjective optimization code was adopted.

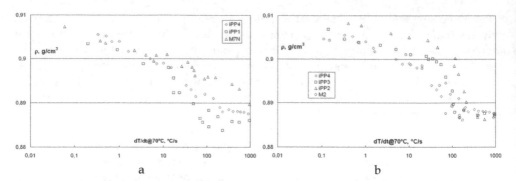

Fig. 17. a.Effect of nucleating agents b.Effect of molecular weight distribution onto the density versus cooling rate behavior.

It should be noticed that **Table 2** reports for the mesomorphic phase a range of values both for the Avrami index n and for X_∞. The uncertainty in those parameters is however not critical for the purpose of the present work. As a matter of fact, a variability of n between 0.4 and 0.5 reflects into very slight changes in the temperature field in which crystallization takes place; consequently, the influence of this parameter is of minor entity. As for X_∞ of the mesomorphic phase, although its variability could turn into larger changes in the crystallization temperature window, its influence is confined to a cooling rate region in which the crystallization of the alpha phase is very little (very high cooling rates), thus not affecting the alpha phase kinetic parameters.

Material	monoclinic						meso					
	K_0, sec^{-1}	T_{max}, °C	D, °C	n	X_∞	K_0*D, °Csec^{-1}	K_0, sec^{-1}	T_{max}, °C	D, °C	n	X_∞	K_0*D, °Csec^{-1}
HPB	1.6	82	28	2	0.55	44.8	1.6	57	19			30.4
M2	1.4	77.3	33	3.0	0.60	46.2	0.6	40	30			18
M12	2.5	79	30	2.0	0.51	75	3.3	42	31			94
M7N	8.0	82	30	2	0.48	240	n.a.	n.a.	n.a.			n.a
M9	2.0	70	36	3.0	0.53	102	2.0	40	36			72
M6	2.4	66	40	3.0	0.54	96	2.0	40	40	0.4 - 0.5	0.45 - 0.55	80
M14	40	72	29	3.0	0.40	1160	1.0	40	34			34
M16	1.8	71	33	3.0	0.50	99.4	1.0	40	34			34
iPP1	1.6	82	28	2	0.55	44.8	1.6	57	19			30.4
iPP2	3.5	73	34	2	0.50	255.5	1.5	40	31			60
iPP3	2.7	70	35	2	0.57	189	0.22	40	40			8.8
iPP4	4.5	85	27	2	0.45	121	0.27	53.5	33.8			9.12

Table 2. Crystallization Kinetics Parameters

Solidification of Polypropylene Under Processing Conditions – Relevance of Cooling Rate, Pressure and
Molecular Parameters

213

As for the Avrami index of the crystalline alpha phase, **Table 2** reports values equal to 2.0 or 3.0, due to a slight round-off with respect to the results obtained via simulation. The Avrami index is here intended as a mere fitting parameter, in line with most of the literature concerning polymer crystallization kinetics (see for instance the review of Pantani et al. – Pantani et al., 2002 and the review of Di Lorenzo et al. - Di Lorenzo & Silvestre, 1999), although its exact physical meaning should indicate the dimensionality of growth (namely 3 or 4 for volume filling depending whether nucleation is predetermined of sporadic). In other words, the Avrami index points out only the sensitivity of the crystallization kinetics to the cooling rate, a larger n leads in fact to a faster dependence of final crystallinity on cooling rate, the curves crossover being always the same, i.e. about one half of the maximum attainable crystallization at an abscissa of K_0*D.

Table 2 shows that differences in materials do not appear to be related in a simple way to kinetic parameters. This may be due to the fact that the set of materials investigated in this work, since representative of iPP's of industrial use, does not cover a wide range of fundamental molecular parameters Mw and Mwd. As a matter of fact, the limited range of the molecular parameters here explored probably does not comply with a complete enlightenment of the role played by each single factor onto the crystallization behaviour.

Nevertheless, some information can be drawn from the table summarizing material kinetics behaviour. For example, the so called "crystallizability", i.e. the product K_0*D instead of the two separate kinetic parameters, allows one to discuss the differences in the non isothermal crystallization behaviour in relationship to the materials investigated in this work. The crystallizability, roughly corresponding to area under the kinetic constant curve versus temperature, has the dimension of a cooling rate, and indicates somehow the ability of the polymer to crystallize (Ziabicki, 1976). A comparison of crystallizability values gives a good insight into the influence of molecular parameters on the crystallization kinetics behaviour. For instance, referring to the monoclinic phase only, it can be observed that the smallest value of K_0*D was obtained for the sample without additives having the highest Mw and narrowest Mwd. The highest values of crystallizability are however observed for nucleated iPP's (M7N and M14). All things considered, it should be however underlined that differences in crystallizability below a factor 2-3 cannot be considered reliably assessed by the crystallization kinetics method, due to the intrinsic errors in the evaluation of both K and D throughout the best fitting procedure. If one looks at **fig. 17a**, reporting density as a function of cooling rate for three polymers having similar features (molecular mass and distribution) except for the presence of nucleating agents, one comes to the conclusion that the presence of nucleants shifts the density cut-off towards larger values of cooling rate; as a matter of fact, the calculated crystallizability of the iPP denominated M7N (strongly nucleated) results larger than the one of iPP1 and iPP4.

Higher values of crystallizability are observed when the molecular weight distribution is broader (see for instance materials iPP2 and iPP3). This behaviour is clearly shown in **fig. 17b**, where four polymers with Mwd ranging from 3.5 (M2) to 26 (iPP2) are reported. The observable shift of density cut-off towards larger cooling rate upon increasing Mwd is correctly accompanied by a parallel increase of crystallizability (see **Table 2**).

On the other hand, no direct and obvious correlation may be found to relate crystallizability to Mw. This apparent inconsistency can be reasonably explained by recalling the already mentioned low variability of molecular weights of the iPP grades investigated in this work, related to their "commercial" nature. Consequently, in the light of crystallization behaviour, all the iPP molecular weights listed in **Table 2** have to be considered rather similar, being their difference in molecular weight not sufficient to develop dissimilar crystallization kinetics. Addition of small amounts of ethylene units in the copolymer does not influence significantly any of the kinetic parameters mentioned above, the small changes of the product $K_0{}^*D$ mainly depending on the differences in Mwd and not upon the ethylene content. No significant differences in the product $K_0{}^*D$ may be argued between materials M6, M9 and M16, although the second couple is copolymerized with ethylene. Also the amount of ethylene used in the copolymerization process does not appear to be relevant. These results are confirmed by density data shown in **fig. 18a**.

On the contrary, the couple of nucleated materials (M7N e M14) that basically differ from the others for the addition of ethylene in the latter, show a large difference in the crystallizability, suggesting a synergetic effect of copolymerization with the addition of a nucleating agent on crystallization kinetics. Although the enhancement of chain mobility, which increases with ethylene content, and nucleation are both factors promoting the crystallization kinetics, the source of the synergy is not simple to interpret. The tacticity index does not seam to have a significant influence on the kinetics of monoclinic alpha phase. **Fig. 18b** shows that the density cut-off of M12, with a lower tacticity, is slightly anticipated with respect to the one of MP6; on the other hand, crystallizability of M12 is somehow lower than the one of M6 (see **Table 2**).

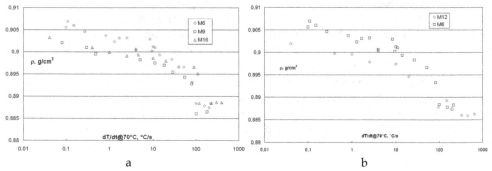

Fig. 18. a.Effect of tacticity b. Effect of ethylene content onto the density cooling rate behavior

Other kinetics parameters of the monoclinic phase are more difficult to be related to molecular parameters. Additionally, their physical meaning is not straightforward with the exception of Avrami index n. This last, in principle, represents the dimensionality of the growth and the kind of nucleation. Experiments, however, rarely well correlate with a value of n in line with the dimensionality of the crystallization process under observation.

Furthermore, the correlation of mesomorphic phase kinetics parameters appears difficult, probably this can be related to the fact that mesomorphic phase determinations are affected by a larger uncertainty due to the broader WAXD peaks characterizing this phase.

With this respect, some recent cooling experiments performed on a nanocalorimeter (De Santis et al., 2007) have shown two distinct crystallization peaks (alpha and mesomorphic phase) appearing in a quite large range of cooling rates, the crystallization of the alpha taking place up to ca 1000 °C/s. The apparent contradiction with the results here presented (alpha phase disappearing above 200-300 °C/s) may be consistently solved if one considers that the amount of alpha phase formed at high cooling rates is of the order of a few percent, hence below the measure limits of WAXD (around 5%). Secondly, being the sample mass undergoing the DSC cooling run in the nanocalorimeter of the order of a few ng, the enhancement of crystallization due to the "surface effect" (high constraints due to the low sample size with respect to the average radius of giration) must be taken into account. Thirdly, the presence of a mesomorphic phase crystallization peak at room temperature justifies the difficulties encountered in iPP amorphization, as confirmed by the present results where a consistent value of the crystallization kinetic constant of the mesomorphic phase at room temperature is shown (see **Table 2**).

Finally, if one considers that, with a few exceptions, the study was executed on a set of materials of industrial interest, a conclusion can be drawn about the fact that crystallization kinetics are mainly influenced only by the presence of nucleating agents. The influence of copolymerization on crystallization kinetics being relevant only if coupled with nucleation.

5. Conclusion

An experimental route for investigating polymer crystallization over a wide range of cooling rates (from 0.01 up to 1000 °C/s) and pressures (from 0.1 to 40 MPa) is illustrated, using a method that recalls the approach adopted in metallurgy for studying structure development in metals. Two typologies of experimental set-up were used; respectively an apparatus for fast cooling of thin films (100 to 200 mm thick) at various cooling rates under atmospheric pressure and a device (based on a on-purpose modified injection moulding machine) for quenching massive samples (about 1-2 cm³) under hydrostatic pressure fields.

In both cases ex-situ characterization experiments were carried out to probe the resulting structure, using techniques like density measurements and Wide Angle X-ray Diffraction (WAXD) patterns. The cooling mechanism and the temperature distribution across the sample thickness were analysed. Results show that the final structure is determined only by the imposed thermal history and pressure.

Experimental results of quiescent crystallization at ambient pressure for various grades of isotactic polypropylene (iPP) are reported, showing the reliability of this experimental approach to assess not only quantitative information but also a qualitative description of the crystallization behaviour. In order to thoroughly describe the crystallization kinetics as a function of molecular and operating parameters, the methodological path followed was the preparation of quenched samples of known cooling histories, calorimetric crystallization isotherms tests, Differential Scanning calorimetry (DSC) cooling ramps, Wide Angle X-ray Diffraction (WAXD) measurements and density determination. The WAXD analysis performed on the quenched iPP samples confirmed that during the fast cooling at least a crystalline structure and a mesomorphic one form. The diffractograms were analysed by a deconvolution procedure, in order to identify the relationship between the cooling history and the distribution of the crystalline phases. The whole body of results (including calorimetric ones) provides a wide basis for the identification of a crystallization model suitable to describe solidification in polymer-processing operations, based on the Kolmogoroff-Avrami-Evans non-isothermal approach.

A systematic investigation about the crystallization kinetics under cooling rates typical of polymer processing for several commercial isotactic polypropylene grades was carried out, aiming to highlight the relevance of a number of molecular parameters, including molecular weight and distribution, tacticity, ethylene units content and nucleating agents.

The approach adopted, although the equations used are clearly empirical, is rather general and it surely represents a development with respect to phenomenological procedures describing relationships between structure and processing conditions. In the intention of the work, the kinetic parameters are the connections among such macroscopic observations.

Furthermore, the chapter provides a large amount of consistent experimental data under non-isothermal conditions (cooling rate range from below 0.1 to above 1000°C/s) for a broad set of commercial iPP's so far not extensively reported in literature.

It should be however underlined that he model provides values of K(T) comparable for the different grades, K(T) being the reciprocal of half-crystallization isothermal time regardless the value of the Avrami index. The most influential factor turned out to be the presence of nucleating agents, which shifts toward larger value the material intrinsic "crystallizability" (represented by area under the "bell-shaped" crystallization kinetics constant vs. temperature curve). In particular, the effect of molecular weight does not appear to be very relevant, due to the limited range of molecular weights available in material grades of a "commercial" nature. On the other hand, an increase in the polydispersity index significantly reflects into a parallel increase in crystallizability. Finally, addition of small amounts of ethylene units in the copolymer does not influence the kinetic parameters.

6. References

Alfonso, G. C. & Ziabicki A. (1995). *Coll Polym Sci*, Vol.273, p. 317, ISSN: 0303-402X.
Avrami M. (1939). *J Chem Phys*, Vol.7, p.1103, ISSN: 0021-9606.
Avrami M. (1940). *J Chem Phys*, Vol.8, p.212, ISSN: 0021-9606.
Avrami M. (1941). *J Chem Phys*, Vol.9, p.177, ISSN: 0021-9606.
Bird, R.B.; Stewart, W.E.; & Lightfoot, E.N. (1960). *Transport Phenomena*, Wiley, New York, ISBN-13: 978-0470115398
Brandgrup, J. & Immergut, E. H (1989). *Polymer Handbook*, John Wiley and Sons, ISBN: 0-471-81244-7
Brucato, V.; Crippa, G.; Piccarolo, S. & Titomanlio, G. (1991a). *Polym Eng Sci*, Vol.31, p.1411, ISSN: 1548-2634.
Brucato, V.; Piccarolo, S. & Titomanlio G. (1991b). *Proceedings of European Regional Meeting of the Polymer Processing Society*, Palermo, p.299.
Brucato, V.; Piccarolo, S. & Titomanlio, G. (1993). *Makromol Chem Macrom Symp*, Vol.68, p.245, ISSN: 1022-1360.
Brucato, V.; Piccarolo, S. & Titomanlio, G. (1998). *Int J Form Proc*, Vol.1, No.1, p.35, ISSN: 1292-7775.
Brucato, V.; Piccarolo, S. & La Carrubba, V. (2000). *Int Pol Proc*, Vol.15, No.1, p.103, ISSN: 0930-777X.
Brucato, V.; Piccarolo, S. & La Carrubba, V. (2002). *Chem Eng Sci*, Vol.57, p.4129, ISSN: 0009-2509.
Brucato V.; Kiflie Z., La Carrubba V., Piccarolo S. (2009). *Adv Pol Techn*, Vol.28, No.2, p.86, ISSN: 1098-2329.
Carslaw, H. S. & Jaeger, J. C. (1986). *Conduction of Heat in Solids*, Oxford Science, London, ISBN-13: 978-0198533689.

Chen J.H., Tsai F.C., Nien Y.H. & Yeh P.H. (2005). *Polymer*, Vol.46, p.5680, ISSN: 0032-3861.

Choi, C. & White, J. L. (2000). *Polym Eng Sci*, Vol.40, No.3, p.645, ISSN: 1548-2634.

Ciofalo, M.; Di Piazza, I. & Brucato, V. (198). *Int J Heat Mass Transfer*, Vol.42, p.1157, ISSN: 0017-9310.

Corradini, P.; Petraccone, V.; De Rosa, C. & Guerra, G (1986). *Macromolecules*, Vol.19, p.2699, ISSN: 0024-9297.

De Rosa C.; Auriemma F. & Resconi L. (2005). *Macromolecules*, Vol.38, p.10080, ISSN: 0024-9297.

De Santis F.; Adamovsky S.; Titomanlio G. & Schick C. (2007). *Macromolecules*, Vol. 40, No.25, p.9026, ISSN: 0024-9297.

Di Lorenzo M.L. & Silvestre C. (1999). *Prog Polym Sci*, Vol.24, p.917, ISSN: 0340-255X.

Ding, Z. & Spruiell J. (1996). *J Polym Sci Part B: Polym Phys*, Vol.34, p.2783, ISSN: 1099-0488.

Douillard, A.; Dumazet, Ph.; Chabert, B. & Guillet, J. (1993). *Polymer*, Vol.34, No.8, p.1702, ISSN: 0032-3861

Eder, G. & Janeschitz-Kriegl, H (1997). *Structure development during processing 5: Crystallization. Material Science and Technology, vol. 18*; Weinheim: H.E.N. Meijer ed.

Elmoumni, A.; Gonzalez-Ruiz, R.A.; Coughlin, E.B. & Winter H.H. (2005). *Macrom. Chem. Phys.*, Vol.206, p.125, ISSN: 1521-3935.

Evans U.R. (1945). *Trans. Faraday Soc.*, Vol.41, p.365, ISSN: 0956-5000.

Fann, D.M.; Huang, S.K. & Lee, J.Y. (1998). *Pol Eng Sci*, Vol.38, No.2, p.265, ISSN: 1548-2634.

Foresta T., Piccarolo S. & Goldbeck-Wood G. (2001). *Polymer*, Vol.42, p.1167, ISSN: 0032-3861.

Geil, P.H.; Anderson, F.R.; Wunderlich, B. & Arakawa, T. (1964). *J Polym Sci A:Polym Chem*, Vol.2, p.3707, ISSN: 1099-0518.

Gerardi, F.; Piccarolo, S.; Martorana, A. & Sapoundjieva, D. (1997). *Macromol. Chem Phys*, Vol.198, p.3979, ISSN: 1521-3935.

Gobbe, G.; Bazin, M.; Gounot, J. & Dehay, G. (1988). *J Polym Sci B: Polym Phys*, Vol.26, p.857, ISSN: 1099-0488.

Goodfellow Catalogue (1996). *Goodfellow*, Cambridge Limited, p.318.

Guerra, G.; Vitagliano, V.; De Rosa, C.; Petraccone, V. & Corradini, P. (1990). *Macromolecules*, Vol.23, p.1539, ISSN: 0024-9297.

He, J. & Zoller, P. (1994). *J Polym Sci B: Polym Phys*, Vol.32, No.6, p.1049, ISSN: 1099-0488.

Isachenko, V. P.; Ossipova, V. A. & Sukomel A. S., (1987). *Heat Transfer*, MIR, Moscow, ISBN: 089875027X.

Kovarskii, A. (1994). *High-Pressure Chemistry and Physics of Polymers*, CRC Press, ISBN: 9780849342394.

La Carrubba, V.; Brucato, V. & Piccarolo, S. (2000). *Polym Eng Sci*, Vol.40, No.11, p.2430, ISSN: 1548-2634.

La Carrubba, V. (2001). *Polymer Solidification under pressure and high cooling rate, Ph.D. Thesis*, CUES, Salerno, ISBN: ISBN: 88-87030-27-8.

La Carrubba, V.; Brucato, V. & Piccarolo, S. (2002a). *J Polym Sci B: Polym Phys* Vol.40, p.153, ISSN: 0024-9297.

La Carrubba, V.; Briatico, F.; Brucato, V. & Piccarolo, S. (2002b). *Polym Bull*, Vol.49, p.159, ISSN: 0170-0839.

La Carrubba, V.; Piccarolo, S. &Brucato, V. (2007). *J Appl Polym Sci*, Vol.104, p.1358, ISSN: 1097-4628.

Leute, U.; Dollhopf, W. &Liska, E. (1976). *Colloid Polym Sci*, Vol.254, No.3, p.237, ISSN: 0303-402X.

Liangbin, L.; Huang, R.; Ai, L.; Fude, N.; Shiming, H.; Chunmei, W.; Yuemao, Z. & Dong W. (2000). *Polymer*, Vol.41, p.6943, ISSN: 0032-3861.

Luikov, A.V. (1980). *Heat and Mass Transfer*, MIR, Moscow, ISBN 13: 9780080166322.

Marigo, A.; Marega, C.; Causin, V. & Ferrari P. (2004). *J Appl Pol Sci*, Vol.91, p.1008, ISSN: 1097-4628.

Martorana, A.; Piccarolo, S. & Scichilone, F. (1997). *Macromol Chem Phys*, Vol.198, p.597, ISSN: 1521-3935.

Nagasawa, S.; Fujimori, A.; Masuko, T. & Iguchi, M. (2005). *Polymer*, Vol.46, p.5241, ISSN: 0032-3861.

Nakamura, K.; Katayama, K. & Amano T. (1973). *J Appl Pol Sci*, Vol.17, p.1031, ISSN: 1097-4628.

Nakamura, K.; Watanabe, T.; Katayama, K. & Amano T. (1972). *J Appl Pol Sci*, Vol.16, p.1077, ISSN: 1097-4628.

Pantani, R.; Speranza, V.; Coccorullo, I. & Titomanlio, G. (2002). *Macrom Symp*, vol.185, p.309, ISSN: 1521-3900.

Pantani, R.; Coccorullo, I.; Speranza, V. & Titomanlio, G. (2005). *Prog Polym Sci*, Vol.30, p.1185, ISSN: 0079-6700.

Piccarolo, S. (1992). *J Macromol Sci B*, Vol.31, No.4, p.501, ISSN: 0022-2348.

Piccarolo, S.; Saiu, M.; Brucato, V. & Titomanlio, G. (1992a). *J Appl Polym Sci*, Vol.46, p.625, ISSN: 1097-4628.

Piccarolo, S.; Alessi, S.; Brucato, V. & Titomanlio, G. (1992b). *Proceedings of Crystallization of Polymers, a NATO Advanced Research Workshop*, Mons, p.475.

Piccarolo, S. & Brucato, V. (1996). *Proceedings of the PPS12-Annual Meeting*, Sorrento, p.663.

Raab, M.; Scudla, J.; & Kolarik, J. (2004). *European Polymer J*, Vol.40, p.1317, ISSN: 0014-3057.

Sakurai, T.; Nozue, V.; Kasahara, T.; Mizunuma, K.; Yamaguchi, N.; Tashiro, K.; Amemiya, Y. (2005). *Polymer*, Vol.46, p.8846, ISSN: 0032-3861.

Schneider, W.; Koppl, A. & Berger, J. (1988). *Int Pol Proc*, Vol.2, p.151, ISSN: 0930-777X.

Strobl, G. (1997). *The Physics of polymers, Concepts for Understanding Their Structures and Behavior*, Springer, New York, ISBN: 978-3-540-25278-8.

Struik, L.C.E. (1978). *Physical ageing in amorphous polymers and other materials*, Elsevier, Amsterdam, ISBN-13: 978-0444416551.

Tchizmakov, M.B.; Kostantinopolskaja, M.B.; Zubov, Yu.A.; Bakeev, N.Ph.; Kotov, N.M. & Belov, G.P. (1976). *Visokomol Soed*, Vol.A18, p.1121.

Titomanlio, G.; Speranza, V. & Brucato, V. (1997). *Int Polym Proc*, Vol.12, No.1, p.45, ISSN: 0930-777X.

Titomanlio, G.; Piccarolo, S. & Levati, G. (1988a). *J Appl Polym Sci*, Vol.35, p.1483, ISSN: 1097-4628.

Titomanlio, G.; Rallis, A. & Piccarolo, S. (1988b). *Polym Eng Sci*, Vol.29, p.209, ISSN: 1548-2634.

Van Krevelen, W. (1972). *Properties of Polymers*, Elsevier, Amsterdam, ISBN: 978-0-08-054819-7.

Wunderlich, B. & Arakaw, T. (1964). *J Polym Sci Part A: Polym Chem*, Vol.2, p.3697, ISSN: 1099-0518.

Wunderlich, B. (1973). *Macromolecular Physics, Vol. 1*, Academic Press, New York, ISBN-13: 978-0127656014.

Wunderlich, B. (1976). *Macromolecular Physics, Vol. 2*, Academic Press: New York, ISBN-13: 978-0127656021.

Wunderlich, B. (1980). *Macromolecular Physics, Vol. 3*, Academic Press: New York, ISBN-13: 978-0127656038.

Wunderlich, B. & Davison T. (1969). *J Polym Sci Part A:Polym Chem*, Vol.7, p.2043, ISSN: 1099-0518.

Ziabicki, A. & Alfonso, G.C. (1994). *Coll Polym Sci*, Vol.272, p.1027, ISSN: 0303-402X.

Zimmermann, H.J. (1993). *J Macromol Sci-Phys*, Vol.B32, p.141, ISSN: 0022-2348.

Ziabicki, A. (1976). *Fundamentals of Fibre Formation*, Wiley, London, ISBN: 0-471-98220-2.

Zuidema, H.; Peters, W.M.P. & Meijer H.E.H. (2001). *Macrom Theory Simul*, Vol.10, p.447, ISSN: 1521-3919.

Zoller, P. (1979). *J Appl Polym Sci*, Vol.23, No.4, p.1051, ISSN: 1097-4628.

11

Charging Property and Charge Trap Parameters in Porous Polypropylene Film Using Thermally Stimulated Current

Fukuzo Yoshida and Masahiko Yoshiura
Osaka Institute of Technology Osaka,
Japan

1. Introduction

The polymeric materials are utilized in industry and an ordinary household with the characteristic that a natural organic material does not have. Research and development are performed actively now because the polymers are materials with a variety of functionality (Imai et al., 2002; Ishii et al., 2009; Ishimoto et al., 2009; Varlow & Li, 2002). By such a background, we aimed at the polymer system piezoelectric material which let it give piezoelectricity as the sensor function. For typical piezoelectric material(Koga & Ohigashi,1985;Lindner etal.,2002), PZT and $BaTiO_3$ are well known until now. In contract, the PVDF of the polymer system piezoelectric material immobilized CF_2 dipolar orientation. Piezoelectric modulus d_{33} of the porous polymer electrets is higher than PVDF, and in a polymer system, polaristation reversal happens.

However, the electrical conduction mechanism of the porous polymer electrets(Cao et al.,1998;Xia et al.,1999) is complicated, and a study is gone ahead with as an important theme for the development. It has been considered that charge carrier traps in a substance play an important role in the charging phenomenon. For this theme, the thermally stimulated current (TSC)(Braünlich,1979;Chen & Kirsh,1981;Ikezaki & Hori,1998;Baba & Ikezaki,1992;Ikezaki & Murata,2006;Oka & Ikezaki,1992;Perlman & Creswell,1971;Yoshida et al.,1998) is one effective measurement. The TSC measurement activates a sample by corona discharge or light and gives a sample heat stimulation by constant heating rate and takes out electric charge in inside of sample to produce a current of the external circuit. This measurement is not a change of state other than a sample, and is extremely high sensitive measurement.

On the other hand, the surface boundary and charging phenomenon are complicated because the inner structure of the sample is not uniform. As a result, it is thought that the signal of the TSC spectrum which measured is a compounds of TSC spectra caused by several charge traps. We developed the evaluation method using the characteristic of the TSC measurement. This evaluation method separates plural TSC spectra precisely and evaluates the information of the trap precisely. As for this separation method (named AEM separation system), it is done computation process on Windows. Above all, an escape frequency factor proposed a directly ratable method (named AEM-v(Yoshida & Maeta,1991)) for the first time even if a waveform condition of the TSC spectrum was bad.

Polymer system piezoelectric material to use for this study is porous polypropylene. The temperature characteristic of the surface charge electric potential was examined directly by the thermally stimulated charge decay (TSCD) other than TSC measurement.

In this study, it applied the AEM separation system which we proposed to the separation of the TSC spectrum and analyzed the property of the trap which imperforate polypropylene and porous polypropylene formed.

2. Materials and methods

An experiment sample is the polypropylene which is one of the four major general- purpose resin. It shows the different polypropylene (PP) of three kinds of properties in Table 1. A polypropylene film as prepared (PP1) and the polypropylene drawn film (PP2) to two axes (Futamura Chemicals) were used as base polymers. The porous film containing pores of several micro meters in radius prepared (PP3) was drawn to the thickness of 75μm.

sample name	type	Thickness (μm)
PP1	solid	50
PP2	solid(drawn to two axis)	50
PP3	porous(drawn to two axis)	75

Table 1. Three PP films used in this experiment.

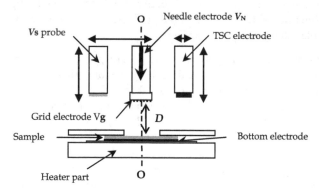

Fig. 1. Electrode arrangement for various measurements in the cryostat.

Figure 1 shows the electrode arrangement that various measurements are possible. The corona electrical charging of the sample removing was performed by a needle electrode to central axis O-O. The corona charge was carried out with voltage V_N of tungsten needle electrode fixed at ±3 kV and the grid voltage V_g at less than ±2 kV for 60 s at 1 atm. The corona discharge was measured under constant humidity after pouring the dry gas of the fixed quantity. Surface potential V_s was measured with surface potential electrometer (Model 344, Trec Japan). The TSC measurement was performed by removing the TSC electrode to the central axis shown in Fig. 1 with a separation of D=1mm from the sample.

The open TSC signal was measured under a vacuum using electrometer (Keithley 610C). As a result of having made shielding on an external circuit, it enabled very sensitive TSC measurement. The TSC observation temperature region reached in the range of 430 K from 250 K using LN₂ cryostat.

Figure 2 shows diagram of the TSC measurement. This figure is an example applying bias electric field (E_b) for present temperature (T_b) and setting time (t_b). This experiment performed charge injection by corona electrical charging not bias electric field. The corona charging processed a sample in the polarity of each positive and negative. After one corona charging process, the TSC spectra were measured two times in succession. It calls the TSC spectrum "1st run TSC" and "2nd run TSC" sequentially. The TSCD measurement is the basically same as TSC. The upper part electrode which showed in Fig. 1 becomes the surface potential prove.

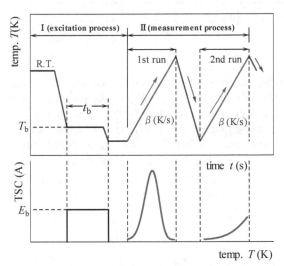

Fig. 2. Diagram of the TSC measurement

3. TSC spectrum analysis

AEM-v is necessary for the construction on the AEM separation system which we are proposed.

3.1 AEM theory

AEM-v is able to evaluate the escape frequency factor v utilizing all the data of an object TSC spectrum. The advantageous property of this method is that both the v and E_t value could be determined. The equation used in AEM-v was derived from an equation of the TSC with constant heating rate β under the condition of first-order slow retrapping is expressed as:

$$I(T) = I_o \exp \left\{ \frac{-E_t}{kT} - \frac{v}{\beta} \int_{T_o}^{T} \exp \left(\frac{-E_t}{kT} \right) dT \right\} \tag{1}$$

The following symbols are used : $I_o = n_o e \mu v \tau A E$ (A), n_o: the carrier density of the filled traps at t =0 (m^{-3}), e: the electric charge (C), μ: the carrier mobility (m^2/V·s), v: an escape frequency factor (s^{-1}), τ: the life time of a free carrier (s), A: the area of the electrode (m^2), E: the applied electric field (V/m), E_t: energy depth of carrier trap(eV), k: the Boltzmann's constant 8.617×10^{-5} (eV/K), β: the heating rate (K/s), T : the absolute temperature (K), T_o : the absolute temperature from which the heating begins after filling of the traps with carrier at the time t.

The theoretical TSC spectrum shown in Fig.3 was calculated by eq.(1) with a TSC maximum m (T_{mo} , I_{mo}) and trap depth E_{to} . The basic formula of AEM-v becomes the eq.(2). The escape frequency factor v is thus expressed by a ratio of I_a to I_b:

$$
v = \frac{\dfrac{\beta k}{E_t}\left\{ \ln\dfrac{I_a}{I_b} + \dfrac{E_t}{k}\left(\dfrac{1}{T_a} - \dfrac{1}{T_b} \right) \right\}}{\displaystyle\sum_{n=0}(-1)^n \dfrac{(n+1)!}{\left(\dfrac{E_t}{kT_b}\right)^{n+2}} \exp\left(\dfrac{-E_t}{kT_b}\right)\left[1 - \left(\dfrac{T_a}{T_b}\right)^{n+2} \exp\left\{ \dfrac{-E_t}{k}\left(\dfrac{T_a - T_b}{T_a T_b} \right) \right\} \right]}
\tag{2}
$$

The integral terms in eq. (1) were integrated by the asymptotic expansion series. As shown in the equation, eq. (2) contains no TSC peak coordinates. Three coordinate points on the TSC spectrum, noted as a (T_a, I_a), b (T_b, I_b) and c (T_c, I_c) as shown in Fig. 3 are used for an application of AEM-v.This equation, therefore, made possible not only the evaluation of v value from a TSC spectrum without a maximum peak, but also the continuous determination of v values from any point on a TSC spectrum.

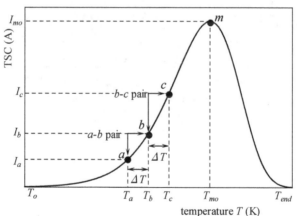

Fig. 3. Coordinate on the TSC spectrum of the AEM-v.

Although two pairs of data points from a TSC spectrum are required for AEM-v, at least three points satisfy the requirement by using one point in common for both pairs. The convenient method to select data by its temperature coordinate, is shown in Fig. 3. In the method, named "moving coordinates method", two pair (a-b and b-c) of points at a same temperature separation (ΔT), as shown in Fig.3, is selected, holding a point at temperature (T_b) in common and inputting their coordinates into the calculation. The E_t and v values at the intermediate temperature (T_b) are evaluated. In other words, for a calculation of v value, E_t value which assumed and a-$b$pair are given in eq.(2). The same calculation is carried out to the other b-c pair at the same time. As a result, it is converged by a computer until v value which calculated in both pairs becomes the same value. Numbers of the E_t and v values were calculated for the coordinates of two pairs of points with ΔT =0.2 K interval, shifting by 0.2 (K) for higher values and plotted continuously on T_b. As the results, this smaller temperature interval of data provides more sensitivity to detect a composite TSC objective.

Three characteristics of AEM-v are shown in Fig. 4. These characteristics applied to a single relaxation TSC theory spectrum of Fig.3. Then, it can calculate the peak temperature T_m of TSC spectrum of to be shown in Fig.4(c) from an E_t and v values at the same temperature T. The part of flat shape means that a target TSC spectrum is caused by single relaxation. The values of three parameters were estimated from the means of the vertical axis of the temperature range indicating the flat shape. It can be understood that the flat shapes of the characteristics in the temperature region of the whole TSC spectrum indicate the signal to be caused by a single trap. Naturally, this peak temperature T_m is temperature indicating the maximum current I_{mo} of the TSC spectrum of Fig. 3.

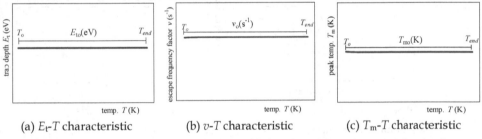

(a) E_t-T characteristic   (b) v-T characteristic   (c) T_m-T characteristic

Fig. 4. Three characteristics of AEM-v which applied to the TSC spectrum of the single trap.

The thermal cleaning method and partial heating are known as experimental separation methods of composite TSC spectrum and only initial rising part of TSC spectrum was used for estimation of E_t value. The initial rise method(Garlick & Gibson,1948) is the only one procedure to apply to data without a peak until now. However, in AEM-v, an application is possible to an omniformity-shaped TSC spectrum.

The initial rising (signal from T_o to T_S) part of a TSC spectrum calculated with arbitrary coefficients (E_{to}, T_{mo} and I_{mo}) was shown in Fig. 5. The attached map of Fig. 5 is an application result of AEM-v. E_t and v values were evaluated by the AEM-v from a part of a TSC spectrum or that without a maximum using three coordinates for the first time. Furthermore, it can be understood that the detection of the peak temperature T_m is possible from the initial part of a TSC spectrum.

The maximum coordinate of TSC spectrum and E_t value is necessary to calculate a theoretical TSC spectrum. In other word, among the maximum coordinate m of TSC spectrum, maximum current I_m is required. AEM-I(Yoshida et al.,1991) which we proposed can calculate maximum current I_m of the TSC spectrum as well as an evaluation of E_t values. It need a maximum temperature T_m in addition to two points, a point and b on the TSC spectrum which were shown in Fig.3 The basic formula of AEM-I is given in the next expression.

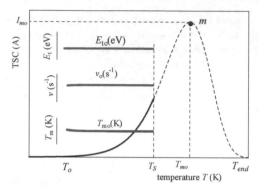

Fig. 5. Three characteristics of AEM-v which applied to the initial rising part of TSC spectrum.

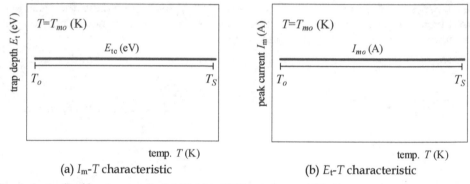

(a) I_m-T characteristic   (b) E_t-T characteristic

Fig. 6. A result of having applied AEM-I to TSC spectrum of Fig.5.

I_m was calculated with two coordinate of a TSC spectrum and T_m obtained from AEM-v. Two characteristics of AEM-I were presented in Fig. 6. Like Fig. 5, the flat shape means the signal was detected from a single trap. It was able to evaluate maximum coordinate and trap depth E_t of the TSC spectrum by AEM-v and AEM-I. As a result, the calculation of the TSC theoretical spectrum is enabled and is shown with a dashed line in Fig. 5. In AEMs, the reconstruction of the whole TSC spectrum is possible from the part of the TSC observed in this way.

$$A_l = -\ln\frac{I_a}{I_b} + \left(\frac{T_b}{T_m}\right)^2 \left[\sum_{n=0}^{\infty} (-1)^n \frac{(n+1)!}{\left(\frac{A_l T_a}{T_b - T_a}\right)^n} \exp\left\{\frac{A_l T_a (T_b - T_m)}{T_m (T_b - T_a)}\right\} \left\{ 1 - \left(\frac{T_a}{T_b}\right)^{n+2} \exp(-A_l) \right\} \right] \quad (3)$$

$$E_t = A_l \frac{kT_b T_a}{T_b - T_a} \quad (4)$$

3.2 AEM separation system

AEM-v can do a judgment whether or not the contribution of the trap is single. The merit of AEM-v is enabled the separation of a compound TSC spectrum without thermal cleaning measurements. We name "AEM separation system" as this separation method and carry it out by computerization.

Then, we explain the procedure of this separation method using the calculated TSC spectrums. A separation object is a compound TSC spectrum formed of two traps.

At first the TSC spectrum is screened by AEM-v. Then, the observed TSC spectrum can visualize by the information from a trap. Because both peaks (P_1 and P_2) are exposed, Fig.7 shows that it is the signal from two traps (C_1 peak to show in Fig.10 and C_2 peak) easily.

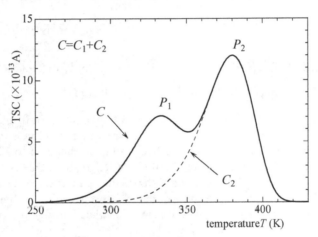

C_1 curve:$E_{to,1}$=0.570(eV),$T_{mo,1}$=330.00(K),$I_{mo,1}$=6.000($\times 10^{-13}$A)
C_2 curve:$E_{to,2}$=0.740(eV),$T_{mo,2}$=380.00(K), $I_{mo,2}$=12.00($\times 10^{-13}$A)

Fig. 7. Compound TSC spectrum consisting of two traps.

Figure 8 is the result that is applied AEM-v to a compound TSC spectrum of Fig. 7. An abscissa of Fig.8 is the temperature region that used in calculation. Figure 8(a) and (b) are trap depth E_t and escape frequency factor v respectively. A flat part is the temperature region that is strong in the contribution of the single trap. This E_t-T characteristic can evaluate E_t values of 0.570eV and 0.740eV. In particular, the large flat part of T_m-T characteristic (Fig.8(c)) leads to a T_m value. The big divergence of the neighborhood of 350 K expresses the place that is strong in contribution of the combined signal. Then, using provided T_m=380.00(K), AEM-I is applied.

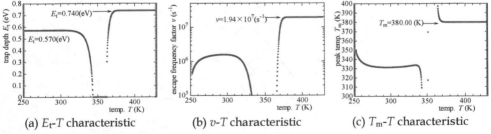

(a) E_t-T characteristic   (b) v-T characteristic   (c) T_m-T characteristic

Fig. 8. Three characteristics that applied AEM-v to Fig.7.

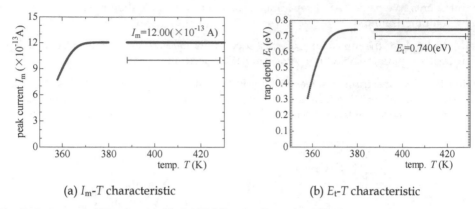

(a) I_m-T characteristic   (b) E_t-T characteristic

Fig. 9. As a result of having applied AEM-I to the P_2 peak of Fig.7.

Figure 9 is two characteristics of AEM-I which applied to P_2 peak of Fig.7 to target separation. I_m value is found from a flat part of Fig. 9(a). In this case, an E_t value is found by a flat part of Fig.9(b) or Fig. 8(a). When the maximum coordinate of the TSC spectrum is exposed, AEM-LH (Maeta & Sakaguchi,1980;Maeta & Yoshida,1989)is applicable. And AEM-LH enables an evaluation of trap depth E_t again, too. At this stage, the AEM separation system fine-tunes the maximum coordinate of the TSC spectrum to raise the flat shape of E_t-T characteristic more fine. The C_2 curve of Fig.7 calculated using maximum coordinate (380.00, 12.00) and E_t (0.740eV) which were detected.

It is the newly exposed TSC spectrum (C_1 curve) which deduct C_2 curve from C curve targeted for separation in Fig. 10. Figure 11 shows the result that applied AEM-I to C_1 curve. Here, AEM-I was applied after determination of the peak temperature T_m (330.00K) by AEM-v. Because two characteristics show flat in all temperature region, this C_1 curve understands that it is the signal caused by a single trap.

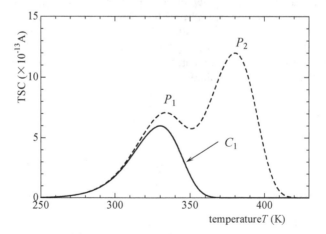

Fig. 10. The C_1 curve exposed from a compound TSC spectrum.

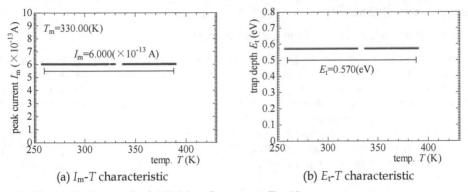

(a) I_m-T characteristic   (b) E_t-T characteristic

Fig. 11. The result that applied AEM-I to C_1 curve in Fig.10.

Figure 12 is an example looking like a single TSC spectrum in an appearance.

Contribution of trap signal (C' curve) is closer than a compound TSC spectrum of Fig. 7. Figure 13 shows the result that applied AEM separation system to Fig. 12.

Three characteristics detect a C' curve is compound contribution in high sensitivity.

This result can evaluate E_t of the outline as 0.740eV directly even if do not separate a TSC spectrum. If a flat part can be detected on a TSC spectrum, the separation is possible.

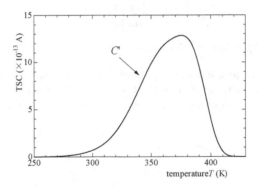

Fig. 12. An example of the TSC spectrum of strong multiplicity.

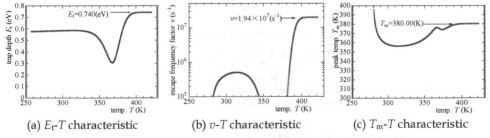

(a) E_t-T characteristic   (b) v-T characteristic   (c) T_m-T characteristic

Fig. 13. Three characteristics that applied AEM-v to Fig.12.

4. Experimental results

4.1 Grid voltage dependence of the surface potential

Figure 14 is surface potential V_s properties of three kinds of PP when changed the grid voltage V_g. The needle voltage V_N is fixed at each ±3kV.The corona discharge condition is charging time t_d 60s at room temperature under 1 atm. In positive corona charge, the maximum charged potential became ca. 1.1kV and ca.1.25kV each in PP1 and PP2. And PP3 of the porous PP film became ca.1.47kV. On the other hand, as for the negative corona charge, PP1 and PP2 became ca.-300V and ca.-690V each, and PP3 became ca.-820V. This result shows that charged surface potential of the positive corona charge is high in all film. This accords with a report that PP of the contact charging is easy to be charged with electricity in a plus.

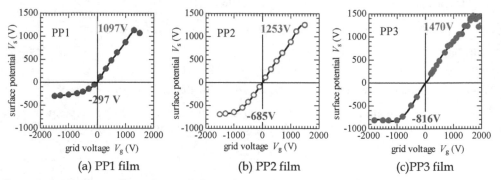

(a) PP1 film   (b) PP2 film   (c)PP3 film

Fig. 14. Grid voltage dependence of the surface potential.

Figure 15 showed Fig.14 in a mass. It is revealed that clear saturation happens in surface potential by the negative corona charge. On the other hand, for positive corona charge, the surface potential shows linear charging characteristics to ca.1kV of the grid voltage.

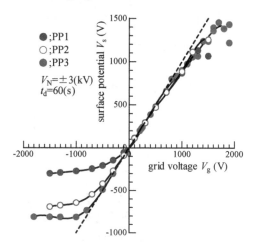

Fig. 15. Comparison of three kinds of PP films.

4.2 Isothermal potential decay

Isothermal decay of the surface potentials were measured during 10^4s as shown in Fig.16. Each characteristic standardized it in surface potential V_{so} at the time of the start of measurement. In positive corona charge, PP1 and PP2 of the solid film are quick in decay, and the decay of ca.4% of initial values is seen in progress for 3h. However, decay is not seen for the negative corona charge. In contrast, even if PP3 passes for 3h, it holds initial potential. In positive corona charge, Time constant of the decay for PP1,PP2 and PP3 were revealed to be 121h, 161h and 806h, respectively.

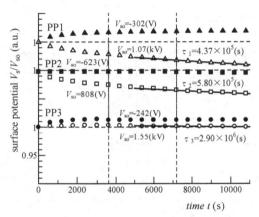

Fig. 16. Isothermal surface potential decay curves for porous and solid PP films at room temperature (▲,■,●:negative charged, △,□,○:positive charged).

4.3 Thermally stimulated charge decay

In generally, temperature condition is important factor in examining the electric-electronic industry material. Therefore, the next performed thermally stimulated charge decay (TSCD) experiment of the PP films. In Fig.17, TSCD characteristics from positively charged PP films were presented. Charge decay of PP1 and PP2 occurred around 390K and 280K, respectively, although PP3 released the charge above 410K. At 430K, PP1, PP2, and PP3 lost 20%, 78%, and 7% of initial surface potential V_{so}, respectively.

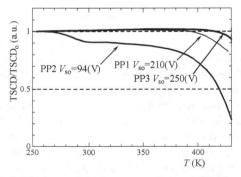

Fig. 17. TSCD characteristics from positively corona charged PP films.

4.4 Thermally stimulated current

The sample which used by TSCD experiment has a TSC experiment successively without exposing a sample to air from a good point of the measuring apparatus. As a result, because it is identical test items, both experimental results compare it directly and can analyze it precisely. TSC spectra measured for PP1, PP2, and PP3 are shown in Fig. 18(a),(b), and (c), respectively.

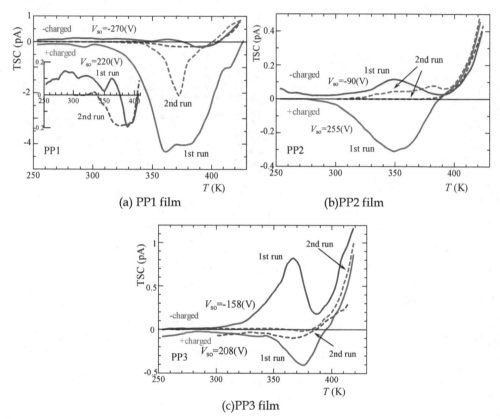

(a) PP1 film   (b)PP2 film

(c)PP3 film

Fig. 18. TSC characteristics for porous PP and solid PP.

In the figures, TSC spectrum recorded after positively charged was presented by red line, beside that after negatively charge was shown by blue line. Solid line represented TSC recorded during initial heating after being charged, though the second heating that performed without charging after the film was rapidly frozen caused TSC spectrum shown by dashed line. The attached map in Fig. 18(a) shows an enlarged picture of negative charged TSC spectrum. A plurality of TSC peaks are observed in a temperature region 375 K from 250 K. It was shown that the magnitude of TSC signals in the base polymer, PP1, charged positively were larger than that charged negatively. The TSC spectrum of PP3 is observed with each charge polarity in ca.370K and ca.375K. In each sample, signal detected in the second run clearly decreased. In every case, however, the increasing current was observed about 400K. As a result of Blank experiment, we regard the increase current at high temperature as a thing by the thermolysis ion.

5. Separation of the actual survey TSC spectrum

Generally, as for the observed TSC, the single trap contribution is rare, and a plurality of signal overlaps in most actual survey TSC spectra thermally. Many insulating materials are easy to catch the temperature distortion from the badness of conduction of heat. In other word, this cannot ignore the influence that a heat cycle history gives a sample. This has been regarded as a cause to disturb an accurate evaluation. The solution must separate the signal contribution of the trap from one TSC measurement. This can be settled by the AEM separation system which it described in Chapter 3. In addition, the AEM separation system can separate thermal noise and the residual current by other causes in a separation process. This chapter uses the actual survey TSC spectrum of Fig.18(b) for an example and explains a separation process.

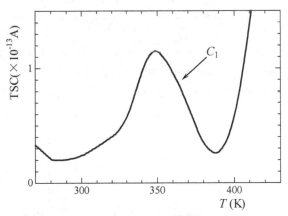

Fig. 19. An example of a TSC spectrum targeted for separation.

Figure 19 is a TSC spectrum targeted for separation and shows the each characteristic in Fig. 20. The maximum coordinate of the TSC spectrum is provided by pushing forward AEM-I and a process from AEM-v. And trap depth E_t evaluates it in AEM-LH. In each characteristic of Fig.20, temperature region that seems to be contributed from a single trap was estimated. Using TSC maximum coordinate (349.00K, $1.150×10^{-13}$A) and trap depth E_t (0.764eV) decided by the screening of the TSC spectrum, P_m peak of Fig. 21(a) was calculated by eq. (1). Figure 21 shows the temporary separation of the TSC spectrum. The C_2 curve of Fig.21(a) is the resultant curve of removal of a P_m peak from actual survey TSC spectrum C_1. Figure 21(b) is the result of application of AEM-LH to C_2 curve. At the stage, the maximum of the TSC spectrum is revealed. And the TSC spectrum which was calculated using this TSC maximum coordinate and trap depth E_t results the P_h peak of Fig. 21(c). The C_3 curve is obtained by removal of P_h peak from C_2 curve. The minus current of the neighborhood of 350K of the C_3 curve shows that P_h peak and agreement of the C_2 curve are incomplete. Therefore it is necessary to revise a P_m peak.

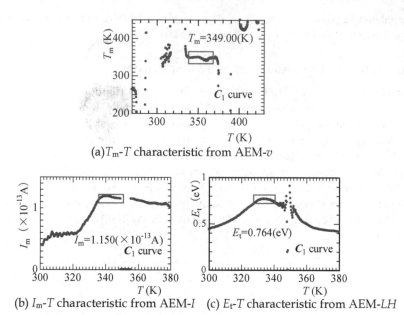

(a)T_m-T characteristic from AEM-v

(b) I_m-T characteristic from AEM-I (c) E_t-T characteristic from AEM-LH

Each evaluation is averaged in a part surrounded with the square of each part.

Fig. 20. Characteristic from each AEM of Fig.19.

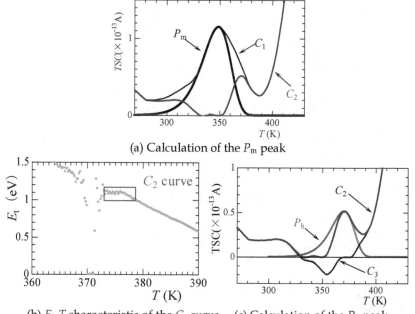

(a) Calculation of the P_m peak

(b) E_t-T characteristic of the C_2 curve   (c) Calculation of the P_h peak

Fig. 21. Temporary separation process of the TSC spectrum.

Figure 22 is a revision process of the P_m peak. The C_4 curve of Fig. 22(a) is obtained by removal of P_h peak from C_1 curve. The E_t-T characteristic of the C_4 curve become Fig.22(b). The P_{mo} peak of Fig.22(c) is the result that P_m peak was revised using an F_t value of Fig.22(b) and the maximum of the TSC spectrum. The P_{mo} peak is the TSC signal which decided. Furthermore, Fig.23 shows the revision process of the P_h peak. The C_5 curve of Fig. 23(a) is obtained by removal of P_{mo} peak from actual survey TSC spectrum C_1. The E_t-T characteristic of the C_5 curve become Fig.23(b). The P_{ho} peak of Fig.23(c) becomes the correction curve of the P_h peak. P_{ho} peak accord with actual survey TSC spectrum C_2 well in comparison with P_h peak of Fig.21(c).The AEM separation system repeats such a calculation process in all temperature region of the measured TSC spectrum. Figure 24 is TSC signal and the residual current C_r that were finally decided.

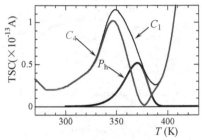

(a) C_4 curve of an exposed low temperature side TSC spectrum

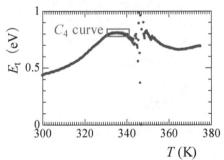

(b) E_t-T characteristic of the C_4 curve

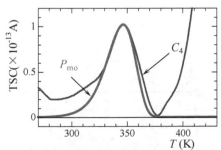

(c)The actual survey TSC spectrum which corrected at P_h peak

Fig. 22. Revision separation process of the TSC spectrum.

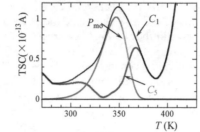

(a) The high temperature side TSC spectrum which was corrected

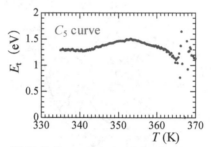

(b) E_t-T characteristic of the C_5 curve

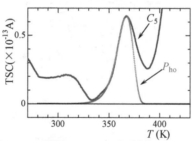

(c) The actual survey TSC spectrum which corrected at P_{mo} peak

Fig. 23. Revision separation process of the TSC spectrum.

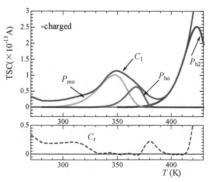

Fig. 24. TSC spectrum separation result of the negative corona charge of the PP2 film.

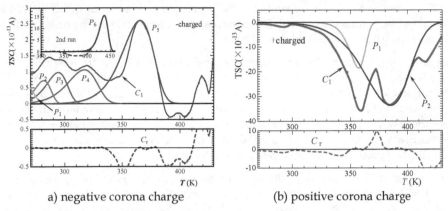

a) negative corona charge

(b) positive corona charge

Fig. 25. Separation result of PP1 film.

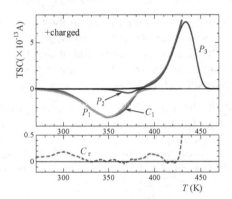

Fig. 26. Separation result of PP2 film due to the positive corona charge.

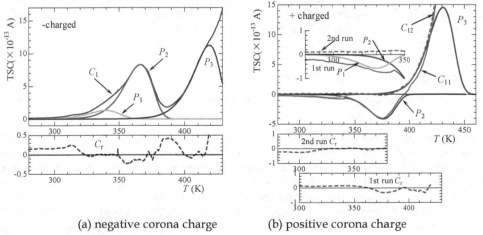

(a) negative corona charge

(b) positive corona charge

Fig. 27. Separation result of PP3 film.

Figure 25-27 are the result that separated a TSC spectrum of three kinds of PP film. In Fig. 25(a) are negative corona charge, Fig. 25(b) are positive corona charge. The residual current C_r after the separation spreads and displays it.

The negative corona charge TSC spectrum was separated at the peaks from P_1 to P_6. Actual survey TSC spectrum C_1 deducts 2nd run TSC and removes the increase current of high temperature region. P_5 peak accords with the high temperature side of the maximum peak of the C_1 curve well, but understand that C_1 curve is distorted in the low temperature side. Because the low temperature side of the C_1 curve accords with a temperature region of inversion current observed in 2nd run TSC, it is considered as influence. The P_6 peak to show in attached map is a separation result of the 2nd run TSC.

On the other hand, in the case of positive corona charge, actual survey TSC spectrum C_1 was separated by two TSC spectra (a P_1 peak and P_2 peak). A P_2 peak and the disagreement in the neighborhood of 370 K of C_1 curve are regarded as the influence of a peak observed in the neighborhood of 370 K of the 2nd run TSC to show in Fig. 18(a).

Then, a result of the PP2 film is shown in the Fig. 26. In the case of negative corona charge, we already showed it at a point of the explanation of the separation process of Chaper 5. For the positive corona charge, the C_1 curve was separated at three peaks from P_1 to P_3.The P_3 peak is the reconstruction of the TSC spectrum from the actual survey C_1 spectrum only for the initial rising part that it explained by AEM-v theory.

Finally Fig. 27 is a separation result of PP3 film. In the 2nd run TSC, the main peaks less than 400 K are cleaned. In negative and positive corona charge, it was divided into the TSC spectrum from three traps. The P_3 peak of the positive corona charge is inversion TSC separated by 2nd run TSC (C_{12} curve). In attached map of Fig.27(b), the enlarged figure of separation result less than 350K is shown. The Cr curve is a result expect the TSC signal.

6. Discussions

A signal for the contribution of the single trap must be separated to evaluate the information of the trap from the measured TSC spectrum. The information of trap to be discussed in this chapter is the result that applied all AEM separation system. Furthermore, we discuss the escape frequency factor v of the trap here and mention the origin of the trap.

6.1 Trap depth E_t and observation temperature T_m

The trap depth E_t values of the TSC signals separated are plotted to the peak temperatures T_m as are shown in Fig. 28.

Generally the trap depth is deepened so that an observed tempearature region of the TSC signal becomes the high temperature. However, the separation result does not necessarily behave like that. The origin of various traps is thought about. In PP1 film to show in Fig. 28(a), there are four traps (from P_1 to P_4 peak) from about 0.59eV to about 0.95eV in less than 350K for negative corona charge. And there is a trap of 0.841eV (P_5 peak) by 400K from 350K. In the case of positive corona charge, two traps (1.61eV of P_1 peak and 0.617eV of P_2 peak) exist in the same temperature region. Only P_6 peak (1.77eV) of the negative corona charge exists when it becomes than 400K.

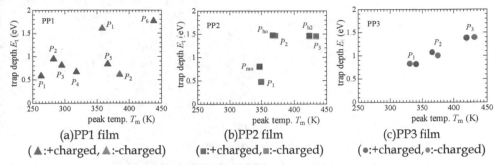

Fig. 28. Correlation of trap depth E_t of three PP film.

In the case of PP2 film to show in Fig. 28(b), three traps exist in the almost same temperature region more than 350 K regardless of polarity of the charge. The depth trap of P_1 peak (0.473eV) is different from P_{mo} peak (0.801eV) greatly. However, as for four trap (P_{ho}, P_2, P_{h2} and P_3 peak), about 1.5eV is evaluated regardless of polarity of the corona charge. When PP2 film compare with PP1 film in the negative corona charge, it understands that the number of charge trapping decreases solid film two axis orients it. Furthermore, the TSC intensity of the P_2 peak of PP1 film for the positive charge is very big. As for the polymeric film which drawn, electrical specification is known to be improved. In the case the solid film has much number of the traps for negative corona charge and it is thought that space charge accumulation that a trap forms is bigger than drawn film for positive corona charge. This does not contradict it about the high insulation that drawing operation of the film gives.

Then, Fig.28(c) is a separation result of the porous film. It is understand that three discrete peaks do not depend on the corona charging polarity. In both P_1 peak, temperature regions less than 350K and both P_2 peak are observed each in the temperature region of 400K from 350K. And there is both P_3 peak in the temperature regions more than 400K. Trap depth E_t of each peak was evaluated as about 0.82eV-0.83eV, about 1.0eV-1.1eV and about 1.4eV from the temperature region sequentially. Each peak regardless of polarity of corona charge understands that it is the same trap depth.

6.2 Escape frequency factor *v* and observation temperature *T*m

Figure 29 are the result that evaluated escape frequency factor v for the observation temperature T_m. The v value of P_1 -P_5 peaks for negative corona charge of solid film (PP1) to show in Fig. 29(a) is about 10^8 s^{-1}-10^{14} s^{-1}. In particular, the v value of P_1 peak for positive corona charge is high with about 10^{20} s^{-1}. With the two axis drawing solid film (PP2) to show in Fig.29(b), P_{ho} and a v values of the P_2 peak are about 10^{17} s^{-1} and P_{h2} and P_3 peak are about 10^{14}-10^{15} s^{-1}. As for this, order of the v value accords well about the observation region of each peak. In contrast, there are approximately 5 orders of differences even if P_{mo} and P_1 peak are the same observation regions.

On the other hand, the v values of the porous film (PP3) accord regardless of corona charge polarity well at both peaks. The v value of each peak is about 10^9-10^{10} s^{-1}of the P_1 peak, about 10^{11}-10^{12} s^{-1}of the P_2 peak and about 10^{13}-10^{14} s^{-1}of the P_3 peak.

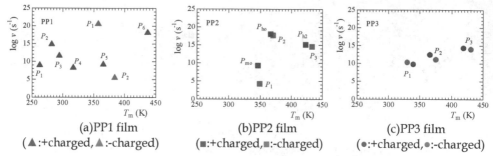

Fig. 29. Correlation of escape frequency factor v of three PP film.

6.3 Magnitude of attenuation of the charged potential

In TSC measurement, the charged potential measured twice of potential (V_{so}) before the start of measurement and residual potential (V_{se}) in the room temperature after the measurement.

sample	polarity	TSC			
		V_{so}(V)	V_{se}(V)	$\Delta V_{s,TSC}$(%)	ave.$\Delta V_{s,TSC}$(%)
PP1	−	-269	-189	-29.7	-33.4
	+	219	138	-37.0	
PP2	−	-89	-2	-97.8	-96.6
	+	272	12.6	-95.4	
PP3	−	-158	-132	-16.5	-12.1
	+	208	192	-7.69	

Table 2. Magnitude of attenuation of the charged potential.

Table 2 shows a result of the magnitude of attenuation of the charged potential. $\Delta V_{s,TSC}$ is a decrement ratio of V_{se} value for the V_{so} value. The ave. $\Delta V_{s,TSC}$ is the mean of $\Delta V_{s,TSC}$ value. A ave. $\Delta V_{s,TSC}$ value of PP1 and PP2 is about 33% and about 97% each. A value of ave.$\Delta V_{s,TSC}$ of the porous film is about 12%. A ave.$\Delta V_{s,TSC}$ value of both solid film understands that two axis drawing film comes to have a bigger degree of the decrement. In the case of positive charge, the TSC experiment shows in particular a decrease in TSC strength of the two axis drawing film. As for this, the decrease in trap charges such as an electronic charge to contribute to charged potential or ionic carrier is thought about. In addition, as for the decrease in ave.$\Delta V_{s,TSC}$, it is thought about the property of the trap such as trap depth and the escape frequency factor having changed. The porous film has less magnitude of attenuation of ave.$\Delta V_{s,TSC}$ than both solid films. Seeing from a point of view of the charged retentivity of charge, the trap which a porous film forms means that it is an important factor in the retention capacity at the high temperature.

6.4 The origin of the trap

We discuss the correlation between trap depth E_t and the escape frequency factor v. In Fig.30, the data (v-E_t correlation) of all provided traps by AEM separation system were plotted. Three domains A, B and C were assumed to three peaks P_1, P_2 and P_3, respectively.

Three domains are shown in the circle of dashed line in Fig.30. There is the trap formed of PP2 film in domain A and C. In the PP1 film, the trap is distributed widely other than P_5 peak in domain A.

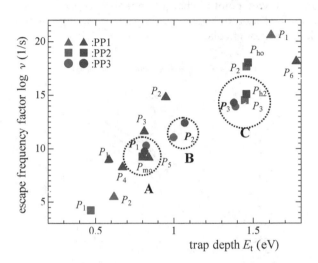

Fig. 30. Correlation of escape frequency factor v and trap depth E_t of three PP film.

The difference in trap distribution of both solid films is regarded as thing by the crystallinity. Table 3 shows correlation of v by E_t of three domains of the porous film. At first, the trap indicating the same correlation as in domain A of the porous film seems to be P_5 of PP1 film and the P_{mo} of PP2 film. Then, the trap of domain C corresponds to P_{h2} and P_3 peak of PP2 film. In order word, in the origin of the trap of the P_1 peak of the porous film, even solid film is formed. And it is thought that a trap of the P_3 peak of the porous film is a trap formed two axis drawing solid film. The dramn structure of the polymer material is complicated, but traps of domain A and domain C is formed in an amorphous part and a crystal part, respectiviely. And it is thought that the trap of the porous film which there is in domain B was formed making polypropylene porous structure. It seems to be possible that this trap is formed in a pores and a boundary of the resin.

Domain	E_t (eV)	v (s^{-1})
A	0.80-0.84	10^9-10^{10}
B	1.0-1.1	10^{11}-10^{12}
C	1.4-1.5	10^{14}-10^{15}

Table 3. Three domain of the PP3 film.

There are P_{ho} peak and P_2 peak formed of PP2 film in the upper part of domain C. The E_t value of both peaks is similar to domain C, but approximately three figures of v value are high. In addition, the trap of the P_1 peak of the PP2 film is located under. The trap (about 0.47eV) of this P_1 peak is considerably lower than trap depth of domain A located near. As a result, it is thought that the property of the trap which PP2 film has with the same two drawing film reduces charging retention capacity in comparison with porous film. When this fact compares the result of the ave. $\Delta V_{s,TSC}$ of Table 2 and TSCD characteristics of Fig.17, it is clear.

The solid film (PP1) which lowest charged quantity, however, is higher than two axis drawn solid film (PP2) from the viewpoint of charging retention capacity in the high temperature. As for the one cause, the action of trap of the P_6 peak of the high-temperature range of the negative corona charge is thought about.

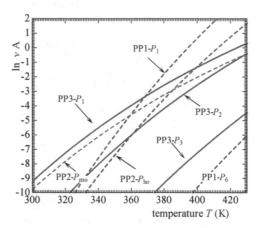

Fig. 31. A detrapping rate of the trapped carrier of the traps in the three domains.

The detrapping rates of the trapped carrier calculated in consideration of Bolzmann factor $A(=\exp(-E_t/kT))$ were presented in Fig. 31. The vA value was calculated in negative corona charging in the observation temperature region of the TSC spectrum using evaluated E_t and v. In the figure, only a P_1 peak (PP1- P_1) of the PP1 film is positive corona charging. The porous film (PP3) and the solid films (PP1 and PP2) are shown with a solid line and a dashed line, respectively. From the viewpoint of charge retention at the high temperature, the vA value for the temperature should be low. At first, it compares the vA value of three peaks of the porous film. The vA value of the P_1 peak (PP3- P_1) is the highest of all, and those of P_2 peak (PP3- P_2) and the P_3 peak (PP3- P_3) followed in order of magnetude of vA value. The detrapping characteristic of P_{mo} peak(PP2- P_{mo}) of the PP2 film resombles that of PP3-P_1. The vA value of P_{ho} peak(PP2- P_{ho}) of the PP2 film increased more rapidly for the temperature than those of the other peaks, and it seems that this trap does not contribute to the charge retention at the high temperature. And it may be said that the P_1 peak (PP1- P_1) of the PP1 film of the positive corona charge is the trap which does not contribute to the charge retention at the high temperature from a reason some as the P_{ho} peak. It is thought that P_{h2} peak and P_3 peak of PP2 film in domain C do not contribute to charge retention from a difference of the film structure with the porous film. As described above, it will be thought that the P_6 peak (PP1- P_6)of PP1 film contributes to charge retention from the result of the vA value.

On the other hand, for the positive corona charge, the trap of P_1 and the P_2 separated, could explain the result. Then trap phenomena more than 430K of both corona charging are suggested when TSCD characteristics and a result of ave. $\Delta V_{s,TSC}$ are added.

As a result of these, it is thought that the formation of the trap of the porous film forms it on an amorphous part and a crystal part as well as a pores and the boundary of the resin. The action of each domain is thought about as follows. It is thought that the trap of all domains takes the increase of the charged quantity. As for the piezoelectricity that a porous film has, it is thought that a trap of domain B participates. And, about the charging retention capacity of the high-temperature range, it is thought that a trap of domain C participates.

7. Conclusions

Using the polypropylene films of the solid state and the porous state, TSC measurements were performed. The next results became clear.

1. The surface potential isothermal decay for the positive corona charge maintained initial potential in the porous film in the progress for 3h, but was the decrement of approximately 4% with both solid films.
2. Both of solid film and porous film, as for the polypropylene, positive corona charge was higher in charged potential maximum than negative corona charge. Porous film showed the highest in the maximum charged potential.
3. From TSCD characteristics, an axis drawn solid film showed the earliest decrement of the charge potential at 430K. The decrement decreased sharply from temperature of about 385K and, for positive corona charge, was approximately 22% of initial value. The porous film had decrement from temperature more than about 410K regardless of corona charging polarity. In the porous film of the positive corona charge, decrement was caused from temperature more than about 410K and, in temperature of 430K, maintained the potential of approximately 93%.
4. As a result of having evaluated a trap by AEM separation method in a TSC spectrum, the trap of the porous film understood that it was distributed over three domains from the property.

    Domain A: E_t: 0.80-0.84eV, v: 10^9-10^{10} s^{-1}
    Domain B: E_t: 1.0-1.1eV, v: 10^{11}-10^{12} s^{-1}
    Domain C: E_t: 1.4-1.5eV, v: 10^{14}-10^{15} s^{-1}

    Because trap which was formed in an amorphous part because domain A existed in all film and domain B were only porous film, it was thought that it was formed in the surface boundary of a pores, and, in domain C, it was thought with the trap of a crystal part formed by extension of the film.
5. It was thought that all traps participated in charging as an action of trap which porous film formed and a trap of domain C participated in the heat resistance of the charging maintenance mainly in particular. In addition, it was thought a trap of domain B acted on piezoelectricity.

8. References

Baba, A. & Ikezaki, K.: "Drawing and annealing effects on thermally stimulated currents in polypropylene films",J. Appl. Phys.Vol. 72, No.5, pp.2057-2059, (1992)

Braünlich, P.: "Thermally Stimulated Relaxation in Solids", Springer-Verlag (1979), ISBN 3-540-09595-0 Springer-Verlag Berlin Heidelberg New York.

Cao, Y., Xia, Z., Li, Q., Shen, L. & Zhou, B.: "Study of Porous Dielectrics as Electret Materials",IEEE Trans. on Dielectrics and Electrical Insulation, Vol.5, No.1,pp.58-62, (1998)

Chen, R. & Kirsh, Y.: "Analysis of Thermally Stimulated Processes", Pergamon Press, Oxford (1981), ISBN 0 08 022930 1.

Garlick, G. F. J. & Gibson, A. F.: "The electron trap mechanism of luminescence in sulphide and silicate phosphors", Proc. Phys. Soc.,, Vol.60, p.574, (1948)

Ikezaki, K. & Hori, T. : "Fundamental Electric Properties of Powder-Formed Materials—Thermally Stimulated Current Spectra of Polymetic Powders—"J. Inst. Electrostatics Jpn., Vol.22, No.2, pp.79-82, (1998) [in Japanese]

Ikezaki, K. & Murata. Y. : "Derivation of Intrinsic Thermally Stimulated Current Spectra of Polymeric Powder Samples"J. Inst. Electrostatics Jpn., Vol.30, No.1, pp.14-19, (2006) [in Japanese]

Imai, T., Hirano, Y., Kojima, S. & Shimizu, T.: "Preparation and Properties of Epoxy-Organically Modified Layered Silicate Nanocomposites", Conf. Rec. 2002 IEEE ISEI, pp.379-383, Boston, USA (2002-4)

Ishii, K., Nagata, K., Osawa, H. & Nanba, N.: "Piezoelectric Properties in Porous Fluoropolymer Having Isolated Voids",Trans. Inst. Electr. Eng. Jpn., Vol.129-A, No.5, pp.373-378, (2009) [in Japanese]

Ishimoto, K., Tanaka, T., Ohki, Y., Sekiguchi. Y. & Murata, Y.: "Thermally Stimulated Current in Low-density Polyethylene/MgO Nanocomposite —On the Mechanism of its Superior Dielectric Properties—",Trans. Inst. Electr. Eng. Jpn., Vol.129-A, No.2, pp.97-102, (2009) [in Japanese]

Koga, K. & Ohigashi, H.: "Piezoelectrisity and related properties of vinylidene fluoride and trifluoroethylene copolymers", J. Appl. Phys., Vol.59, No.6, pp.2142-2150, (1985)

Lindner, M., Bauer-Gogonea, S., Bauer, S., Paajanen, M. & Raukola, J.: "Dielectric barrier microdischarges:Mechanism for the charging of cellular piezoelectric polymers", J. Appl. Phys., Vol.91, No.8, pp.5283-5287, (2002)

Maeta, S. & Sakaguchi, K.: "On the Determination of Trap Depth from Thermally Stimulated Currents", Jpn. J. Appl. Phys., Vol.19, No.3, pp.519-526, (1980)

Maeta, S. & Yoshida, F.: "On the Determination of Trap Depth from Thermally Stimulated Currents II", Jpn. J. Appl. Phys., Vol.28, No.9, pp.1712-1717, (1989)

Oka, K. & Ikezaki, K.: "Effect of Etching Treatment on Thermally Stimulated Current in Spherulitic Polypropylene", Jpn. J. Appl. Phys., Vol.31, No.4, pp.1097-1101, (1992)

Perlman, M. M & Creswell, R.: "Thermal Current Study of the Effect of Humidity on Charge Storage in Mylar", J. Appl. Phys., Vol.42, No.2, pp.531-533, (1971)

Varlow, B. R. & Li, K.: "Non-linear Characteristics of Filled Resins under Alternating Field", 2002 Annu. Rep. CEIDP, pp.52-55, Cancun, Mexico (2002-10)

Xia, Z., Gerhard-Multhaupt, R., Nunstler, W. K., Wedel, A. & Danz, R.: "High surface-charge stability of porous polytetrafluoroethylene electret films at room and elevated temperatures", J. Phys. D: Appl. Phys., Vol.32, pp.L83-85, (1999)

Yoshida, F., Kamitani, Y., Maeta, S., Yoshiura, M. & Ohta, T.: "Thermally Stimulated Currents in Polyaniline Film and their Analyses",Trans. Inst. Electr. Eng. Jpn., Vol.118-A, No.9, pp.1035-1042, (1998) [in Japanese]

Yoshida, F. & Maeta, S.: "Proposal of Asymptotic Estimation v Method Evaluating Escape Frequency Factor from a Partial Thermally Stimulated Current Curve Directly", Trans. Inst. Electr. Eng. Jpn., Vol.111-A, No.4, pp.323-331, (1991) [in Japanese]

Yoshida, F., Tanaka, M. & Maeta, S.: "Proposal of Asymptotic Estimation I Method with High Sensitivity to Thermally Stimulated Current Curve and its Application to New Analysis",Trans. Inst. Electr. Eng. Jpn., Vol.111-A, No.2, pp.104-110, (1991) [in Japanese]

12

Tailoring of Morphology and Mechanical Properties of Isotactic Polypropylene by Processing

K. Schneider[1], L. Häussler[1] and S.V. Roth[2]

[1]*Leibniz-Institut für Polymerforschung Dresden,*
[2]*DESY, Hamburg,*
Germany

1. Introduction

The deformation behaviour of semi-crystalline materials is mainly determined by the behaviour of the two components – the crystalline and the amorphous phase with their characteristic temperature-dependent mechanical behaviour and sometimes their anisotropy. So the crystalline phase is elastically with a rather high modulus. Above a certain stress the crystallites break down into smaller fragments. Aligned chains enable recrystallisation. The mobility in the amorphous phase depends on the difference between the ambient temperature and the temperature characteristic of the glass transition, which is the dominant relaxation process in the temperature range under investigation. On the other side the amorphous phase is constrained within the crystalline one. So it shows to some extent stress relaxation or frozen stress. Both phases are connected via anchor molecules, bridging the phase boundaries. Those molecules are mainly responsible for stress transfer between the phases.

The actual morphology and so the subsequent interaction of the different phases within a samples are mainly determined by processing and the thermal history of the material. For instance, in an injection moulded plate the properties can be extremely dependent on the position and direction of a specimen, changing from relatively brittle to highly stretchable (Schneider, 2010).

The whole stress-strain curves of semi-crystalline materials generally show three characteristic regions. Although the initial region prior to the yield point apparently behaves elastically, stress relaxation due to rearrangements in the amorphous phase can be observed also here. The mobility in the amorphous phase depends on the difference between the ambient temperature and the temperature characteristic of the glass transition, which is the dominant relaxation process in the temperature range under investigation. In the case of confined stretched amorphous regions between crystalline phases the glass transition temperature can be changed significantly. After the yield point typically local necking occurs with high local strain while the overall strain remains moderate. In the case of dog-bone specimens then the neck propagates over the whole specimen during constant load. In the true stress-strain diagram necking is a fast local deformation from the yield strain to the strain of the fully yielded specimen. In the third step during further elongation strain hardening occurs, until finally the specimen fails. Here the strain hardening modulus can be a relevant parameter for the long term stability against creep (Kurelec, 2005).

Already in his first model of structural changes during deformation Peterlin (Peterlin, 1971) discussed spherulites consisting of lamellae separated by amorphous regions. Due to tie molecules stress is transferred between the lamellae inducing deformation and finally disintegration and rearrangement in the form of fibrillae. Later numerous authors modified and enhanced Peterlin's approach to describe their experiments. The discussion gained new impetus after the first online structure investigation during deformation by Zuo et al. (Zuo, 2005).

Breese and Beaucage (Breese, 2004, 2008) gave an overview of older models. Finally, they describe mechanical behaviour with a model refining the models of Weeks and Porter (Weeks, 1974) and Gibson, Davies and Ward (Gibson, 1978) in the sense that the effects of the orientation process on the modulus of the non-fibrous gel component are incorporated into the model by allowing a transition to fibres.

To describe the deformation behaviour of semi-crystalline polymers Strobl et al. (Hong, 2004, 2006) separated different mechanisms of stress transfer with respect to amorphous and crystalline units. They distinguished between four phases: onset of local block sliding (1), collective motion slightly below the yield point (2), disassociation of crystalline blocks and transformation into fibrils (3) and the start of disentangling (4).

In most models the common process of cavitation in polymers during deformation is not yet incorporated. Many systems show a characteristic whitening during plastic deformation due to the formation of voids or cavities with a typical length scale growing up to the wavelength of visible light, i.e. some hundreds of nanometres (Pawlak, 2010; Men, 2004). It is obvious that these large-scale structures are not accessible with a conventional small angle X-ray scattering setup. But at least the beginning of this process should be well detectable by X-ray scattering due to the strong difference in electron density between the polymer and the voids in the relevant angular range accessible by USAXS (Ultra-SAXS (Lode, 1998; Gehrke, 1995; Roth, 2006)).

In previous studies Davies et al. (Davies, 2004) demonstrated a continuous generation of the voids in the plastic phase starting at the yield point, which led to a conspicuous increase in the scattering power. Furthermore, a change in the diffuse scattering profile indicates a monotone change in the size and in the shape of the voids.

Stribeck et al. (Stribeck, 2008) described nanostructure evolution in Polypropylene during online mechanical testing with simultaneous SAXS and refined details of the interaction between the different phases including cavitation.

As an example the properties of isotactic Polypropylene (iPP) samples, produced by compression and injection moulding as well as after hot stretching, respectively, were investigated under stepwise loading. Simultaneously the structural changes were characterised by Synchrotron-SAXS and –WAXS and discussed together with the mechanical properties. It was found that there is a highly preoriented structure within the injection moulded specimen, which strongly influences the further temperature-dependent deformation behaviour. The influence of temperature and pre-orientation on the deformation behaviour is described extending the present models of plastic deformation. The detailed knowledge of the structure and the microscopical behaviour of the material enables the tailoring of the processing conditions for material with certain mechanical behaviour.

Recently we published an overview about our first investigations about the structural changes in iPP during deformation using SAXS, WAXS, DSC and SEM-micrographs (Schneider, 2010, 2011). The present work will complement and round up the investigations described in those papers.

2. Experimental

2.1 General procedure

In order to investigate structural modifications during deformation X-ray scattering experiments were performed on samples mounted in a miniaturised tensile rig placed in the synchrotron X-ray beam. The synchrotron radiation is necessary to have sufficient intensity to get high time resolution. A general description of the equipment was given by Davies et al. (Davies, 2004).

The actually used experimental arrangement and the specimen geometry are illustrated in figure 1.

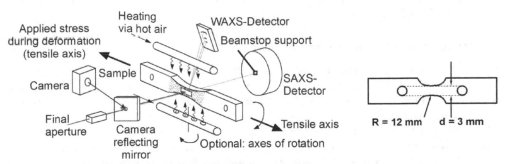

Fig. 1. Sketch of the experimental arrangement for SAXS and WAXS during deformation (left) and waisted specimen (mini-dumbbell) mainly used for simultaneous structure and mechanical investigation (right), the dimensions of the specimen can be scaled.

To investigate local strain-dependent properties, small waisted specimens were used in order to concentrate the stress in the centre of the specimen. By using a relatively large radius of curvature (12 mm) compared to the specimen width (3 mm) the stress state in the middle of the specimen is in good approximation maintained constant. The strain was determined optically by observing the deformation of a grid pattern applied on the specimen surface. The grid pattern was applied using a self-made flexible ink and a mesh size of 0.35 mm. Alternatively, also a classical image correlation analysis for strain estimation can be used. A comparison of stress-strain curves of the waisted specimens with standardised dog-bone specimens shows a good consistency: The curve progression from the beginning to the yield point as well as during strain hardening (that is where the yielded parallel region of the dog-bone specimen is deformed in a uniform way) are identical. For the range of neck formation and propagation in the dog-bone specimen an estimation of true stress and strain via global strain measurement are not possible. Only a local strain measurement solves this problem. In the present context the specimens are approximated as incompressible. This allows us to calculate the true stress σ_t from the measured force F, initial cross section A and tensile strain ε_t as $\sigma_t = F * (1 + \varepsilon_t) / A$.

In order to keep the beam in a fixed position relative to the gauge length of the waisted sample throughout the measurement, both grips were moved simultaneously in opposite direction.

To get simultaneous SAXS and WAXS patterns, the WAXS could be monitored only in a limited range using a tilted detector. By using a horizontal tensile direction the vertically arranged WAXS detector monitors mainly the equatorial scattering of the sample. To follow also quick changes exclusively in the crystalline phase WAXS measurements were performed alternatively with the WAXS detector directly in the beam behind the sample and a sampling rate of 1 s.

For temperature-dependent tensile and scattering experiments a small heating device blows locally preheated air at the sample.

Besides the global scanning of samples the experimental setup generally also permits spatially resolved pattern recording, e. g. around a crack tip, if it is employed in a microfocus beamline. The described arrangement was successfully used investigating semi-crystalline polymers during deformation (Schneider, 2006, 2010, 2011) and fracture (Schneider, 2008, 2009).

2.2 Measurements at synchrotron beamlines

WAXS and SAXS measurements during deformation were performed at the synchrotron beamlines BW4 and P03 at HASYLAB in Hamburg, Germany (Roth, 2006). At BW4 the wavelength of the X-ray beam was 0.13808 nm, the beam diameter was about 400 µm. The SAXS images were collected by a two-dimensional MarCCD-detector (2048 x 2048 pixels of 79.1 x 79.1 µm^2). The sample-to-detector distance was set to 4080 mm. The WAXS images were collected by a two-dimensional PILATUS 100K-detector (487 x 195 pixels of 172 x 172 µm^2). The position of the WAXS detector was estimated using certain reference reflexes of iPP. By a special procedure the position of the tilted WAXS-detector was determined to have a tilt angle of 22.7°. The distance between the sample and the point of normal incidence was 247 mm. Exposure times were chosen in the range of 5 to 60 s per pattern. The frame rate, determined by exposure and data storage, was 15 to 70 s per pattern.

Additionally, WAXS measurements were performed at the synchrotron beamline P03 at HASYLAB (for a beamline overview see Roth et al., 2011) with a wavelength of 0.0941 nm and a beam size of about 15 µm diameter, using a PILATUS 300K-detektor (487 x 619 pixels of 172 x 172 µm^2). The detector was placed vertically in the beam in a distance of 144.9 mm, the sampling rate was 1 s. To prevent sample damage due to the high intensity of the used X-ray beams at P03 during the measurements normally slow scans along or across the sample were performed.

For the discussion all SAXS- and WAXS-2D-patterns are shown with vertical tensile direction. For specimens with fibre symmetry this means that the fibre axis and so the scattering vector s_3 is also vertical, the scattering vector s_{12} is always horizontal.

2.3 Supplementary investigations of the material properties

For the investigation of the general temperature-dependent stress-strain-behaviour and the preparation of specimens for dynamic mechanical analysis (DMA) of highly stretched specimens short dogbone-specimen (parallel length 25 mm, specimen width 10 mm, thickness 4 mm, strain rate 80 mm/min) were measured with a testing machine Zwick 1456 from Zwick, Germany with heating chamber and optical strain measuring system.

DMA was performed by a rheometer ARES G2 from TA-instruments in torsion mode with a heating rate of 5 K/min and a frequency of 1 rad/s (6.28 Hz) as well with a dynamic mechanical spectrometer EPLEXOR 150N from GABO QUALIMETER Testanlagen GmbH, Germany in tensile mode with a heating rate of 3 K/min and a frequency of 1 Hz.

For qualitative comparison of crystallinity and the melting behaviour of samples stretched under certain conditions some differential scanning calorimeter (DSC) measurements were performed on drawn samples using a DSC Q 1000 from TA-instruments. The specimens were measured between -80 and 230 °C with a heating rate of 20 K/min. Before each scan the sample was equilibrated at constant temperature for 300 s. In the case of heating into the melting region a heating rate of 10 K/min and a subsequent cooling rate of 80 K/min were used.

As qualitative check of the discussed structures some SEM images of the drawn samples were taken on a Zeiss Gemini Ultra plus SEM.

2.4 Material and sample preparation

The investigations were done on isotactic polypropylene (iPP), grade HD 120 MO from Borealis. Both injection and compression moulded plates were used. The melt temperature in injection moulding was 245 °C, the mould temperature was 40 °C. The mould has a film gate with a width of the half of plates thickness.

Compression moulded plates were used, produced by heating injection moulded plates to 210 °C for 5 minutes in a vacuum press. Then the plates were cooled down to room temperature at 15 K/min. The specimen used for scattering have always a thickness of about 1 mm. Mini-dumbbell specimens were produced by CNC milling. For the investigation of the general temperature-dependent stretching behaviour from the plates with 4 mm thickness short dumbbell specimen (parallel length 25 mm, specimen width 10 mm, thickness 4 mm) were produced also by CNC milling.

3. Evaluation of scattering data

3.1 SAXS data evaluation

The evaluation of SAXS images is strongly related to the approach for materials with fibre symmetry developed by Stribeck (Stribeck, 2007). Data processing was realised with the software package pv-wave from Visual Numerics.

The images were normalised with respect to the incident flux and blind areas were masked. Considering the sample absorption the instrument background was subtracted. By translation and rotation the images were aligned in that way that the tensile direction is vertical and the beam position in the middle of the patterns. Finally, the patterns were averaged with respect to the four quadrants of the detector. The blind area of the beam stop is interpolated assuming a Guinier-type behaviour of intensity in this range.

Generally, SAXS realises a projection of the specimen structure into the reciprocal space, where it produces a slice of the Fouriertransform (FT) of the structure whose amplitude is measured. A single back-transformation would deliver a projection of the autocorrelation function of the structure.

Complete information about the specimen would be available only by tomographic methods with a stepwise rotation of the sample (see e.g. Schroer, 2006) or using inherent symmetry properties of the sample. Under the assumption of fibre symmetry of the stretched specimen around the tensile axis, from the slices through the squared FT-structure the three-dimensional squared FT-structure in reciprocal space can be reconstructed and hence also the projection of the squared FT-structure in reciprocal space. The Fourier back-transformation of the latter delivers slices through the autocorrelation function of the initial structure. Stribeck pointed out that the chord distribution function (CDF) as Laplace transform of the autocorrelation function can be computed from the scattering intensity $I(s)$ simply by multiplying $I(s)$ by the factor $L(s) = -4\pi^2(s_{12}^2 + s_3^2)$ prior to the Fourier back-transformation. Here s is the scattering vector with the component s_{12} in transversal direction and s_3 in fibre direction.

The interpretation of the CDF is straightforward (Stribeck, 2001), since it has been defined by the Laplacian of Vonk's multidimensional correlation function (Vonk, 1979). It presents autocorrelations of surfaces from the scattering entities in that way that positive values characterise distances between surfaces of opposite direction, negative values distances between surfaces of the same direction, see figure 2.

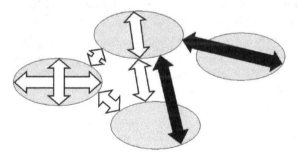

Fig. 2. Information of CDF's: In any particular structure positive values of the CDF represent correlations of interfaces with opposite direction (light arrows); negative values represent correlation of interfaces with the same direction (dark arrows).

Hence positive peaks in the vicinity of the origin characterise size distributions of the primary domains. In the case of semi-crystalline materials this can be crystallites as well as amorphous regions in-between. If cavitation occurs it will be superimposed by the size of the cavities. Also a relatively small amount of cavities will be visible because the difference in electron density is here much higher than between crystalline and amorphous phase within the polymer.

Following negative peaks characterise distances between adjacent repetition units ("long periods").

Positive peaks at greater distances describe size and orientation of domains (from the beginning of the first domain to the end of the second one). In well oriented systems also peaks of higher order can be observed.

For detailed discussion the CDF's can be presented as contour plots or as density plots in the plane.

3.2 WAXS data handling

Unfortunately, the PILATUS detector has some dark regions. After masking the beam stop those regions were reconstructed using the symmetry properties of the pattern.

For the qualitative discussion in the present case the individual images were not normalised with respect to the incident flux and not corrected with respect to background scattering.

4. Results

4.1 Mechanical behaviour of injection moulded iPP

The temperature-dependent stress-strain-behaviour of iPP is shown in figure 3.

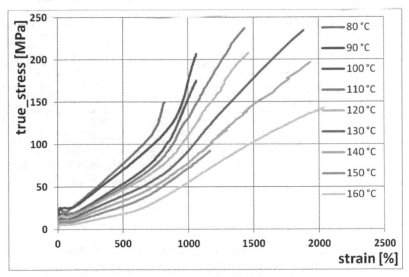

Fig. 3. Temperature-dependent stress-strain-behaviour

The general behaviour is characterized by a decrease of the initial modulus as well as the yield stress with temperature. Yielding of the material is finished at a strain of about 600 ... 900 %, afterwards there is a strain hardening. The strain hardening modulus (slope of the true stress vs. strain curve in the region of strain hardening) decreases with the temperature.

The stress at failure initially increases with the temperature. At temperatures above 130 °C the stress at failure again decreases. The strain at failure increases steadily to finally more than 2000 %.

There is some whitening of the specimens, mainly due to cavitation. It starts at lower temperatures above the yielding point. With increasing temperatures the initiation of whitening shifts to higher strains. At temperatures above 110 °C it happens in the strain hardening region, with increasing temperature nearer to failure.

The small decay in the curves above the yielding point is due to the fact, that these experiments were performed with tensile bars with parallel gauge length and a distance of the optical marks of about 10 mm. Using samples with waisted geometry and strongly localized optical strain measurement this decay can be prevented to the greatest extent (Schneider, 2010).

4.2 Crystallite identification

According to Bragg's law the positions of WAXS reflexes refer to the distance between crystalline planes within the crystallites. In the compression moulded as well as in the injection moulded plates a couple of crystalline reflexes could be identified. The reflexes with the highest intensity within the relevant angular region are summarised in table 1.

Reflex	Intensity (qualitatively)	d / nm	Scattering angle 2 Θ / deg	
			P03	BW4
1 1 0	strong	0.6269	8.608	12.64
0 4 0	strong	0.5240	10.303	15.13
1 3 0	strong	0.4783	11.291	16.59
1 1 1	strong	0.4170	12.957	19.05
0 4 1	strong	0.4058	13.316	19.58

Table 1. Crystalline reflexes of iPP, which were used for the calibration of the detector distance as well as for the further discussion of crystallite orientation and changes during deformation.

4.3 Elastic crystallite deformation

Loading of the samples below the yield point causes a certain deformation of Debye-Scherrer patterns of the crystallites. According to Bragg's law a scattering signal refers to the spacing between lattice planes. The shift inversely reflects the strain vertical to the corresponding plane. In the present context the longitudinal tension as well as the transversal compression could be followed. Due to the stiffness of the crystallites deformations only in the range of about 1 % could be found, which are difficult to be resolved with the pixel size of the used detector. By unloading the samples the crystallite reflexes shift again to the initial positions. Similarly, also the thermal expansion of the crystals could be found only qualitatively.

This behaviour we reported yet in the past (Schneider, 2010), where we compared the mismatch between crystalline and global strain by a factor in the order of 10 indicating the different stiffness of the crystallites compared to that of the whole specimen due to the relatively soft amorphous regions.

In the present context the shift of the observed (h k 0)-reflexes in equatorial direction (perpendicular to tensile direction) display the transversal crystalline strain of the crystallites aligned in tensile direction.

Cyclic loading and unloading gives an estimation of the elastic crystalline strain also during plastic deformation. Figure 4 shows transversal crystalline strain of the (1 1 0)-reflex vs. the overall tensile strain for 3 consecutive loading cycles. The first strain amplitude was 0.12, the remaining strain after unloading 0.04. The following cycles were conducted within the overall elastic region. The shift of other (h k 0)-reflexes yield the same result.

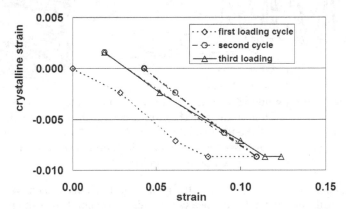

Fig. 4. Transversal crystalline strain vs. optically measured tensile strain of semi-crystalline iPP specimen estimated via the shift of the (1 1 0)-reflex at room temperature.

In a range of up to about 8 % strain the crystallites deform transversally by about -0.8 %. The mismatch between crystalline and global strain indicates the different stiffness of the crystallites compared to the whole specimen due to the relatively soft amorphous regions. During further loading the strain of the crystallites remains constant. This means that further deformation is realised on the micro-scale only by the amorphous phase and by additional shear dislocation of the crystallites.

4.4 Crystallite orientation of the samples and reorientation during drawing

In a preliminary investigation at BW4 stepwise loading at different temperature was performed. The stress-strain-diagrams captured during these measurements are presented in detail recently (Schneider, 2010, 2011), they are comparable to the mechanical behaviour of the preliminary tests. The stress-strain-level is slightly lower due to the generally lower strain rate. For the flat samples under investigation the transmission was 0.83 in the unstretched state. During stretching and simultaneous thinning the transmission generally increased to 0.96.

The images were always transformed to reference coordinate systems with respect to the scattering angles (scattering angle 2Θ and azimuthal angle φ or scattering vectors s_{12} and s_3). For further discussion the scattering intensity was projected over a selected range of φ on the 2Θ-axis. Figure 5 shows some characteristic equatorial cuts together with the transformed pattern.

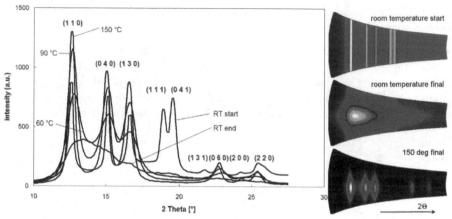

Fig. 5. Equatorial scattering intensity of undeformed iPP at room temperature and during strain hardening at elevated temperatures. The curves always show a projection of the angular range of ±2 ° around the equatorial direction. Right: WAXS-patterns, azimuthal angle vs. scattering angle 2Θ at three characteristic points: undeformed as well as immediately before failure at room temperature and 150 °C.

During deformation the WAXS patterns change in a quite characteristic way, see figure 6. Initially the intensity of the crystalline peaks is relatively constant over the azimuthal angle pointing to a homogeneous distribution of crystallite directions. At deformations below the yield point slight changes in the intensity of the peaks are reversible.

Above the yield point the deformation behaviour is strongly temperature-dependent. Generally in the investigated equatorial direction only the (h k 0)-peaks remain and the scattering intensity concentrates azimuthally in equatorial direction. This is generally an indication of orientation of crystallites with the c-axis (chain direction) in tensile direction. But while the peaks sharpen with increasing temperature, which is an indication of growing crystallites, they decrease strongly and broaden at temperatures in the range immediately above room temperature. This indicates a gradual disruption of crystallites into smaller fragments. Finally, at room temperature the distinct peaks disappear but give a broad halo in the angular range of the previous peaks. This indicates a rough orientation of the chains or very small crystallite fragments in tensile direction, probably in the form of fibrils. However, the thermal mobility of the chains seems to be all in all insufficient to establish new crystallites.

After these recent investigations with simultaneous SAXS and WAXS measurements with limited range of the WAXS-detector (Schneider, 2010) some representative WAXS measurements were repeated at elevated temperatures with whole WAXS-range and certain temperature-strain-program. The samples were heated to 110 °C in the tensile rig. At this temperature the samples were stretched to about 300 %. Afterwards they were heated stepwise to temperatures below the melting temperature and cooled again up to 80 °C, both with and without load. The individual behaviour is described in detail.

The investigated injection moulded samples showed initially mostly homogeneous WAXS pattern with Debye-Scherrer patterns related to the reflexes mentioned in table 1. The samples are to a large extent crystallised. Heating the samples to about 110 °C didn't change the WAXS pattern.

During stretching the samples to about 300 % the pattern changed drastically, see figure 6. The intensity of the rings concentrates to certain positions, characteristic for well aligned crystallites in stretching direction.

Fig. 6. Changing of the WAXS pattern during deformation at 110 °C (strain from left to right: 0 %, 70 %, 85 %, 300 %) of a waisted specimen (mini-dumbbell). The stretching direction is vertical.

Under the condition of fibre symmetry signals in the pattern vertical to the tensile direction report highly oriented lamellae or lamellar fragments in tensile direction. This final pattern remains mainly constant also during further heating to about 160 °C (below the melting point) under load as well as unloaded.

Further heating of the samples very close to the melting point creates a sharpening and broadening of the crystalline reflexes in azimuthal direction: Apparently the thermal stress is released, the mobility of the unloaded crystallites becomes higher and the orientation of it becomes a little bit less, see figure 7.

Fig. 7. Sharpening of the WAXS reflexes during deformation approaching the melting point (temperature from left to right: 110, 160, 165, 170 °C) of a waisted specimen (mini-dumbbell). The stretching direction is vertical.

Immediately above the last position in figure 7 the sample melts and all WAXS-reflexes disappear.

4.5 Crystallisation behaviour during subsequent cooling

The azimuthal width of the standard reflexes of the stretched iPP – if present - remains mainly constant also during following crystallisation processes by cooling. Cooling to about

140 °C doesn't change the pattern. During further cooling different routes of crystallisation can be observed, see figure 8.

Fig. 8. Pattern during recrystallisation of a stretched iPP sample at temperatures below 140 °C: Isotropic pattern after crystallisation from the melt, a slight preferential orientation may be observed (creation of crystallites with a wide range of orientation); crystallisation of the common stretched lamellar fragments; generation of a meridional double-reflex (daughter lamellae); creation of a new series of (h k 1)-crystallites (from left to right). The stretching direction is vertical.

While the generally present reflexes of the lamellar fragments remain nearly unchanged, notable is the formation of a new meridional double-reflex with different strength. This reflex is discussed in the literature often as the formation of epitactic daughter lamellae (Kumaraswamy, 2000). Otherwise in some cases a whole system of reflexes in the plane of the (h k 1)-reflexes is found, which is described in the literature as nano-oriented crystals (Okada, 2010). The appearance of the latter depends strongly on a small temperature window below the melting point, to which the stretched sample must be heated.

4.6 Structural changes across a sample – influence of a temperature gradient

In the present investigations there had been a certain temperature gradient across the samples. So the initial temperature of the sample before cooling was different across the sample. As a consequence finally all the different structures discussed above could be found side by side in the same sample.

4.7 SAXS during stretching

From the set of successive scattering patterns during the deformation for the following discussion only patterns and the corresponding CDF's were chosen where characteristic changes could be seen.

The respective results of SAXS from stretching at room temperature are shown in figure 9. The regular circular form of the CDF from the beginning is nearly unchanged to about 6 % strain also during unloading. It represents the randomly distributed crystallites, which were deformed elastically, as discussed in the section about WAXS above. The diameter of the inner ring in the CDF indicates a lamellar thickness of 9.8 nm. The isotropic ring in the negative direction represents the long period of 17.7 nm. The following second positive and negative reflexes indicate a certain correlation with the third neighbours. Missing higher-order reflexes indicate that there are surprisingly no remarkable correlations above the third neighbours. During deformation below the yield point (loading, unloading, relaxation and repeated loading) this general situation is unchanged.

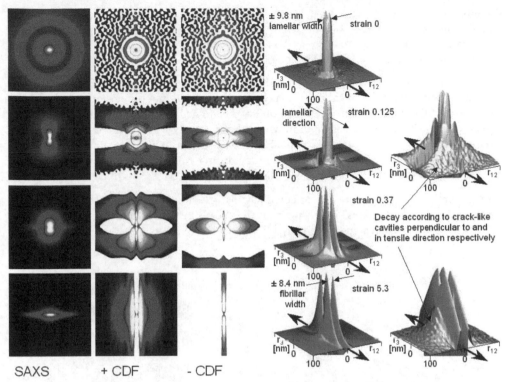

SAXS   + CDF   - CDF

Fig. 9. Deformation of iPP at room temperature, from left to right: SAXS patterns, positive and negative CDF's (always log scale, pseudo-colour) as well as a surface plot of the CDF's (linear scale and for two strains also log scale) at different strains ε = 0.0, 0.125, 0.37 and 5.3 (from top). Each square of the pattern covers a range – 0.12 nm^{-1} < s_1, s_3 < 0.12 nm^{-1}, each square of the 2D-CDF covers a range of – 100 nm < r_{12}, r_3 < 100 nm, fibre direction always vertical. The stretching (fibre) direction is indicated in the surface plots by arrows.

Beyond the yield point the situation changes drastically. First of all it is noticeable that the total scattering intensity as well as the extrapolated intensity I_0 increase dramatically suggesting the activity of new strong scatterers – presumably cavities, which exhibit a strong scattering contrast to iPP due to their extremely low electron density. They start as small cracks transversal to the tensile direction and finally deform to long drawn cavities between fibrils. According to their successive growth and hence their broadly distributed dimensions they hardly show distinct scattering signals, but overlap with the signals arising from the lamellae or their fragments. The growing cavities are also responsible for the whitening of the cold drawn samples (see also Pawlak, 2010). In contrast to the scanning experiments with a microbeam reported by Roth (Roth, 2003), which allows to resolve certain individual voids, here over a large ensemble of cavities is averaged. Within the CDF's these cavities are reflected by the broad initial decay in fibre direction, at higher strains perpendicular to fibre direction, see figure 9, CDF's in log scale on the right.

Yet at a strain of 12.5 %, the beginning of the yielding region, the structure becomes anisotropic. While thickness of the lamellae remains constant, they align perpendicular to the tensile direction. The correlation to the next neighbours in transversal direction is lost and concentrated within a 4-point-pattern. This indicates occasionally an internal shear deformation. Under the external stress the lamellae are sheared and finally break down into certain blocks as described by Strobl (Hong, 2004).

A new transversal correlation length is established at about 37 % strain. By shearing the lamellar blocks break down to a size which later represents the dimensions of the fibrils. According to the relatively low internal mobility the aligned chains are not able to form new crystallites. Their mean dimension in transversal direction is with 8.4 nm clearly below the lamellar thickness. This correlates with the WAXS results, which also indicate the final disappearance of the lamellar transversal order during cold drawing. Ultimately, at high strains, this transversal correlation dominates the whole CDF.

The general changes within the oriented chains – within crystallites as well as non-crystallized stretched chains – are shown in the sketch in figure 10, deformation at room temperature is shown in steps a), b), c), f) and g).

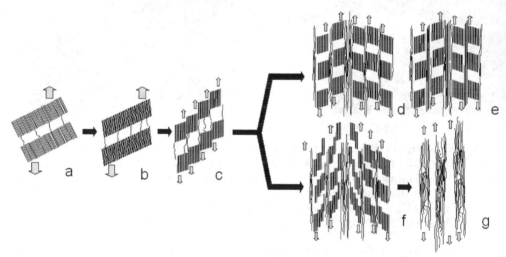

Fig. 10. Sketch of the transformations of the oriented chains during deformation, in-between are additional amorphous chains: a) lamellae with some tie-molecules, b) elastic shear-deformation of lamellae under small load and reorientation with respect to the load, c) fracture of lamellae into smaller blocks due to local stress concentration caused by tie molecules, d) stretched aligned, but not re-crystallised chains between the blocks during stretching at higher temperatures, e) some of the fibrillar arranged molecules crystallise, final stage in the case of hot stretched iPP, f) further dissolution of the blocks creating more extended chains at room temperature, g) finally, there are several strands of extended chains, not crystallised, with some amorphous regions in between, final stage in the case of cold stretched iPP.

With increasing temperature the deformation behaviour totally changes. The patterns and the CDF's of the stretching of iPP at 130 °C are shown in figure 11. Here the averaged as well as the extrapolated maximum intensity I_0 does not change dramatically above the yield point. Instead soon the lamellae align in tensile direction increasing their correlation in this direction, indicated by higher order of reflexes in the CDF's. The transversal displacement of the reflexes suggests that the aligned lamellae are shifted against each other. The situation does not change generally also at high strains. This interpretation is also supported by the WAXS results described above. The deformation is shown schematically in figure 10 a) – e).

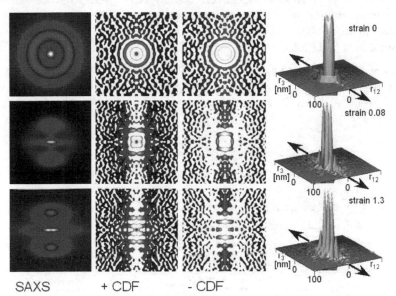

Fig. 11. Deformation of iPP at at 130 °C, from left to right: SAXS patterns, positive and negative CDF's (always log scale, pseudo-colour) as well as a surface plot of the CDF's (linear scale) at different strains ε = 0.0, 0.08 and 1.3 (from top). Each square of the pattern covers a range – 0.12 nm^{-1} < s_1, s_3 < 0.12 nm^{-1}, each square of the 2D-CDF covers a range of -100 nm < r_{12}, r_3 < 100 nm, fibre direction always vertical. The stretching (fibre) direction is indicated in the surface plot by arrows.

With increasing temperature also the whitening is remarkably reduced indicating that due to the higher mobility of the amorphous phase nearly cavitation is negligible as compared to the extent observed at room temperature.

Over the whole deformation range stress relaxation is observed as soon as the tensile rig was stopped. However since this can be ascribed to the amorphous phase, SAXS results remain mostly unaffected.

4.8 Influence of processing on the samples morphology

The influence of injection moulding on a preliminary orientation of the crystallites and their deformation we described recently (Schneider, 2010). It was shown, that injection moulded specimens in injection direction as well as the compression moulded and quickly cooled specimens are highly stretchable. By contrast the specimens transversal to the injection direction fail very soon. Here in some cases even crazing could be observed before failure.

This points out that there will be a strong structural anisotropy due to the processing history (shear stress as well as cooling rate).

4.9 DSC investigation of unstretched and stretched specimens

To check the thermal behaviour respectively melting of different stretched samples some DSC measurements were performed. In a first test samples were pre-stretched at different temperatures to about 500 %. The heating peaks of the first heating curves of the stretched samples are shown in fig. 12.

It seems to be possible to split each curve into 2 individual melting peaks whose positions shift with the stretching temperature to higher values. Furthermore, while the high-temperature peak is nearly constant the low-temperature peak increases with stretching temperature.

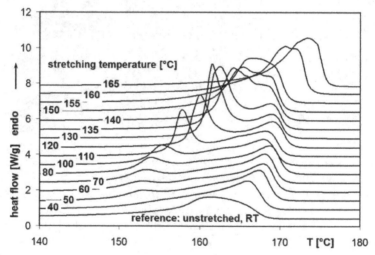

Fig. 12. DSC-measurements: first heating curves of compression moulded iPP-specimens stretched at different temperatures. For clarification the curves are vertically shifted.

To check, whether there are two individually melting crystalline species in a second series samples, stretched to about 500 % at 110 °C, were heated to different temperatures behind the first peak in the heating curve. Afterwards they were cooled to room temperature before they were heated over the melting point in a second run, see figure 13.

The position of the first peak flutters a little bit due to the different specimen taken from a stretched sample. Heating in the first run well above the melting temperature – 175 or 200 C – the second heating curve is similar to the first one with a slight double peak. If the first heating is stopped at 160 °C, the second heating curve has the narrower double peak, shifted to higher temperatures. Surprisingly in the case of the interrupted first heating at 165 °C the second heating peak is shifted to remarkable higher values and only a unique peak is observed.

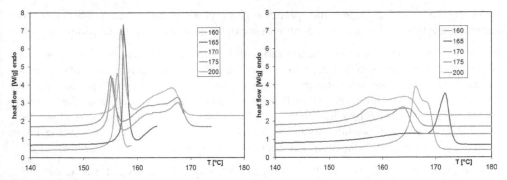

Fig. 13. DSC-measurements: first and second heating curves of compression moulded iPP-specimens stretched at 110 °C to about 500 %. The first heating was stopped at different temperatures after the first peak in the heating curve, see the label within the diagram. For clarification the curves are vertically shifted.

4.10 DMA investigation of unstretched and stretched specimens

The general different interaction of amorphous and crystalline regions in the initial (spherulitic) as well the highly stretched (fibrillar) state was investigated with DMA. The stretched material was produced by hot-stretching of small tensile samples from plates with a thickness of 4 mm and 10 mm width at different temperatures.

With shear and tensile load two independent measurements were performed, see figure 14. Under shear load the crystallites (fibrils) and the amorphous regions are more or less in series, under tensile load they are parallel. Accordingly the combination of the moduli is quite different.

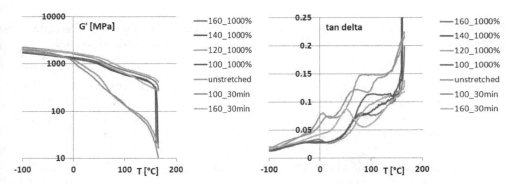

Fig. 14. Measurement under torsion: G′ (storage modulus) and loss tangent of differently treated iPP: injection moulded, annealed and hot-stretched samples. Parameters of the sample preparation are shown in the legend

At low temperatures the storage modulus is almost the same: Crystalline as well as amorphous regions behave stiff. At the common glass transition temperature at about 0 °C there is a small peak in the loss tangent and a decrease in the storage modulus due to the onset of the mobility of the amorphous regions of the unstretched (annealed or not annealed) sample. This content is nearly missing in the case of the stretched samples.

In the range between room temperature and melting temperature there is a continuous drop in the storage modulus of the unstretched samples due to the increasing mobility of the amorphous phase with temperature. The amorphous phase is constrained between the crystalline one. Therefore, here is such a high temperature dependence of the mobility. Annealing at 100 °C enables some post- or re-crystallisation causing an increase in modulus and a decrease in the loss-tan. Annealing at 160 °C, i.e. shortly below the melting point, supports the further crystallization. The relatively strong drop down of G' remains.

The situation is totally different after high stretching. The signal of the undisturbed amorphous phase nearly totally disappears; the constrained amorphous phase enables only very slight molecular motion between the aligned crystallites. Annealing at increasing temperatures reduces this remaining mobility.

The behaviour under tensile load is somewhat different, see figure 15.

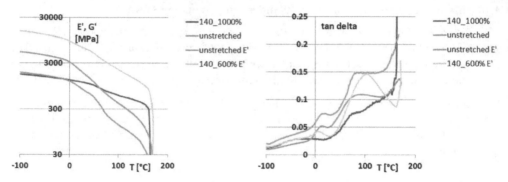

Fig. 15. Measurement under tension: E' (storage modulus) and loss tangent of differently treated iPP in comparison to the torsion measurement G': injection moulded and hot-stretched

Besides the generally higher values of the tensile modulus with respect to the shear modulus according to $E = 2 (1 + \mu) * G$, the crystalline and the amorphous phase after the stretching are more or less parallel. By this a noticeable increase in E', small contribution of the glass transition near 0 °C and a further softening around 100 °C appears.

4.11 Micrographs and SEM of stretched specimens

Micrographs under polarized light of iPP crystallised within a rheometer as well a stretched sample are shown in figure 16. Here the change from spherulitic to fibrillar structure is clearly visible. SEM images were taken to verify the described general deformation behaviour. Figures 17 to 20 show the compression moulded unstretched reference specimens and specimens stretched at room temperature, 90 °C and 150 °C. The specimens were cut in tensile direction, and vapour deposited with platinum. Stretching direction and thus s_3-axis are always vertical in the figures 17 to 20, the s_1-axis horizontal.

Fig. 16. iPP crystallised within a rheometer without shear and cold stretched sample with fibrillar strands (stretching direction diagonal, see arrow)

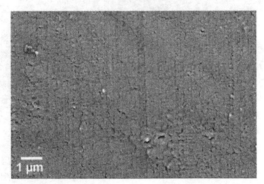

Fig. 17. Scanning electron micrograph of the unstretched reference iPP sample.

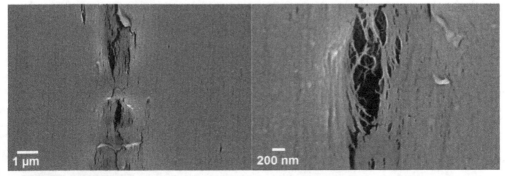

Fig. 18. Scanning electron micrograph of the iPP sample stretched at room temperature. Stretching direction vertical.

Fig. 19. Scanning electron micrograph of the iPP sample stretched at 90 °C. Stretching direction vertical

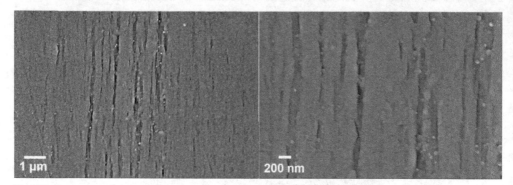

Fig. 20. Scanning electron micrograph of the iPP sample stretched at 150 °C. Stretching direction vertical.

Figure 17 to 20 show the voids in the stretched samples the formation of which was discussed above. Their size decreases with increasing stretching temperature. In the sample stretched at 150 °C they are not found. The lamellae and fibrillae are not visible at the present magnification, only a slight texture in stretching direction can be seen. A similar trend was reported recently by Pawlak (Pawlak, 2010), who found voids only at stretching temperatures below 60 °C, at higher temperatures only deformed spherulites and fibrils.

5. Discussion

Stretching semi-crystalline iPP at temperatures changes the morphology totally. Below about 80 °C the crystallites become destroyed, the stretched fragments oriented in stretching direction. This causes the wide peak in the WAXS pattern without a fine structure referring to the individual crystalline parameters. A strong cavitation appears because the transversal strength of the lamellae is less than that of the stretched amorphous phase, which one transfers stress to the lamellae. The relatively stiff but irregularized structure fails at comparable low stresses.

Stretching above about 80 °C causes a destruction of the initial morphology without cavitation. The crystalline/lamellar fragments were oriented via the stress transferred by the tie molecules in stretching direction. Due to the high degree of crystallinity and mainly the transferred stress the amorphous regions between the oriented crystallites are in confined geometry and stretched. The stress of the amorphous phase is in balance to a load of the crystallites, which are slightly deformed on this way. The total degree of crystallinity changes only marginal. The frozen stresses can relax only approaching the melting temperature. There also the load of the crystallites goes down and the orientation in stretching direction becomes worse.

By aligning of crystallites or lamellar fragments by stretching as nuclei for subsequent crystallisation the whole morphology of a later fully crystallised material can become highly oriented. By this way the mechanical properties can become strongly improved. The accessible orientation by shearing a melt during processing is much lower than this route. This procedure to improve mechanical properties is normally used in hot drawing of melt spun fibres. But the present investigations show, that it should also be possible for macroscopic respectively 3-dimensional structures. It is much easier than the compression procedures of undercooled melt to get highly oriented material (Okada, 2010). The process is strongly dependent on the temperature regime. In the case of too strong heating all crystallites can melt and the sample will crystallise later on in a spherulitic or in the presence of nucleating agents and with sufficient quick cooling in an unoriented manner.

The splitting of the melting peak of stretched iPP seems to be initially due to the frozen internal stresses. The first peak appears if the stresses are finally released. This is, dependent on the stretching temperature, between 150 and 165 °C. The second heating shows again a double melting peak between 165 and 173 °C. The first part is due to the remaining lamellar fragments, the second due to the new established fibrils. If the first heating was to 165 °C, only a unique melting peak of the fibrils above 170 °C is found. It seems that this is also the region where the multitude of new crystalline reflexes is observed. This discussion of the material behaviour corrects our first conclusions about the melting behaviour of the different phases in stretched iPP, which we published recently (Schneider, 2010).

To describe the temperature-dependent stress-strain behaviour of semi-crystalline materials we established a model on the basis of the described structural elements. It describes the interaction of the temperature-dependent mobility of the amorphous phase, the initially relative stable crystalline phase, the step-wise re-arrangement of the crystalline phase with respect to orientation and transformation via crystallite fragments and extended chains into fibrils. It will be soon published separately.

6. Conclusion

The experiments revealed the continuously changing interplay of elastic and plastic deformation and energy dissipation (the transformations of amorphous and crystalline phases) during deformation of semi-crystalline polymers using the example of iPP. Due to the high number of parallel processes, here the use of structure characterisation by X-ray scattering techniques was used in small and wide angle range and complemented by mechanical tests, DMA, DSC and SEM. The combination of these methods is likely to provide a well-founded basis for understanding material structure. DSC is an essential extension to characterise crystallisation behaviour.

It was found that there is a strong temperature-dependent interaction of deformations within the amorphous phase and reorientations as well as transformations of crystalline units. They are strongly determined by the molecular structure as well as the processing-dependent initial morphology of the samples. According to the temperature-dependent mobility within the amorphous phase the stress transfer to the crystalline phase and so the changes within this phase are also strongly temperature-dependent.

On the other side by re-melting, stretching and re-crystallizing semi-crystalline material it is possible to get materials with highly improved mechanical behaviour due to well-defined and strongly oriented re-crystallisation. Of cause for using this general technique the parameters of treatment must be optimized.

The mechanisms found and discussed here in the case of iPP seem to be very universal. It might also apply to other polymers. This will be checked in the future by additional investigations.

The detailed understanding of the multiple deformation and structure-establishing processes is an essential base for the development of semi-crystalline materials with well-defined properties.

7. Acknowledgement

The authors are grateful to HASYLAB for beamtime within the projects II-20060086 and I-20100280 and A. Timmann, J. Perlich and M. A. Kashem for local support, and N. Stribeck (University of Hamburg) for the support in the course of data evaluation. They thank their colleagues from the IPF for support, in particular V. Körber for the construction of the tensile rig, W. Jenschke for the software support for the tensile rig, R. Boldt for SEM micrographs, R. Vogel and R. Jurk for DMA-measurements, and D. Krause for sample preparation.

8. References

Breese, D.R. & Beaucage, G. (2004). A review of modeling approaches for oriented semi-crystalline polymers, *Current Opinion in Solid State and Materials Science*, Vol. 8 (2004), pp. 439-448

Breese, D.R. & Beaucage, G. (2008). Modeling the mechanical properties of highly oriented polymer films: A fiber/gel composite theory approach, *Journal of Polymer Science: Part B: Polymer Physics*, Vol. 46, No. 6 (2008), pp. 607-618

Davies, R. J. et al. (2004). The use of Synchrotron X-ray Scattering coupled with in situ Mechanical Testing for studying Deformation and Structural change in Isotactic Polypropylene, *Colloid. Polym. Sci.*, Vol. 282 (2004), pp.854-866

Gehrke, R. et al. (1995). An ultrasmall angle scattering instrument for the DORIS-III bypass, *Rev. Sci. Instrum.* Vol. 66 (1995), pp. 1354-1356

Gibson, A.G. at al. (1978). Dynamic mechanical behaviour and longitudinal crystal thickness measurements on ultra-high modulus linear polyethylene: a quantitative model for the elastic modulus, *Polymer*, Vol. 19 (1978), pp. 683-693

Hong, K. et al. (2004). A model treating tensile deformation of semi-crystalline polymers: Quasi-static stress-strain relationship and viscous stress determined for a sample of polyethylene. *Macromolecules,* Vol. 37, 2004, pp.10165-73 and Model treatment of tensile deformation of semicrystalline polymers: Static elastic moduli and creep parameters derived for a sample of polyethylene, *Macromolecules,* Vol. 37 (2004), pp.19174-79

Hong, K. & Strobl, G. (2006). Network stretching during tensile drawing of polyethylene: A study using X-ray scattering and microscopy, *Macromolecules* Vol.39 (2006), pp.268-273

Kumaraswamy, G. et al. (2000). Shear-Enhanced Crystallization in Isotactic Polypropylene: 2. Analysis of the formation of the Oriented 'Skin', *Polymer* Vol.41 (2000), pp.8931-8940

Kurelec, L. et al. (2005). Strain hardening modulus as a measure of environmental stress crack resistance of high density polyethylene, *Polymer,* Vol. 46 (2005), pp. 6369-6379

Lode, U. et al. (1998). Development of crazes in polycarbonate, investigated by ultra small angle X-ray scattering of synchrotron radiation, *Macromol. Rapid Commun.* Vol. 19 (1998), pp.35-39

Men, Y.M. et al. (2004). Synchrotron Ultrasmall-Angle X-ray Scattering Studies on Tensile Deformation of Poly(1-butene), *Macromolecules* Vol. 37 (2004), pp. 9481-9488

Okada, K.N. et al. (2010). Elongational crystallization of isotactic polypropylene forms nano-oriented crystals with ultra-high performance, *Polymer Journal* Vol. 42 (2010), pp.464-473

Pawlak, A. & Galeski, A. (2010). Cavitation and morphological changes in polypropylene deformed at elevated temperatures, *Journal of Polymer Science: Part B: Polymer Physics* Vol.48 (2010), pp.1271-1280

Peterlin, A. (1971). Molecular model of drawing polyethylene and polypropylene, *J. Mat. Sci.* Vol.6 (1971), pp.490-508

Roth, S.V. et al. (2003). Fatigue behaviour of industrial polymers - a microbeam small-angle X-ray scattering investigation, *J. Appl. Cryst.* Vol.36 (2003), pp.684-688

Roth, S.V. et al. (2006). Small-angle options of the upgraded ultrasmall-angle x-ray scattering beamline BW4 at HASYLAB, *Rev. Sci. Instrum.* Vol.77 (2006), 085106

Roth, S.V. et al. (2011). In situ observation of cluster formation during nanoparticle solution casting on a colloidal film, *J. Phys. Condens. Matter* Vol. 23 (2011), 254208

Schneider, K. et al. (2006). The Study of Cavitation in HDPE Using Time Resolved Synchrotron X-ray Scattering During Tensile Deformation, *Macromolecular Symposia* Vol.236 (2006), 241-248

Schneider, K. & Schöne, A. (2008). Online-structure characterisation of polymers during deformation and relaxation by Synchrotron-SAXS and WAXS, In: *Reinforced Elastomers: Fracture Mechanics, Statistical Physics and Numerical Simulations*; Kaliske, M.; Heinrich, G. & Verron, E. (Eds.); EUROMECH Colloquium 502, Dresden, 2008 pp.79-81

Schneider, K. et al. (2009). Investigation of changes in crystalline and amorphous structure during deformation of nano-reinforced semi-crystalline polymers by space-resolved synchrotron SAXS and WAXS, *Procedia Engineering* Vol.1 (2009), pp.159-162

Schneider, K. (2010). Investigation of Structural Changes in Semi-Crystalline Polymers During Deformation by Synchrotron X-Ray Scattering, *Journal of Polymer Science: Part B: Polymer Physics,* Vol.48 (2010), pp.1574-1586

Schneider, K. et al. (2011). Online structure investigation during deformation and fracture using synchrotron radiation, *Proceedings of 13. Problemseminar "Deformation und Bruchverhalten von Kunststoffen"*, CD-ROM, ISBN 978-3-86829-400-2, Halle-Merseburg, 29.6.-1.7.2011, pp. 122-130

Schroer, C.G. et al. (2006). Mapping the local nanostructure inside a specimen by tomographic small-angle x-ray scattering, *Appl. Phys. Lett.* Vol.88 (2006), 164102

Stribeck, N. (2001). Extraction of domain structure information from small-angle X-ray patterns of bulk materials, *J. Appl. Cryst.* Vol.34 (2001), pp.496-503

Stribeck, N. (2007). *X-Ray Scattering of Soft Matter*, ISBN 978-3-540-69855-5, Springer. Heidelberg, 2007

Stribeck, N. et al. (2008). Nanostructure Evolution in Polypropylene During Mechanical Testing, *Macromol. Chem. Phys.* Vol. 209 (2008), pp.1992-2002

Vonk, C.G. (1979). A small angle X-ray scattering study of polyethylene fibres, using the two-dimensional correlation function, *Colloid Polym. Sci.* Vol. 257 (1979), pp.1021-1032

Weeks, N.E. & Porter, R.S. (1974). Mechanical properties of ultra-oriented polyethylene, *Journal of Polymer Science: Part B: Polymer Physics* Vol.12 (1974), pp.635-643

Zuo, F. et al. (2005). An in Situ X-ray Structural Study of Olefin Block and Random Copolymers under Uniaxial Deformation. *Macromolecules*, Vol. 38 (2005), 3883

13

Structure of Polypropylene Fibres Coloured with Organic Pigments

Jan Broda
University of Bielsko-Biala
Poland

1. Introduction

Polypropylene is known as a versatile and valuable fibre-forming polymer material, which is widely used for the production of medical and hygienic products, carpets and floor coverings, apparel and household textiles, filtering media, agro and geotextiles, automotive interior and many other technical textiles. The wide range of goods comprises a variety of products including mono- and multifilaments, staple fibres, tapes and fibrillated fibres as well as spun-bonded and melt-blown nonwovens.

For the formation of polypropylene textiles different techniques were developed. The common method used for the formation of mono- and multifilaments is classical melt spinning.

In melt spinning, polypropylene in powder or pellet form is heated above the melting point and then is extruded through fine orifices of a spinneret into the air. Below spinneret in the air thin streams are intensively cooled and subjected to an intense stretching. During cooling the liquid streams solidify. Finally, solidified filaments are taken by final take-up device.

Formation of fibres with good properties requires proper selection of formation parameters. Changes of the particular parameters have a great impact on the fibres' structure.

The significant changes of the fibres' structure were observed right from the beginning of their production (Natta, 1961; Ross, 1965; Sheehan & Cole, 1964). It was stated that inside the fibres both, less ordered mesophase and well ordered crystalline phase can be formed.

Mesophase reveals intermediate order between amorphous and crystalline phases. In the first studies it was labelled as smectic (Natta & Corradini, 1960) or paracrystalline (Miller, 1960). Further studies revealed that mesophase is made up of bundles of parallel chains, which maintain typical for all polymorphic forms of polypropylene three-fold helical conformation. Bundles are terminated in the direction of the chain axis by helix reversals or other conformational defects (Androsch et al., 2010). In the bundles long range ordering maintains only along the chain axes, whereas in lateral packing a large amount of disorder is present (Natta & Corradini, 1960). The mesophase is formed by quenching of the molten polypropylene (Miller, 1960; Wyckoff, 1962) or by deformation of the crystalline structure (Saraf & Porter, 1988; Qiu, 2007). As for the fibres, the mesophase was observed in fibres taken at low take-up velocity (Spruiell & White, 1975; Jinan et al., 1989, Bond & Spruiell, 2001) in fibres intensively cooled in water with addition of ice or in the mixture of dry ice and acetone (Sheehan & Cole, 1964; Choi & White, 1998; Yu & White, 1999; Choi & White, 2000), in fibres extruded at the high extrusion temperature (Dees & Spruiell, 1974) and fibres extruded from polypropylene with low molecular weight (Lu & Spruiell, 1987).

The content of the mesophase significantly decreases in fibres taken at other parameters. At certain conditions the mesophase completely disappears and inside fibres only two-phase structure, crystalline and amorphous, is formed. The high crystalline content was obtained in fibres extruded at low extrusion temperature and taken at high take-up velocities (Spruiell & White, 1975; Bond & Spruiell, 2001).

Investigations of the crystalline structure revealed that inside the fibres the crystalline phase is usually built from α crystals. The α form is one of the three known polymorphic forms of polypropylene (Brückner et al.,1991; Lotz et al., 1996). It can be easily obtained by crystallization of polymer melts or solutions. It is the most stable and the most often encountered form in different polypropylene products.

The structure of the α form was early established. Natta and Corradini (Natta & Corradini, 1960) proposed a model based on the monoclinic unit cell. According to this model, polypropylene helices of the same hand are arranged in layers parallel to the *ac* plane. Layers of isochiral helices alternate with layers formed from helices of the opposite hand. The model was commonly accepted and confirmed by subsequent studies of Mencik (Mencik, 1972), Hikosaka (Hikosaka & Seto, 1973) and Corradini (Corradini et al., 1980).

During investigations of the fibres' structure in few cases, except commonly encountered α form, crystals of β form were observed. The β form occurs rarely in the polypropylene products and its formation requires special crystallization conditions. It was revealed that the β form arises during crystallization of a sheared melt (Leugering & Kirsch, 1973), crystallization in a temperature gradient (Lovinger et al., 1977) or during crystallization in the presence of special additives (Tjong et al., 1996; Varga, J. et al. 1999; Li & Cheung, 1999; Varga, 2002). The β form was obtained in the case of fibres containing efficient β nucleating agents and extruded at appropriately selected spinning parameters (Yu & White, 2001; Takahashi, 2002).

The β form exhibits complex structure and during the years different models characterizing this structure were proposed. Finally, the arrangement of the β form was discovered only several decades after its initial observation. It was shown that the β form consists of the characteristic left-handed or right-handed helices arranged in the original packing schemes. Lotz (Lotz at al, 1994) and Meille (Meille at al., 1994) proposed a model of the frustrated structure based on a trigonal cell containing three isochiral helices.

The structure of fibres is formed during the crystallisation, which occurs below spinneret orifices, in the cooling zone, by solidification of the extruded stream. The crystallization rate is comparable to the cooling rate of the polypropylene in the spinning line. Therefore, the formation of crystalline structure in the melt-spinning process depends strongly on the spinning conditions and polymer characteristics.

During the formation of fibres the crystallization takes place in non-isothermal conditions in the field of high tensile stress. Cooling rate and tensile stress strongly affect polypropylene crystallization.

During formation at low take-up velocities the cooling rate plays the dominant role. At low take-up velocity the cooling rate and the applied tensile stress are low. The crystallization proceeds at relatively high temperatures and in the as-spun fibres monoclinic α crystals are formed. When take-up velocity is increased the cooling rate is enhanced. Simultaneously, the crystallization temperature decreases. By crystallization in such conditions the structure with high mesophase content is formed. Further increase of take-up velocity leads to an increase of the tensile stress. Consequently, the molecular orientation of polypropylene chains increases significantly. At higher molecular orientation the crystallisation temperature increases. The crystallisation rate rapidly increases and the crystal structure with highly oriented monoclinic crystals is formed.

2. Dyeing of polypropylene fibres

Polypropylene fibres are highly crystalline and show extremely low wettability. Due to the aliphatic structure of the polypropylene chain fibres have no polar groups and no dye sites capable of reacting permanently with dye molecules. Therefore, polypropylene fibres cannot be coloured by bath methods commonly used for the coloration of other natural and synthetic fibres. Over the years many efforts were undertaken for improving the fibres' dyeability. Various methods include copolymerizing with other monomers, grafting of dye sites, adding dyeable polymers before fibres spinning, adding dyeable filaments before fibre processing, dissolving or dispersing additives of low molecular weight in the polymer melt, treating to modify the surface of fibres after extrusion or adding halogen compounds (Zhu & Yang, 2007). All mentioned techniques have not found broader commercial application. The mainstream technique for the coloration polypropylene fibres consists of using pigments. Despite of its disadvantages, low flexibility and suitability only for a large-scale manufacturing coloration with pigments has a great importance. The most coloured polypropylene textiles are made from fibres coloured this way.

Pigments used for the fibres coloration must be finely dispersed and stable to the thermal conditions and environment applied in fibres formation. Most of the pigments reveal tendency to agglomerate and form large aggregates, which clog the spinneret orifices and spoil the fibres properties. To avoid spinning problems, fine dispersed pigment concentrates, so called masterbatches, are commonly used. Pigment concentrates are made from previously dispersed pigments mixed with polypropylene resin at high pigment concentration. The concentrates are usually mixed in a proper ratio with polymer granulate or powder. Some concentrates are injected directly into the polypropylene melt during fibre formation.

For the coloration of polypropylene fibres inorganic and organic pigments are used. Most inorganic pigments reveal low to moderate colour strength combined with good to excellent thermal stability, lightfastness and weather resistance. Inorganic pigments are easily dispersible. For polypropylene coloration inorganic pigments for black and white colours are mostly applied. For this purpose carbon black and titanium dioxide white pigment based on the rutile modification are used.

Coloration effects are usually achieved with organic pigments. Organic pigments provide high colour strength and high light stability. Some pigments heave negative influence on the efficiency of light stabilizers and limited heat resistance. By reason of above mentioned negative influence only selected groups of pigments can be used. In the literature the following groups of pigments are usually mentioned: azo pigments, isoidolinones, perylene, anthraquinone, quinacridone and phthalocyanine.

During mass coloration pigments are mixed physically with the polypropylene melt in the barrel of the extruder.

It is well known that impurities and foreign substances present in crystallizing polymer strongly affect its crystallisation. Impurities provide a foreign surface, what reduces free energy of the formation of primary nuclei and significantly reduces their critical dimensions. As a result of the formation of heterogeneous nuclei the nucleation density in the crystallizing melt significantly increases. Consequently, the crystallisation temperature is higher and the crystallisation rate rapidly increases.

On the basis of investigations of nucleating ability of many compounds Binsbergen revealed that good nucleating agents are insoluble in the polypropylene melt or crystallize earlier at higher temperature prior polypropylene crystallisation (Binsbergen, 1970). Pigments have such desired properties and fulfil requirements for good nucleating agents of polypropylene. The most pigments form stable crystals, which are insoluble in the polypropylene melt. The rough surface of pigments crystals enables the epitaxial growth of polypropylene crystals. The various geometries of the contact surface can lead to the formation of the different polypropylene modifications.

3. Nucleating ability of phthalocyanine and quinacridone pigments

The nucleating ability toward polypropylene crystallisation for some pigments was investigated and the efficient ability of quinacridone and phthlacyanine was revealed (Broda, 2003a). The investigation were performed in non-isothermal conditions by polarizing microscopy and differential scanning calorimetry (DSC).

Phthalocyanine and quinacridone belong to organic pigments, which are commonly used for the coloration of polypropylene fibres. The chemical formulae of pigments are presented in Figure 1.

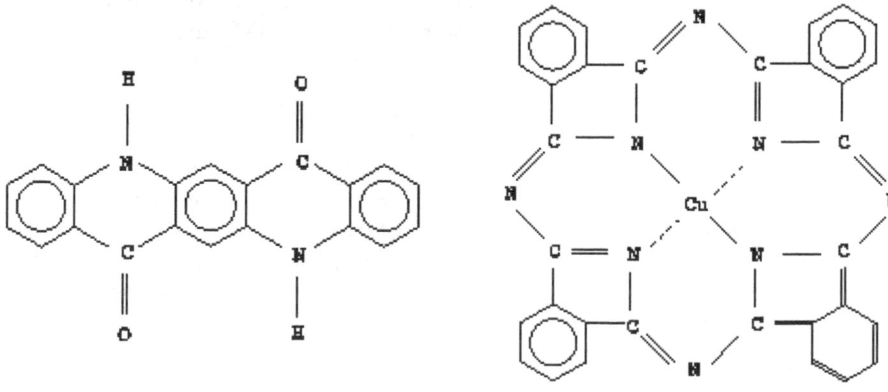

a)   b)

Fig. 1. Chemical formulae of: a) quinacridone, b)phthalocyanine blue

The blue phthalocyanine pigment (Pigment Blue 15, C.I.74160) is built of a tetrabenzoporphyrazine nucleus with a central copper atom. The molecule assumes a planar conformation and possesses a square shape with a side length of 1.3 nm. The red quinacridone pigment (Pigment Violet 19, C.I.73900) belongs to deeply colored pigments characterized by a relatively small molecular size. The molecule of dimensions 1.406 x 0.52 nm is formed from five heterocyclic rings. The molecule is planar with no significant departure of the carbonyl groups from a molecular plane defined by all nonhydrogen atoms.

Both pigments added to quiescent melt accelerate polypropylene crystallization. In the presence of pigments the crystallization temperature moves toward higher temperature. The increase of the crystallization temperature for phthalocyanine and quinacridone pigment is 13 K and 14.5 K, respectively (Broda et al., 2007). Such increase of the crystallization temperature is very high and comparable with the increase observed for effective nucleating agents. In the presence of pigments the nucleation density significantly increases. Consequently, the overall crystallization rate is enhanced and, as a result, fine spherulitic structure is formed.

The nucleating ability of both pigments results from their crystalline structure. Both pigments form fine dispersed crystals, which have very high thermal stability. The degradation temperature of phthalocyanine and quinacridone pigments is higher then 400 and 500 °C, respectively, what considerably exceeds the melting temperature of polypropylene.

The surface structure of pigments crystals enables the epitaxial growth of the polypropylene.

In crystals of the quinacridone pigment planar molecules are arranged in parallel stacks, with the molecule tilted to the stacking direction. The neighboring stacks adopt a herringbone arrangement. In crystals each molecule is bonded through hydrogen bonds to four adjacent molecules. Very strong intermolecular hydrogen bonds combined with strong van der Waals' forces ensure quinacridone pigments the high heat and chemical resistance.

Seven different crystalline forms of quinacridone are known (Filho & Oliveira 1992; Potts et al. 1994, Lincke, 2000). Most synthetic methods lead to the formation of an unstable α form. Subsequent treatments lead to the more stable β and γ forms, most commonly used as commercial pigments.

In the investigations the γ form of the quinacridone pigment was used. For years this modification has been known as a very efficient nucleating agent for the β form of polypropylene (Leugering, 1967; Moos & Tilger, 1981). Stocker and co-workers showed that the β form of polypropylene grows epitaxially on the surface of the γ crystals of the quinacridone pigment (Stocker et al., 1998). The epitaxy involves the (110) plane of the trigonal unit cell of the polypropylene, which contacts the bc surface of the γ crystals of the quinacridone.

Hydrogen atoms of stacked benzene rings form on the surface bc of the γ quinacridone crystals a parallel array of bulges and grooves. The spacing between grooves 0.65 nm is close to the axis repeat distance of the polypropylene helix. The arrangement of polypropylene chains on the pigment surface perpendicularly to the parallel grooves ensures nearly perfect matching of the above-mentioned dimensions and in this way enables the epitaxial growth of β crystals.

The formation of the β form is confirmed by WAXS measurements. In the WAXS pattern of polypropylene crystallized by addition of the quinacridone pigment characteristic β peaks are observed (Fig.2).

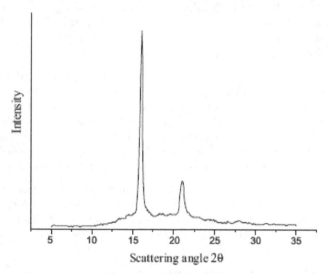

Fig. 2. WAXS pattern of polypropylene crystallized in the presence quinacridone pigment.

In crystals of the copper phthalocyanine fairly rigid molecules can be packed in different arrangements giving rise to different polymorphs. In literature ten different polymorphic forms of copper phthalocyanine are described (Erk & Hengelsberg, 2003). From all polymorphic forms commercial interest exhibits the first recognized α form and the most thermodynamically stable β form.

Each polymorphic form of phthalocyanine is built from molecules arranged in uniform stacks with rings tilted with respect to the stacking direction. For the α and β forms the molecules are tilted in stacks with respect to the stacking direction by 25° and 46°, respectively.

The interactions between molecules within stacks are mainly defined by π-π interactions. The interplanar distance between adjacent molecules is consistent with a van der Waals bond and equals 0.34 nm. In the β modification neighboring stacks are arranged in a herringbone style. In the α form the herringbone interactions are not present.

Crystals of phthalocyanine tend to form needles or rods parallel to the stacking direction. The side faces of crystals are mainly covered by aromatic hydrogen atoms, while the basal faces expose the π system and the copper atom. The lateral surfaces exhibit nonpolar character, while the basal surfaces have relatively polar character.

The aromatic hydrogen atoms occurring on the lateral surfaces of the pigments crystals are arranged in parallel rows. The shallow nonpolar grooves formed between such rows force polypropylene molecules to assume a stretched conformation over some distance, making the nucleation much easier (Binsbergen, 1970).

The investigated α form of phthalocyanine promotes mostly formation of the α form of polypropylene. On the wide-angle X-ray scattering (WAXS) pattern for polypropylene crystallized in the presence of phthalcoyanine pigment, crystalline peaks characteristic for the α form are observed (Fig.3).

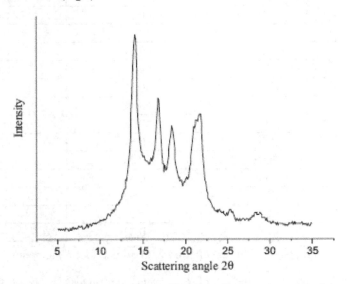

Fig. 3. WAXS pattern of polypropylene crystallized in the presence phthalocyanine pigment.

4. Structure of coloured fibres

The investigations of morphology and structure of coloured fibres were carried out. The morphology of fibres was investigated for samples sputtered with gold by scanning electron microscopy (SEM). The fibres structure was studied by the wide angle X-ray scattering (WAXS) and small angle X-ray scattering (SAXS) methods.

During formation of fibres coloured with pigments the structure containing crystalline, mesophase and amorphous phases is formed. Content of particular phases changes across a broad range depending on formation parameters.

4.1 Structure of fibres coloured with quinacridone pigment

Table 1 presents structural parameters determined on the basis of WAXS patterns for fibres coloured with quinacridone pigment.

For fibres taken at very low velocity 100 m/min the high crystalline structure without mesophase is formed. The structure consists mainly from β crystals with admixture of a small amount of α crystals.

The content of β form is usually characterised by the K value, which is determined as a ratio of the intensity of the $(300)_\beta$ peak to the sum of intensities of the $(110)_\alpha$, $(040)_\alpha$, $(130)_\alpha$ and $(300)_\beta$ peaks on the WAXS patterns (Turner Jones, 1964).

Extrusion temperature [°C]	Take-up velocity [m/min]	Crystallinity index	K value
210	100	0.54	0.95
	200	0.52	0.47
	300	0.51	0.09
	400	0.51	0.03
	880	0.51	0.03
	1050	0.52	0.03
	1350	0.52	-
250	100	0.51	0.93
	200	0.51	0.33
	300	0.50	0.08
	400	0.51	-
	880	0.52	-
	1050	0.52	-
	1350	0.52	-

Table 1. Structural parameters of fibres coloured with quinacridone pigment

In fibres the relatively high amount of β crystals is formed already at low pigment concentration. For fibres containing 0.1% of pigment the K value achieves relatively high value of 0.76. With the increase of the pigment concentration until 0.5% the β form content successively increases. For the pigment concentration of 0.3% and 0.5% the K value grows to 0.89 and 0.95, respectively. Similar high K value is observed for different extrusion temperature, 210, 225 and 250 °C (Broda, 2004c).

The K value observed in fibres coloured with quinacridone pigment is comparable to the value obtained during crystallization of polypropylene melt in quiescent conditions in the presence of very effective β nucleating agents.

The appearance of β crystals in the coloured fibres taken at low velocity is a result of the ability of quinacridone pigment to nucleate β modification of polypropylene and appropriate crystallisation conditions. During formation of fibres at low take-up velocity the crystallisation occurs at low cooling rate and low molecular orientation. In these conditions, during solidification of fibres, pigment takes part in the nucleation process. The crystallisation proceeds on heterogeneous nuclei produced with the participation of pigments. As a result of the crystallisation fine spherulitic structure is formed (Fig.4) (Broda, 2003c).

The low cooling rate favours the formation of the β form of polypropylene (Huang et al., 1995). The crystallization process starts at the relatively high temperature, above the lower critical temperature for the formation of the β modification (Lovinger et al., 1977). As a result of the epitaxial growth on the surface of quinacridone crystals, the crystals of β modification are formed. At this temperature, the growth rate of β crystals exceeds the growth rate of α crystals (Varga, 1989). The β nuclei formed on the surface of pigment crystals quickly grow forming the high amount of β crystals.

a)
b)

Fig. 4. Spherulitic structure of fibres coloured with quinacridone taken at very low velocity: a) gravity spun ; b)100 m/min.

In the fibres taken at higher velocities the crystalline structure changes. With the increase of the take-up velocity the content of the β form rapidly decreases. For fibres extruded at 210°C taken at 200 m/min and 300 m/min the K value decreases to 0.47 and 0.09, respectively. For the fibres extruded at higher melt temperature 250 °C the decrease of the β form content is even higher. For fibres taken at 200 m/min and 300 m/min the K value drops to 0.33 and 0.08. In the same time the content of the α form increases and the crystallinity index does not change significantly.

The increase in take-up velocity and the extrusion temperature results in the increase of the cooling rate. Consequently, the crystallization temperature moves toward lower values. The crystallization starts at lower temperature and then quickly, by further cooling, moves below the critical temperature for the formation of β crystals. Then, the smaller part of the material crystallizes at conditions favourable for the formation of the β form. At the beginning of the crystallization process β crystals are formed. During crystallization temperature decreases into the range below the critical temperature and remaining crystallisable material crystallizes at a lower temperature, forming α crystals.

In temperature below the low critical temperature the growth rate of α crystals is higher than the growth rate of β crystals (Fillon et al., 1993). Then α nuclei grow quickly, while the growth of the β nuclei is strongly constrained.

With the increase in take-up velocity, the temperature is lower and lower, and less and less material can crystallize at a temperature above the critical temperature for the formation of β crystals. With the increasing take-up velocity more and more material crystallize at lower temperatures and β form content decreases.

For the fibres taken at medium velocities, from 400 to 1050 m/min only the minimal content of β crystals is formed. The crystalline structure is built almost exclusively from the α crystals. In the fibres produced without pigments at the same conditions the high mesophase content was observed (Broda, 2004a). This fact suggests that at medium take-up velocities the quinacridone pigment induces the formation of α crystals. The investigations of Rybnikar (Rybnikar, 1991) and Mathieu (Mathieu et al., 2002) showed that γ quinacridone reveals a versatile nucleating ability and may induce either α or β modification of polypropylene. The versatile nucleating ability of the quinacridone pigment results from the fact that the spacing between the grooves on the *bc* surface of γ crystal is similar as well to the interchain distance of the isochiral helices in the (010) plane of the α form of polypropylene. By the arrangement of polypropylene helices parallel to the grooves the nearly perfect match between above-mentioned dimensions may be achieved and the epitaxial growth of α crystals may be initiated.

By further increase of the take-up velocity the structure with the high content of well oriented α crystals is formed. It is know that during formation of fibres at high velocities the molecular orientation has a great influence on the polymer crystallization. Under high molecular orientation certain chain segments become aligned. Bundles of aligned segments form so called row nuclei, which initiate lamellae growth in the perpendicular direction. At high velocities a big number of row nuclei is formed. In this condition the heterogeneous nucleation on pigments loses its importance. The nucleation proceeds without the participation of pigments. Crystals of pigment do not participate in the formation of row nuclei and do not disturb their formation. Relatively small amount of pigment does not affect the mobility of polypropylene chains and does not influence the growth of polypropylene crystals. The growing polypropylene crystals push out pigment outside crystals to the amorphous regions.

The crystallisation in coloured fibres occurs at the same rate and at the same temperature as in non-coloured fibres. The lamellar crystals formed on the row nuclei alternate with the amorphous areas forming fibrillar structure (Broda, 2004b). The long period for coloured fibres determined on Small Angle X-Ray Scattering (SAXS) measurements equals 10.6 nm and has the similar value as in the case of non-coloured fibres (Broda, 2003b).

4.2 Structure of fibres coloured with phthalocyanine pigment

The values of the crystallinity index and K value for fibres coloured with phthalocyanine pigment calculated on the basis of WAXS measurements are presented in Table 2.

Extrusion temperature [°C]	Take-up velocity [m/min]	Crystallinity index	K value
210	100	0.57	0.31
	200	0.56	0.14
	300	0.53	-
	400	0.51	-
	880	0.42	-
	1050	0.43	-
	1350	0.52	-
250	100	0.54	0.29
	200	0.52	0.09
	300	0.44	-
	400	0.40	-
	880	0.36	-
	1050	0.31	-
	1350	0.36	-

Table 2. Structural parameters of fibres coloured with phthalocyanine pigment

For fibres taken at low take-up velocity 100 m/min the high crystalline structure is formed. The cristallinity index reaches high value of 0.57. The crystalline structure is built mainly from α crystals with an addition of β crystals. The content of the β crystals characterised by the K value equals 0.31.

Similarly as for fibres coloured with quinacridone at low take-up velocity, phthalocyanine participates in the nucleation process. The crystallisation starts on heterogenous nuclei formed on the crystals surface. The numerous nuclei grow forming fine spherulitic structure.

Phthalocynine is known as efficient nucleating agent of the α form of polypropylene. The high crystalline structure of the fibres results from its high nucleation ability. Appearance of the small number of β crystals is surprising.

The ability to nucleate β crystals can be explained by the surface geometry of phthalocyanine crystals. For α modification of the phthalocyanine the distance between grooves on the lateral surfaces equals 1.19 nm (Honigmann et al. 1965). This dimension is comparable with the spacing of 1.1 nm between helices of the same hand in the trigonal cell of the β form of polypropylene.

The compatibility of these dimensions is responsible for a good nucleating ability toward the β form of several calcium dicarboxylates (Li et al., 2002) and may also explain the formation of a certain amount of β crystals in fibres coloured with phthalocyanine.

With the increase of the take-up the content of β crystals rapidly decreases. For fibres taken at take-up velocity 200 m/min and extruded at 210 °C and 250 °C the K value drops to 0.14 and 0,09, respectively. At higher velocities β crystals are not observed.

At higher cooling rates the crystallization conditions for formation of the β form are less favorable. On the surface of the phthalocyanine pigment only α nuclei are formed. The growth of nuclei leads to formation of the high crystalline structure built only from α crystals.

Similarly to fibres coloured with quinacridone at high take-up velocity above 1000 m/min, pigment does not participate in the nucleation process. In these conditions, under high molecular orientation numerous row nuclei are formed. Row nuclei quickly grow forming well oriented lamellar α crystals. In fibres the fibrillar structure is formed (Fig.5)

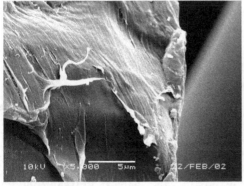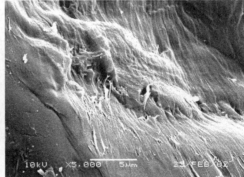

Fig. 5. Fibrillar structure in fibres coloured with phthalocyanine taken at 1350 m/min.

4.3 Structure of fibres coloured with the mixture of quinaridone and phthalocyanine pigments

Mixing of pigments is the common procedure for achieving different colour effects. By mixing of qunacridone and pthtalocyanine two pigments with different efficient nucleating ability toward polypropylene crystallisation are introduced. One can expect that by mixing of both pigments a competition between formation of α and β nuclei will be observed. Such competition, together with a different growth rate of both forms, should lead to formation of a structure with different constitution.

Table 3 presents structural parameters determined on the WAXS measurements for fibres coloured with the 1:1 mixture of quinacridone and phthalocyanine.

Take-up velocity [m/min]	Crystallinity index -	K value -
100	0.53	0.75
200	0.53	0.52
300	0.53	0.14
400	0.51	0.05
880	0.52	0.04
1050	0.51	-
1350	0.52	-

Table 3. Structural parameters of fibres coloured with the mixture of quinaridone and phthalocyanine pigments

At the lowest velocity 100 m/min inside fibres the crystalline structure consisting of both polymorphic modifications is formed. The K value equals 0.75. The content of β crystals is much higher than the content of α crystals. Nevertheless, the content of the β form is much lower in comparison to fibres coloured with quinacridone alone, but significantly greater in comparison to fibres coloured with phthalocyanine (Broda, 2003d).

In fibres taken at very low velocities pigments participate in the formation of the crystallization nuclei. By coloration with the mixture of pigments, both pigments reveal their nucleating activity. On the surface of the quinacridone the β nuclei, while on the surface of phthalocyanine crystals the α nuclei, are formed.

Different content of both modifications in the fibres may be caused by a different number of produced nuclei and/or different rate of crystals growth.

By even proportion of both pigments and their similar dispersion one can assume that the number of the quinacridone crystals equals the number of the phthalocyanine crystals. Both pigments reveal their nucleating activity at similar temperature and formation of α and β nuclei occurs at similar conditions. Taken into account the above mentioned statements, there is no indication of significantly greater number of the β nuclei. The higher content of the β crystals in the fibres coloured with the mixture of pigments have to result from the higher rate of crystals growth.

During formation of fibres at low velocity the crystallization occurs at high temperature. The crystallization temperature exceeds the critical temperature for the formation of the β form. In temperature above critical, the growth rate of the β form crystals exceeds the growth rate of the α crystals (Lovinger et al., 1977). Due to higher growth rate, the β crystals grow quicker and even by equal number of both nuclei form the structure with the high content of the β form crystals.

For fibres formed at higher velocities the crystallization conditions prefer the formation of the α form crystals. Similarly as for fibres coloured with quinacridone, the content of the β form rapidly decreases.

Similarly as for fibres coloured with particular pigments at high velocities, the mechanism of the nucleation changes. The crystallization proceeds without pigments on row nuclei formed under high molecular orientation. As a result, in all fibres the structure consisting of the α modification characterized by the same crystallinity index and the same lamellar thickness is produced.

5. Influence of processing on the structure of polypropylene fibres coloured with quinacridone pigment

To achieve final properties, as-spun fibres obtained by the spinning line are usually submitted to further processing. During processing the fibres structure formed by spinning undergoes further transformation. Heat stabilisation and drawing are of a great importance. Both treatments strongly affect the structure of the fibres and their final properties.

During heat stabilisation of as-spun fibres coloured with quinacridone pigment the less thermodynamically stable β crystals transform into α crystals.

In the case of fibres extruded at the lowest velocity exhibiting the highest amount of β crystals the first changes of the content of β form are observed at heating at 140°C. The amount of β crystals decreases already after 3 minutes of stabilization. For longer time decrease of the β form content is more pronounced. The K value decreases from 0.95 for not stabilized fibres to 0.63 for fibres stabilized for 5 minutes. After a very long time the high amount of β crystals remains unchanged. For the fibres stabilized at higher temperatures, in the range 140–148°C, the β form content gradually decreases with the increment of the stabilization temperature. The content of the β form decreases quickly during the first minutes of heating. For longer times of heating (up to 10 min) further changes of the β form content are less meaningful. After certain time the β→α transition stops. Despite the long time of stabilization the content of β crystals does not change. During stabilization at 150°C a rapid drop of the β form content is observed already after 3 minutes of heating. In this time the K value drops to 0.1. Then, after 10 minutes the β crystals disappear (Broda, 2004d).

The transition of the polymorphic forms of polypropylene was observed many times. It was stated that the β→α transition is not reversible and occurs in the temperature range of 130 to 150°C (Forgacs et al., 1981; Varga, 1995). For years several mechanisms of the transition have been proposed. Garbarczyk suggested that the β→α transition proceeds in a solid state in three intermediate stages based on rotations and translations of polypropylene chains (Garbarczyk, 1985, 1989). Due to considerable differences in unit cells of both modifications, Samuels stated that the β→α transition must take via the liquid phase (Samuels & Yee,

1972). The investigations of Vleeshouwers confirmed that the transition is connected to the melting of β crystals (Vleeshouwers, 1997). Zhou proved that the melting of β crystals occurs partially, starting from the less perfect crystals possessing the lowest melting temperature (Zhou et al., 1986). Rybnikar suggested that new α crystals are formed on the lateral faces of the remaining thickened β lamellae by regular, probably epitaxial, overgrowth (Rybnikar, 1991). Varga and Fillon stated that the transition can be attributed to the formation of α nuclei within the β crystals during secondary crystallization at temperatures below 100°C (Varga, 1986; Fillon et al., 1993).

Taking into consideration the above mentioned statements, one can conclude that the changes of the β form content in fibres with quinacridone results from the melting of β crystals and their recrystallisation into α crystals. At temperatures near 140°C only a small fraction of the less perfect β crystals is melted. After the melting of such crystallites the transition stops in spite of the long time of stabilization. At higher temperatures, closer to the melting point of β crystals the greater part of the β lamellae is melted and the transition proceeds further. At 150°C, the temperature above the melting temperature of β crystals, all β lamellae melt and recrystallize, forming α crystals. At this temperature the β→α transition is completed within few minutes.

The changes, which occur in the fibres' structure during drawing depend on drawing temperature. During drawing at room temperature the crystalline structure with the high content of β crystals transforms into mesophase. The significant changes are observed already at draw ratio of 2. At this draw ratio the β form content drops to half of the value for undrawn fibres. With the increase of the draw ratio the transition from crystalline structure to mesophase proceeds further and the content of the β form gradually decreases. At draw ratio of 7 all crystalline structure transforms into mesophase and β crystals disappear.

During drawing at 120°C the gradual transition from β form to α form is observed. As a consequence the β form content in fibres gradually decreases. At draw ratio of 2 the K-value drops to 0.45. At higher draw ratios of 3 and 5 it drops to 0.07.

The transition of β crystals into mesophase generated by drawing at room temperature results from pulling polypropylene chains from the lamellar crystals (Ran et al. 2001). As a result, partial destruction of the crystals is observed. Due to limited molecular mobility at room temperature, chains pulled from the crystals aggregate into bundles with no specific arrangement of helical hands. Such bundles, representing a collection of helical segments with a random assembly of helical hands, form the mesophase. At low draw ratios only a few chains are pulled out from the crystals. At higher draw ratios the number of pulled chains increases. Consequently, the β form content in drawn fibres gradually decreases, while the mesophase content increases. At higher draw ratios, the regular lamellar structure of β crystals is completely destroyed and the β form disappears.

During drawing at 120°C chains pulled out from polypropylene crystals posses much higher mobility and may form assemblies ready to crystallize into well ordered α form crystals.

6. Conclusions

Organic pigments commonly used for the coloration of polypropylene fibres exhibit nucleating ability toward polypropylene crystallisation. The rough surface of pigments' crystals enables epitaxial growth of polypropylene. The various geometry of the surface of pigments' crystals ensures matching of polypropylene chains in different way, what leads to the formation of different polymorphic forms of polypropylene. The red quinacridone pigment promotes mainly formation of the β form crystals. On the contrary the blue phthalocynine pigment enables growth of α crystals.

During fibres coloration pigments are added to the polypropylene granulate and mixed with the polypropylene melt. Presence of insoluble pigments' crystals effect the polypropylene crystallisation. The influence of pigments on the structure of fibres reveals at the lowest take-up velocity. Then, pigments participate in the nucleation process. For fibres coloured with quinacridone almost only β crystals are formed. For fibres coloured with phathalocynine the formation of the high crystalline structure containing α crystals with the addition of the small amount of β crystals is observed. For fibres coloured with the mixture of both pigments the structure with α and β crystals is produced.

For higher velocities the cooling rate increases and the crystallisation conditions favour formation of the α form crystals. As a result in fibres coloured with pigments the β form content rapidly decreases.

By further increment of the take-up velocity the nucleation ability of pigments does not influence the polypropylene crystallisation. Inside fibres under high molecular orientation numerous row nuclei are formed. The crystallisation proceeds on row nuclei without pigments.

By processing of fibres with high content of β form crystals transitions: β→α and β→mesophase are observed. The direction and range of transition depends on the processing parameters.

7. References

Androsch. R. et al. (2010). Mesophases in polyethylene, polypropylene, and poly(1-butene). *Polymer*, Vol.51, Issue 21, pp. 4639-4662.

Binsbergen, F. L. (1970). Heterogeneous nucleation in crystallization of polyolefins. I. Chemical and physical nature of nucleating agents. *Polymer* , Vol.11, Issue 5, pp. 253-267.

Bond, E.B. & Spruiell, J.E. (2001). Melt spinning of metallocene catalyzed polypropylenes. I. On-line measurements and their interpretation. *J.Appl.Polym.Sci.,*Vol. 82, Issue 13, pp. 3223-3236.

Broda, J. (2003a). Nucleating activity of the quinacridone and phthalocyanine pigments in polypropylene crystallization. *J.Appl.Polym.Sci.*, Vol.90, Issue 14, pp. 3957-3964.

Broda, J. (2003b). Polymorphism in Polypropylene Fibres. *J.Appl.Polym.Sci.*, Vol. 89 , Issue 12, pp. 3364-3370.

Broda, J. (2003c). Morphology of the noncoloured and coloured polypropylene fibers. *Polymer*, Vol.44, Issue 5, pp. 1619-1629.

Broda, J. (2003d). Structure of polypropylene fibres coloured with a mixture of pigments with different nucleating ability. *Polymer* Vol.44 , Issue 22, pp 6943-6949.

Broda, J. (2004a). WAXS Investigations of The Mass Coloured Polypropylene Fibers. *Fibres and Textiles in Eastern Europe,* Vol.11, pp. 95

Broda, J. (2004b). SEM Studies of Polypropylene Fibres Coloured with Quinacridone Pigment, *Microscopy and Analysis*, Vol. 91, Issue 5, pp. 5-6.

Broda, J. (2004c). Polymorphic composition of colored polypropylene fibers. *Crystal Growth & Design*, Vol.4, Issue 6, pp. 1277- 1282.

Broda, J. (2004d). Influence of Processing on Structure of β-Nucleated Poly(propylene) Fibers. *J. Appl. Polym. Sci.*, Vol.91, Issue 3, pp. 1413-1418.

Broda, J. (2007). The influence of additives on the structure of polypropylene fibres. *Dyes & Pigment*, Vol.74, Issue 3, pp. 508-511.

Brückner, S. et al. (1991). Polymorphism in isotactic polypropylene. *Prog Polym. Sci.* Vol.16, Issue 2-3, pp. 361-404.

Choi, C.H. & White, J.L. (1998). Comparative study of structure development in melt spinning polyolefin fibers. *Intern.Polym.Proc.*, Vol.13, Issue 1, pp. 78-87.

Choi, D. & White, J.L. (2000). Structure development in melt spinning syndiotactic polypropylene and comparison to isotactic polypropylene. *Intern.Polym.Proc.*, Vol.15, Issue 4, pp. 398-405.

Corradini, P. et al. (1980). Structural variations in crystalline isotactic polypropylene (alpha-form) as a function of thermal treatments. *Gazz.Chim.Ital.*, Vol.110, Issue 7-8, pp.413-418.

Dees, J.R. & Spruiell, J.E. (1974). Structure development during melt spinning of linear polyethylene fibers. *J.Appl.Polym.Sci.*, Vol. 18, Issue , Issue 4, pp.1053-1078.

Erk, P. & Hengelsberg, H. (2003). Phthalocyanine Dyes and Pigment, In *The Porphyrin Handbook*; Kadish, K. M., Smith, K. M., Gulard, R., pp 106-149, Academic Press, ISBN 10: 0-12-393201-7, Amsterdam.

Filho, D. S. & Oliveira, C. M. F. (1992). Crystalline modifications of linear trans-quinacridone pigments. *J. Mater. Sci.*, Vol. 27, pp. 5101-5107.

Fillon, B. et al. (1993). Self-nucleation and recrystallization of polymers. Isotactic polypropylene, β phase: β-α conversion and β-α growth transitions. *J. Polym.Sci. Part B. Polym. Phys.*, Vol. 31, Issue 10, 1407-1424.

Forgacs, P. et al. (1981). Study of the beta-alpha solid-solid transition of isotactic polypopylene by synchrotron radiation. *Polymer Bulletin*, Vol. 6, Issue 1-2, pp. 127-133.

Garbarczyk, J. (1985). A study on the mechanism of polymorphic transition beta-alpha in isotactic polypropylene. *Makromol Chem.* , Vol. 186, Issue 10, pp.2145-2151.

Garbarczyk, J. et al (1989). Influence of additives on the structure and properties of polymers .4. study of phase-transition in isotactic polypropylene by synchrotron. *Polym. Commun.*, Vol. 30, Issue 5, pp.153-157.

Hikosaka, M. & Seto, T. (1973). Order of molecular chains in isotactic polypropylene crystals. *Polym.J.*, Vol.5, Issue 2, pp.111-127.

Honigmann, B. Et al.(1965). Beziehungen zwischen den strukturen der modifikationen des platin- und kupferphthalocyanins und einiger chlorderivate . *Z. Kristallogr.* Vol.122, Issue 3-4, pp. 185-205.

Huang, M. R. et al. (1995). β nucleators and β crystalline form of isotactic polypropylene. *J. Appl. Polym. Sci.*, Vol 56, Issue 10, pp. 1323-1337.

Jinan, C.; et al. (1989). Nonisothermal orientation-induced crystallization in melt spinning of polypropylene. *J. Appl.Polym. Sci.*, Vol. 37, Issue 9, pp. 2683-2697.

Leugering, H. J. (1967). Einfluß der kristallstruktur und der überstruktur auf einige eigenschaften von polypropylen. *Makromol. Chem.* Vol. 109, Issue 1, pp.204-216.

Leugering, H. J.& Kirsch, G.(1973). Effect of crystallization from oriented melts on crystal-structure of isotactic polypropylene . *Angew. Makromol. Chem.*, Vol.33, Issue OCT, pp. 17-23.

Li, J. X. & Cheung, W. L. (1999). Conversion of growth and recrystallisation of β-phase in doped iPP. *Polymer*, Vol.40, Issue 8, pp. 2085-2088.

Li, X. et al. (2002). Calcium dicarboxylates nucleation of β-polypropylene. *J.Appl.Polym.Sci.* , Vol. 86, Issue 3, pp. 633-638.

Lincke, G. (2000). A review of thirty years of research on quinacridones. X-ray crystallography and crystal engineering. *Dyes Pigm.*, Vol.44, Issue 2, pp.101-122.

Lotz, B. et al. (1996). Structure and morphology of poly(propylenes): A molecular analysis *Polymer*, Vol. 37, Issue 22, pp. 4979-4992.

Lotz, B. et al. (1994). An original crystal-structure of polymers with ternary helices *Comptes rendus de l academie des sciences serie II*, Vol.319. Issue 2, pp.187-192.

Lovinger, A. J. et al. (1977). Studies on the α and β forms of isotactic olypropylene by crystallization in a temperature gradient. *J. Polym. Sci.Polym. Phys. Ed.*, Vol. 15, Issue 4, pp. 641-656.

Lu, F.M. & Spruiell, J.E. (1987). The influence of resin characteristics on the high speed melt spinning of isotactic polypropylene. I. Effect of molecular weight and its distribution on structure and mechanical properties of as-spun filaments *J.Appl.Polym.Sci.* , Vol.34, Issue 4, pp.1521-1539.

Mathieu, C. et al. (2002). Specificity and versatility of nucleating agents toward isotactic polypropylene crystal phases *J. Polym.Sci. Part B. Polym. Phys.* , Vol.40, Issue 22, pp.2504-2515.

Meille, S.V. et al.(1994). Structure of beta-isotactic polypropylene - a long-standing structural puzzle. *Macromolecules*, Vol.27, Issue 9, pp.2615-2622.

Mencik, Z. (1972). Crystal-structure of isotactic polypropylene. *J.Macromol.Sci., Part B.*, Vol. 6, Issue 1, pp.101-115.

Moos, K. H. & Tilger, B. (1981). Nukleierung und Polymorphie in isotaktischem Polypropylen. *Angew. Makromol.Chem.*, Vol.94, Issue 1, pp. 213-225.

Natta, G. & Corradini, P. (1960). Structure and properties of isotactic polypropylene. *Nuovo Cimento Suppl.*, Vol. 15, No. 10, pp. 40- 51.

Miller, R.L. (1960). On the existence of near-range order in isotactic polypropylenes *Polymer*, Vol.1, Issue 2, pp. 135-143.

Potts, G. D.; et al. (1994). The crystal structure of quinacridone: An archetypal pigment. *J. Chem. Soc., Chem. Commun.*, Vol.40, Issue Pt01, pp. 2565-2566.

Qiu, J. et al. (2007). Deformation-induced highly oriented and stable mesomorphic phase in quenched isotactic. *Polymer*, Vol. 48, Issue 23, pp.6934-6947.

Ran, S. et al. (2001). Structural and morphological studies of isotactic polypropylene fibers during heat/draw deformation by in-situ synchrotron SAXS/WAXD. *Macromolecules*, Vol.34, Issue 8, pp. 2569-2578.

Ross, S.E. (1965). Some observations concerning behavior of polypropylene polymers and fibers. *J.Appl.Polym.Sci.*, Vol.9, Issue 8, pp. 2729-2748.

Rybnikar, F. (1991). Transition of β to α phase in isotactic polypropylene. *J. Macromol. Sci.-Phys.*, Vol.30, Issue 3, pp. 201-223.

Samuels, R. J. & Yee, R.Y. (1972). Characterization of structure and organization of beta-form crystals in type-iii and type-iv beta-isotactic polypropylene spherulites *J. Polym. Sci. Part A-2 Polymer Phys.*, Vol.10, Issue 3, pp. 385-432.

Saraf, R.F. & Porter, R.S. (1988). A deformation induced order-disorder transition in isotactic polypropylene. *J.Polym.Eng.Sci.*, Vol. 28, Issue 13, pp. 842-851.

Sheehan, W.C. & Cole, T.B. (1964). Production of super-tenacity polypropylene filaments. *J.Appl.Polym.Sci.*, Vol. 8, Issue 5, 2359-2388.

Spruiell, J.E.& White, J.L. (1975). Structure development during polymer processing - studies of melt spinning of polyethylene and polypropylene fibers. *Polym.Eng.Sci.*, Vol. 15, Issue 9, pp.660-667.

Stocker, W. et al. (1998). Epitaxial Crystallization and AFM Investigation of a Frustrated Polymer Structure: Isotactic Poly(propylene), β Phase. *Macromolecules*. Vol.31, Issue 3, pp. 807-814.

Takahashi, T. (2002). Crystal modification in polypropylene fibers containing β-form nucleating agent. *Sen Gakkaishi,* Vol.58, No. 10, pp.357-364.

Tjong, S. C. et al. (1996). Morphological behaviour and instrumented dart impact properties of beta-crystalline-phase *Polymer*, Vol.37, Issue 12, pp. 2309-2316.

Turner Jones, A.; et al. (1964). Crystalline forms of polypropylene *Makromol.Chem.* , Vol. 75, Issue 1, pp.134-154.

Varga, J. (1986). Melting memory effect of the beta-modification of polypropylene. *J. Therm. Anal.* Vol.31, Issue 1, pp.165-172.

Varga, J. (1989). β-Modification of polypropylene and its two-component systems. *J.Therm. Anal.*, Vol.35, Number 6, pp.1891-1912.

Varga J. (1995). Crystallization, melting and supermolecular structure of isotactic polypropylene, In: *Polypropylene: Structure, Blends and Composities; Structure and Morphology Copolymers and Blends Composites,* Karger-Kocsis, J., pp. 56-115 , Chapman & Hall, ISBN 9780412614309, London.

Varga, J. et al. (1999). Highly active thermally stable beta-nucleating agents for isotactic polypropylene. *J. Appl. Polym. Sci.* , Vol.74, Issue 10, pp. 2357-2368.

Varga, J. (2002). Beta-modification of isotactic polypropylene: Preparation, structure, processing, properties, and application. *J. Macromol. Sci., Phys. B.*, Vol. B41, Issue 4-6, pp.1121-1171.

Vleeshouwers, S. (1997). Simultaneous in-situ WAXS/SAXS and DSC study of the recrystallization and melting behaviour of the alpha and beta form of iPP. *Polymer*, Vol.38, Issue 13, 3213- 3221.

Yu, Y.; White, J.L. (1999). Structure development in melt spinning polypropylene-EPM blends and dynamically vulcanized polyolefin TPEs. *Intern.Polym.Proc.*, Vol.14, Issue 2, pp. 159-167.

Yu, Y.; White, J.L. (2001). Comparison of structure development in quiescent crystallization, die extrusion and melt spinning of isotactic polypropylene and its compounds containing fillers and nucleating agents. *Polym.Eng.Sci.*, Vol. 41, Issue 7, pp.1292-1298.

Wyckoff, H.W. (1962). X-ray and related studies of quenched, drawn, and annealed polypropylene. *J.Polym.Sci.*, Vol.52, Issue 173, pp.83-114.

Zhou, G. et al. (1986). Studies on the beta-form of isotactic polypropylene .1. characterization of the beta-form and study of the beta-alpha-transition during heating by wide angle X-ray-diffraction. *Makromol. Chem.*, Vol. 187, Issue 3, pp. 633-642.

Zhu, M.F. & Yang, H.H. (2007) Polypropylene fibers, In: *Handbook of fiber chemistry*, Lewin M., pp.139-260, CRC Press, ISBN 10 08247-2565-4, Boca Raton.

Permissions

The contributors of this book come from diverse backgrounds, making this book a truly international effort. This book will bring forth new frontiers with its revolutionizing research information and detailed analysis of the nascent developments around the world.

We would like to thank Dr. Fatih Doğan, for lending his expertise to make the book truly unique. He has played a crucial role in the development of this book. Without his invaluable contribution this book wouldn't have been possible. He has made vital efforts to compile up to date information on the varied aspects of this subject to make this book a valuable addition to the collection of many professionals and students.

This book was conceptualized with the vision of imparting up-to-date information and advanced data in this field. To ensure the same, a matchless editorial board was set up. Every individual on the board went through rigorous rounds of assessment to prove their worth. After which they invested a large part of their time researching and compiling the most relevant data for our readers. Conferences and sessions were held from time to time between the editorial board and the contributing authors to present the data in the most comprehensible form. The editorial team has worked tirelessly to provide valuable and valid information to help people across the globe.

Every chapter published in this book has been scrutinized by our experts. Their significance has been extensively debated. The topics covered herein carry significant findings which will fuel the growth of the discipline. They may even be implemented as practical applications or may be referred to as a beginning point for another development. Chapters in this book were first published by InTech; hereby published with permission under the Creative Commons Attribution License or equivalent.

The editorial board has been involved in producing this book since its inception. They have spent rigorous hours researching and exploring the diverse topics which have resulted in the successful publishing of this book. They have passed on their knowledge of decades through this book. To expedite this challenging task, the publisher supported the team at every step. A small team of assistant editors was also appointed to further simplify the editing procedure and attain best results for the readers.

Our editorial team has been hand-picked from every corner of the world. Their multi-ethnicity adds dynamic inputs to the discussions which result in innovative outcomes. These outcomes are then further discussed with the researchers and contributors who give their valuable feedback and opinion regarding the same. The feedback is then collaborated with the researches and they are edited in a comprehensive manner to aid the understanding of the subject.

Apart from the editorial board, the designing team has also invested a significant amount of their time in understanding the subject and creating the most relevant covers. They scrutinized every image to scout for the most suitable representation of the subject and create an appropriate cover for the book.

The publishing team has been involved in this book since its early stages. They were actively engaged in every process, be it collecting the data, connecting with the contributors or procuring relevant information. The team has been an ardent support to the editorial, designing and production team. Their endless efforts to recruit the best for this project, has resulted in the accomplishment of this book. They are a veteran in the field of academics and their pool of knowledge is as vast as their experience in printing. Their expertise and guidance has proved useful at every step. Their uncompromising quality standards have made this book an exceptional effort. Their encouragement from time to time has been an inspiration for everyone.

The publisher and the editorial board hope that this book will prove to be a valuable piece of knowledge for researchers, students, practitioners and scholars across the globe.

List of Contributors

Azza M. Mazrouaa
Petrochemical Department, Polymer Laboratory, Egypt
Egyptian Petroleum Research Institute, Nasr City, Cairo, Egypt

Alireza Aslani
Nanobiotechnology Research Center, Baqiyatallah University Medical of Science, Tehran,
Department of Chemistry, Faculty of Basic Science, Jundi Shapur University of Technology,
Dizful, Islamic Republic of Iran

Yan Wu, Dingguo Zhou, Yang Zhang and Zhihui Wu
Nanjing Forestry University, China

Siqun Wang
University of Tenneessee, USA

Yosuke Nishitani
Kogakuin University, Japan

Chiharu Ishii
Hosei University, Japan

Takeshi Kitano
Tomas Bata University in Zlin, Czech Republic

Maciej Jaroszewski, Janina Pospieszna, Jan Ziaja and Mariusz Ozimek
Wrocław University of Technology, Institute of Electrical Engineering Fundamentals, Poland

Karim Shelesh-Nezhad, Hamed Orang and Mahdi Motallebi
University of Tabriz, Iran

Kamil Şirin
Celal Bayar University, Faculty of Science and Arts, Department of Chemistry, Manisa,

Mehmet Balcan
Ege University, Faculty of Science, Department of Chemistry, İzmir,

Fatih Doğan
Çanakkale Onsekiz Mart University, Faculty of Education, Secondary Science and Mathematics
Education, Çanakkale, Turkey

Jia Ma and Ton Peijs
Queen Mary University of London, UK

Diene Ndiaye, Bouya Diop, Papa Alioune Fall and Abdou Karim Farota
Universite Gaston Berger de Saint-Louis, Senegal

Coumba Thiandoume and Adams Tidjani
Universite Cheikh Anta Diop de Dakar, Senegal

Valerio Brucato and Vincenzo La Carrubba
Dipartimento di Ingegneria Chimica, Gestionale, Informatica, Meccanica Università di Palermo, Palermo, Italy

Fukuzo Yoshida and Masahiko Yoshiura
Osaka Institute of Technology Osaka, Japan

K. Schneider and L. Häussler
Leibniz-Institut für Polymerforschung Dresden,

S.V. Roth
DESY, Hamburg, Germany

Jan Broda
University of Bielsko-Biala, Poland